Asia and Australia Regional Meeting of Polymer Processing Society (PPS2016)

Chengdu, China
11 - 14 October 2016

ISBN: 978-1-5108-4340-0

Printed from e-media with permission by:

Curran Associates, Inc.
57 Morehouse Lane
Red Hook, NY 12571

CURRAN ASSOCIATES INC. proceedings .com

Some format issues inherent in the e-media version may also appear in this print version.

Printed by Curran Associates, Inc. (2017)

For permission requests, please contact Polymer Processing Society
at the address below.

Polymer Processing Society
810 E. 10th St.
Lawrence, KS 66044
USA

Phone: +1-800-627-0326
+1-785-865-9403
Fax: +1-785-843-6153

pps@allenpress.com

Additional copies of this publication are available from:

Curran Associates, Inc.
57 Morehouse Lane
Red Hook, NY 12571 USA
Phone: 845-758-0400
Fax: 845-758-2633
Email: curran@proceedings.com
Web: www.proceedings.com

TABLE OF CONTENTS

RHEOLOGY AND RHEOMETRY

MODIFICATION OF RHEOLOGICAL PROPERTIES FOR POLYMER MELT 1
M. Yamaguchi

RHEOLOGICAL BEHAVIOR OF HYPERBRANCHED COPOLYMER BASED NANO-SIO_2 FUNCTIONAL MONOMER 5
N. Lai, L. Tang, Z. Ye

INTERACTION BETWEEN SODIUM DODECYL BENZENE SULFONATE AND HYPERBRANCHED POLYMER BASED ON DENDRITIC STRUCTURE 10
N. Lai, Y. Zhang, Z. Ye

ANALYSIS OF THE VISCOELASTIC PROPERTIES OF CARBON NANOTUBE REINFORCED POLYPROPYLENE COMPOSITES 14
J. Huang, D. Rodrigue, L. Sun

EFFECT OF FIBRE ORIENTATION AND CONCENTRATION ON THE VISCOSITY OF POLYMER COMPOSITES 19
D.V. Perumal, S.N. Bhattacharya, R.K. Gupta, F.S. Costa

EFFECT OF EXTRUSION TEMPERATURE ON THE ELONGATIONAL PROPERTIES OF POLYPHENYLENE SULFIDE USING MELT SPINNING TECHNIQUE 24
C. Tan, Y. Yang, Y. Zhang, H. He, J. Ning

EXTRUSION

PREPARATION AND CHARACTERIZATION OF PPO/PA66/CARBON BLACK/SISAL FIBERS ELECTROMAGNETIC SHIELDING COMPOSITE 29
H. He, S. Cheng, G. He, Y. Xing

THE STUDY ON THE STRESS-PHOTOOXIDATION AGING OF POLYETHYLENE CABLE SHEATH UNDER DIFFERENT PROCESSING CONDITIONS 34
X. Ren

INJECTION MOLDING

INFLUENCES OF RESIN COMPONENTS AND INJECTION RATES ON GENERATION OF TIGER-STRIPED FLOW MARKS IN INJECTION MOLDING 39
S. Owada, H. Yokoi

THE FUNCTION OF DETECTING BACKFLOW PHENOMENON ON INJECTION MOLDING MACHINE AND THE EVALUATION 44
S. Takatsugi

APPLICATION PROGRESS OF IN-MOLD DECORATION 49
Y. Wang, J. Chen

IMPROVEMENT OF MOLDING QUALITY OF LIGHT GUIDE PLATE USING MULTI-COATING SURFACE TREATMENT FOR INJECTION MOLDING SCREW 54
R. Jian, W. Yang, Z. Jiao, Y. Ding, P. Xie

BLOW MOLDING, ROTATIONAL MOLDING AND THERMOFORMING

PRODUCTION AND CHARACTERIZATION OF ROTOMOLDED HYBRID POLYMER COMPOSITES BASED ON NATURAL FIBERS 59
F.E. Hanana, D. Rodrigue

FILMS, FIBERS, AND FOAMS

DYNAMIC MECHANICAL ANALYSIS AND DYNAMIC FATIGUE OF BEAD FOAMS 64
J.H. Keller, V. Altstadt

THE EVALUATION OF STRETCHABILITY AND ITS APPLICATIONS FOR BIAXIALLY ORIENTED POLYPROPYLENE FILM 69
T. Kanai, T. Takebe

NEW TRENDS IN PREPARATION AND CHARACTERIZATION OF NANOFIBRE FILTERS 74
W. Sambaer, M. Zatloukal, D. Kimmer, T. Sedlacek, P. Saha

EFFECT OF LONG-CHAIN BRANCHES ON FABRICATION OF HIGH EXPANSION MICROCELLULAR INJECTION MOLDING OF POLYPROPYLENE FOAMS 79
L. Wang, S. Ishihara, A. Minato, M. Ohshima

NEEDLELESS MELT ELECTROSPINNING FOR NANOFIBER PREPARATION AND ITS SCALING-UP 84
H. Li, J. Tan, X. Ma, H. Chen, W. Yang

IMPROVING THE TRANSPARENCY OF THE HIGHLY-ORIENTED MELT-CRYSTALLIZED POLYETHYLENES: TOWARD HIGH-IMPACT WINDOW APPLICATION 88
N/A

FIBER STRUCTURE DEVELOPMENT AND CRYSTALLIZATION BEHAVIORS OF PHBH IN HIGH-SPEED MELT SPINNING PROCESS 89
Q. Qin, W. Takarada, T. Kikutani

PREPARATION AND MAGNETIC PROPERTY OF ELECTROSPRAYED PES/SPES MAGNETIC MICROSPHERES BY IN-SITU REACTION 94
T. Yu, M. Peng, G. Zhang, S. Long, X. Wang, J. Yang

PREPARATION AND CHARACTERISTIC OF INFLAMMATORY RESPONSE TYPE GUIDED TISSUE REGENERATION MEMBRANE 98
J. Ye, R. Shi, L. Zhang

POLYMER BLENDS AND COMPOSITES

MODELING AND EXPERIMENTAL VALIDATION OF ANOMALOUS MOISTURE ABSORPTION OF MICRO AND NANOCOMPOSITE LAMINATES 99
Y.K. Hamidi, G.E. Guloglu, J.P. Anderson, M.C. Altan

BLOWN FILM EXTRUSION AND COMPRESSION MOULDING OF HIGH DENSITY POLYETHYLENE/MULTI-WALLED CARBON NANOTUBE COMPOSITES: STRUCTURE, MECHANICAL AND ELECTRICAL PROPERTIES 104
D. Xiang, E. Harkin-Jones

DEVELOPMENT OF PLA-PBAT AND PLA-PBSA BIO-BLENDS: EFFECTS OF PROCESSING TYPE AND PLA CRYSTALLINITY ON MECHANICAL PROPERTIES 109
H. Oguz, C. Dogan, D. Kara, Z.T. Ozen, D. Ovali, M. Nofar

ENHANCED PERFORMANCE OF POLYARYLENE ETHER NITRILES BASED COMPOSITE PROTON EXCHANGE MEMBRANES WITH GRAPHENE OXIDE 114
M. Feng, Y. Huang, X. Liu

TRIBOLOGICAL PROPERTIES OF CARBON AND ARAMID FIBER HYBRID REINFORCED POLYAMIDE 66 COMPOSITES 119
T. Shinohara, Y. Nishitani, T. Kitano

GEAR STRENGTH OF HEMP FIBER REINFORCED PLANTS-DERIVED POLYAMIDE1010 BIOMASS COMPOSITES 124
Y. Nishitani, J. Mukaida, Y. Yamada, T. Yamanaka, T. Kajiyama, T. Kitano

PREPARATION OF DYNAMICALLY VULCANIZED SEBS/PP TPVS WITH HIGH SOLVENT-RESISTANCE AND GOOD TENSILE PROPERTIES 129
Y. Wu, B. Shentu, Z. Weng

EFFECTS OF COUPLING AGENTS ON THE THERMAL AND MECHANICAL PERFORMANCE OF GLASS FIBER REINFORCED PA6T/66 COMPOSITES 133
X. Luo, Q. Zhang, M. Zhang, G. Zhang, X. Wang, S. Long, J. Yang

ANTIOXIDATIVE MODIFICATION FOR SEMI-AROMATIC POLYAMIDE PA6T/66 138
W. Peng, X. Luo, M. Zhang, S. Long, X. Wang, J. Yang

EFFECTS OF INJECTION MOLDING PROCESS PARAMETERS ON THE MECHANICAL PERFORMANCE OF LONG GLASS FIBER REINFORCED PPS COMPOSITES 143
X. Zhang, G. Zhang, X. Wang, S. Long, J. Yang

PREPARATION AND MECHANICAL PROPERTIES OF GLASS FIBER CLOTH REINFORCED POLY (ARYLENE SULFIDE SUOFONE) LAMINATE VIA SOLUTION IMPREGNATION 147
J. Yang, S. Long, G. Zhang, X. Wang, J. Yang

AN UNIVERSAL METHOD TO FABRICATE POLYMER COMPOSITES WITH HIGH THERMAL CONDUCTIVITY AT LOW FILLER LOADING 151
J. Huang, Y. Zhu, W. Jiang, X. Nie

THE INTUMESCENT FLAME RETARDANT (IFR) CONTENT-GRADIENT DISTRIBUTION IN POLY (BUTYLENE TEREPHTHALATE) 155
N/A

CONTROLLING THE CRYSTALLIZATION AND CROSSLINKING OF PHTHALONITRILE END-CAPPED POLYARYLENE ETHER NITRILE 156
L. Tong, R. Wei, X. Liu

FACILE FABRICATION OF 3D ORDERED LAYER-BY-LAYER GRAPHENE OXIDE/COPPER PHTHALOCYANINE FILMS WITH HIGH DIELECTRIC PROPERTIES 161
Z. Wang, R. Wei, X. Liu

MORPHOLOGY AND STRUCTURE DEVELOPMENT

CRYSTALLIZATION COEXISTING WITH CROSSLINKING IN PHTHALONITRILE END-CAPPED POLYARYLENE ETHER NITRILE / COPPER PHTHALOCYANINE NANOCOMPOSITE 166
R. Wei, L. Tong, Y. You, X. Liu

THE CAPACITIVE PERFORMANCE RESEARCH BASED ON POLYPYRROLE/CARBON NANOTUBES/GRAPHENE TERNARY COMPOSITES 167
X. Zhou, W. Wang, S. Wu

CRYSTALLIZATION OF POLYMERS IN PROCESSING CONDITIONS: AN OVERVIEW 172
J.M. Haudin, S.A.E. Boyer

SUPER-TOUGHENED POLYLACTIDE PREPARED BY BLENDING WITH BIOBASED COPOLYESTERS 173
X. Hu, L. Zhang

THERMAL INSULATING NANOCOMPOSITE FOAMS WITH BIMODAL CELLULAR STRUCTURE 176
P. Gong, S. Zhai, M.P. Tran, P. Buahom, G. Wang, C.B. Park

A COMPARISON BETWEEN SEVERAL COMMERCIAL POLYMER HOLLOW FIBER MEMBRANES FOR GAS SEPARATION 181
X.Y. Chen, S. Kaliaguine, D. Rodrigue

STUDY ON DIFFERENT HIGH-DENSITY POLYETHYLENES DURING STRESS AND PHOTO-OXIDATIVE AGING 186
X. Ren, D. Luo

PROCESS MODELING AND SIMULATION

THE DEVELOPMENT OF THE SYNTHESIS TECHNIQUE 191
N/A

NUMERICAL SIMULATIONS FOR INJECTION MOLDING OF POSTCONSUMER WASTE 192
F. Desplentere, S. Deceur

POLYMER NANOCOMPOSITES (PNCS) FILLED WITH VARIOUS KINDS OF NANOPARTICLES (NPS) 197
N/A

FATIGUE DESIGN METHODOLOGY FOR SHORT FIBRE REINFORCED POLYMERS CONSIDERING FIBRE ORIENTATION AND WELD LINES 200
A. Primetzhofer, A. Mosenbacher, G. Pinter

FRONTIER PROCESSING TECHNOLOGY AND MACHINERY

SPECIAL MATERIALS AND TECHNOLOGIES FOR FUSED FILAMENT FABRICATION 205
M. Spork, J. Gonzalez-Gutierrez, C. Kukla, S. Schuschnigg, C. Holzer

SELECTIVE LASER IRRADIATION TO IMMISCIBLE POLYMER BLEND FOR DISPERSION CONTROL 210
T. Saito, T. Kawaguchi, I. Satoh

ENVIRONMENTALLY FRIENDLY POLYMERS AND POLYMER RECYCLING

HYDROPHOBIC NANOCOATING OF CELLULOSE BY SOLVENTLESS MECHANICAL MILLING 214
Y. Huang, M. Wu, M. Zhao, S. Kuga
A WATER-SOLUBLE BRANCHED POLYMER FOR ENHANCED OIL RECOVERY 220
N. Lai, X. Guo
DEVELOPMENT AND CHARACTERIZATION OF THE POLYMER BASED ON MODIFIED CHITOSAN 225
Y. Liu, N. Lai, Z. Ye
RESEARCH OF FLOODING POLYMER WITH A CERTAIN DEGREE OF SURFACE ACTIVITY 230
N. Lai, S. Li
RADIATION EFFECT OF POLY (BUTYLENE SUCCINATE) 235
Y. Lu, J. Liu, W. Wang, Y. Jiao, X. Zeng
DEGRADATION OF BISPHENOL A POLYCARBONATE ON OUTDOOR EXPOSURE IN WESTERN AREAS OF CHINA 242
W. Gao, M. Yang, S. Han, L. Jiang, G. Li, Y. Dan
Author Index

Modification of Rheological Properties for Polymer Melt

Masayuki Yamaguchi [a,*]

[a] *School of Materials Science, Japan Advanced Institute of Science and Technology*

Abstract. Modification techniques of rheological responses under elongational flow for polymer melts are introduced. One is the addition of a weak gel just beyond sol-gel transition point. When chain segments of a weak gel are miscible with matrix polymer chains, the strain-hardening behavior in elongational viscosity is enhanced greatly by a small amount of the gel. Another method is to add flexible nanofibers. Interdigitation of nanofibers provides marked elastic properties especially under elongational flow, which is attributed to bending force of deformed fibers and frictional force between fibers. Furthermore, novel techniques to enhance the force to stretch a molten polymer at non-isothermal condition, defined as the drawdown force, are also demonstrated using crystalline polyolefins. After passing through a long die, an extrudate has high level of molecular orientation with fewer entanglement couplings. This situation leads to prompt crystallization by enhanced diffusion capability of molecules onto the crystalline surface. Then the crystallization is responsible for a sudden increase in elongational stress. Consequently, a long die provides a high level of molecular orientation. The drawdown force enhancement becomes prominent for a polymer melt with long relaxation time and high crystallization temperature, because of the pronounced chain orientation. Moreover, the addition of a nucleating agent, which enhances the crystallization temperature, provides a high level of the drawdown force and its sensitivity to the die length. Finally, the addition of a small amount of an acrylate polymer with low molecular weight is found to enhance the drawdown force of polypropylene. The deformed droplets of the acrylate polymer act as the rigid fibers near the crystallization temperature of polypropylene. As a result, the excess deformation is generated in a matrix between the fibers, leading to high drawdown force.

Keywords: Elongational Flow; Capillary Extrusion; Drawdown Force; Crystallization.
PACS: 61.25.Hq

INTRODUCTION

Rheological properties under elongational flow are considered to be significantly important for some processing operations, in which free surface deforms during processing, such as blown-film processing, T-die extrusion, thermoforming, blow-molding, and foaming. The relationship between rheological responses and processability has been explained mainly by the transient elongational viscosity [1-3]. According to them, the strain-hardening behavior in elongational viscosity plays an important role in polymer processing. Since most crystalline polymers with linear structure do not show strain-hardening, various efforts have been carried out. Incorporation of long-chain branches is the most well-known technique, although the shear modification behavior should be considered seriously [4-6]. Here, various methods to modify the rheological responses under elongational flow are demonstrated.

CRITICAL GEL

It has been found that blending a small amount of a weak gel near the sol-gel transition point, in which chain segments in the gel are miscible with a matrix polymer, enhances melt elasticity drastically, whereas it has less/no influence on the steady-state shear viscosity [7-9]. Figure 1 exemplifies the growth curves of uniaxial elongational viscosity at various strain rates for blends of linear isotactic polypropylene (PP) and gEHDM that is the gel fraction of a crosslinked terpolymer composed of ethylene, 1-hexene, and ethylidene norbornene with a low density of crosslink points. The precursor of gEHDM, *i.e.*, uncrosslinked terpolymer (EHDM), is

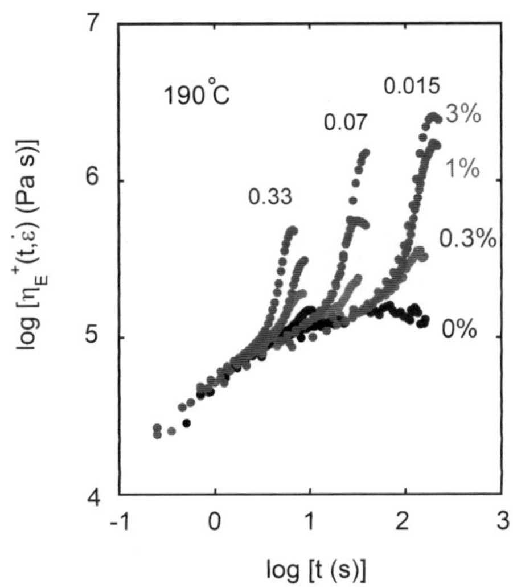

Figure 1 Transient elongational viscosity for PP containing various amounts of a weak gel composed of ethylene, 1-hexene, and ethylidene norbornene.

miscible with PP in a molten state because of the high 1-hexene content, 67.5 mol% [7]. The strain-hardening behavior in the elongational viscosity, which hardly takes place for linear PP, is clearly detected in the curves for the blends, whereas the linear region in the growth curves is the same as that for the pure PP. The steady-state shear viscosity of the blend is, on the other hand, almost similar to that of the pure PP. These results prove that the gEHDM acts as a processing modifier of PP to enhance the melt elasticity remaining the shear viscosity.

FLEXIBLE NANOFIBERS

Surface modified polytetrafluoroethylene PTFE has been commercialized as a processing modifier for PP. Although PTFE is in the solid state at mixing temperature, it is deformed into fibrous shape with a diameter of approximately two hundred nanometers because of the complicated structure growth due to weak cohesive strength in PTFE crystals [10]. Addition of flexible PTFE nanofibers affects the rheological properties of PP to a great extent [11,12]. Figure 2 shows the drawdown force, defined as the force required for stretching an extruded strand from a capillary rheometer, of the blends prepared by a twin-screw extruder. The drawdown force has been known to have a close relation with elongational viscosity [13]. As seen in the figure, the drawdown force increases rapidly with the PTFE content. On the contrary, steady-state shear viscosity does not change greatly. The enhanced drawdown force is found to be attributed to the marked strain-hardening in elongational viscosity [12]. This phenomenon is significantly different from a polymer composite with rigid fibers. As revealed by Mewis and Mezner [14], elongational viscosity is greatly enhanced by an addition of rigid fibers with large aspect ratio. The high level of elongational stress is ascribed to the localized shear deformation between fibers during elongational flow [15,16]. Therefore, strain-hardening behavior is not detected for the blend systems.

The mechanism of the enhanced elongational stress of the present system containing flexible nanofibers should be different from that of a polymer composite with rigid fibers. Since the fibers are flexible, network structure appears in the molten PP owing to interdigitation of fibers. The deformation of the network during flow with keeping the interfibrous interaction provides the elastic response by bending force of each fiber and frictional force between fibers. Similar phenomena were detected observed for other composite systems with flexible nanofibers, such as poly(lactic acid) containing poly(butylene terephthalate) PBT nanofibers [17] and PP containing PBT nanofibers [18,19].

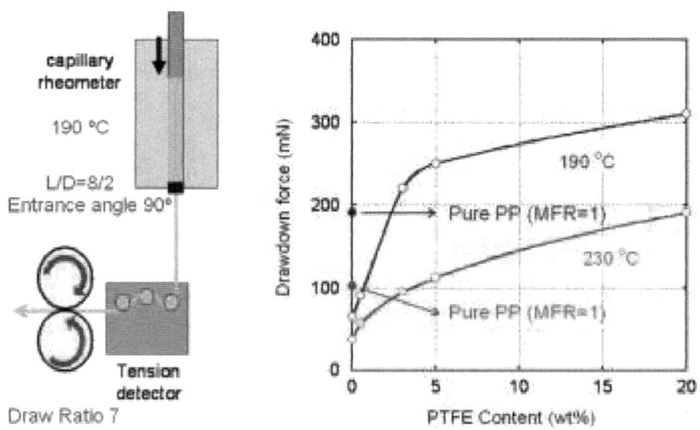

Figure 2 Set-up of drawdown force measurements and the results of PP containing PTFE fibers.

EFFECT OF DIE GEOMETRY

Drawdown force of PP and HDPE evaluated by dies with various lengths is shown in Figure 3 [20,21]. As seen in the figure, the drawdown force increases with the die length. This phenomenon is detected in the wide range of shear rates and draw ratios. Furthermore, our previous study revealed that it is pronounced when a molten material has a long characteristic time. The enhanced drawdown force is detected with the marked molecular orientation to the flow direction [20], which can be characterized by the 2D-WAXD pattern as exemplified in Figure 4.

These results suggest that a melt passing through a long die has a low density of entanglement couplings because of the orientation to the flow direction by the applied shear stress in the die. This situation becomes prominent beyond the shear strain at stress-overshooting, and has an impact also on the swell ratio as discussed by Tanner [22]. The reduced interaction of entanglement couplings leads to the rapid crystallization during stretching, owing to the enhanced chain diffusivity onto the surface of crystalline nuclei. In other words, polymer chains in a molten state do not have to "reptate" in the entanglement network to reach to the crystalline surface. This is a similar phenomenon to the flow-induced crystallization at injection-molding.

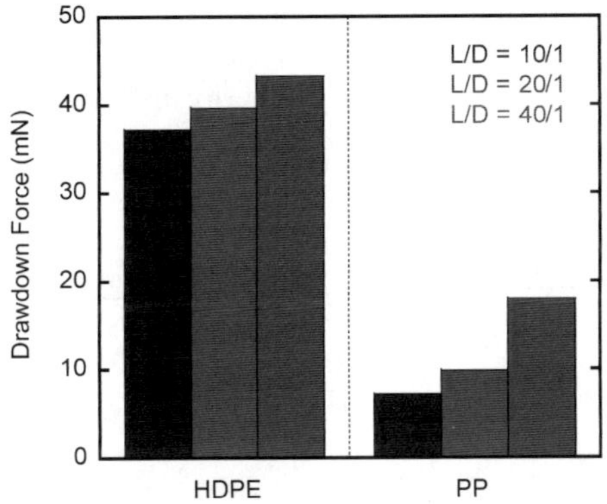

Figure 3 Drawdown force for HDPE and PP evaluated by dies with various lengths.

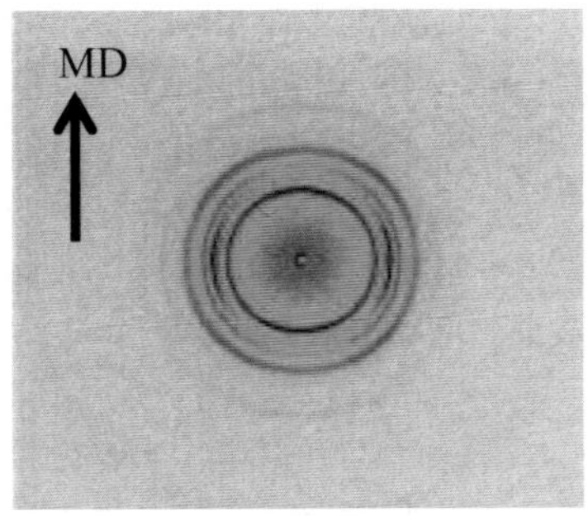

Figure 4 Exemplified 2D diffraction pattern of the PP strand extruded from a die with L/D = 20/1 at 190 °C. The strand was irradiated by X-ray beam.

Since the prompt crystallization enhances the drawdown force, the nucleating agent, which enhances the crystallization temperature effectively [23], also affects the drawdown force. In fact, the drawdown force and its die-length-sensitivity are pronounced by the addition of the nucleating agent as shown in Figure 5 [20]. Consequently, the drawdown force of the sample containing the nucleating agent is almost twice as high as that of the pure PP when using the long die. A similar phenomenon is detected also for linear low-density polyethylene (LLDPE) containing a small amount of HDPE [21]. In this system, HDPE acts as a nucleating agent for LLDPE and enhances the crystallization temperature and the drawdown force.

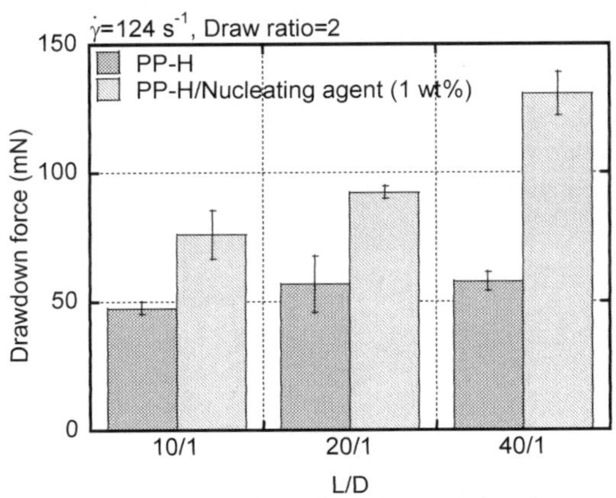

Figure 5. Drawdown force for PP containing the nucleating agent evaluated by dies with various lengths. The data should be compared with Figure 3

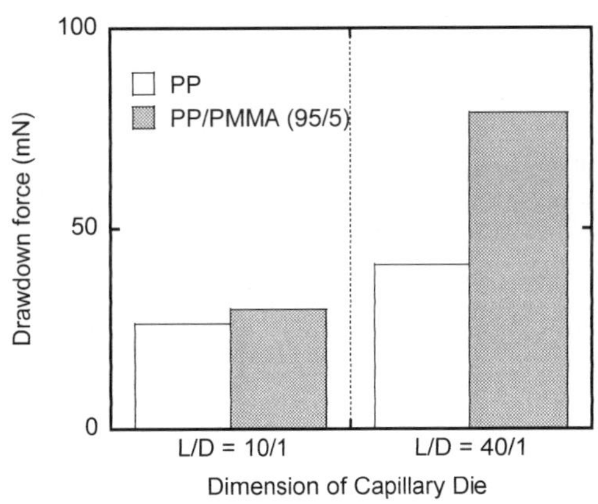

Figure 6. Drawdown force for PP containing PMMA evaluated by dies with various lengths.
The data should be compared with Figure 3

Finally, the addition of a small amount of PMMA with low viscosity also enhances the drawdown force of PP, whereas the shear viscosity and oscillatory shear moduli are almost unchanged by the PMMA addition, except for the shoulder peak in the storage modulus in the low frequency region, which is ascribed to the interfacial tension. As well known, PMMA is immiscible with PP and thus deformed to the flow direction in the die, which is enhanced in a long die. After passing through the die, the extruded strand is cooled down quickly. Then the deformed PMMA droplets behave like rigid fibers because the glass transition temperature is higher than the crystallization temperature for PP at the quench condition. In other words, solidification of PMMA occurs prior to the PP crystallization. This situation is confirmed by DSC measurements performed at a fast cooling rate. Of course, the sensitivity to the die length is more pronounced for the blends as shown in Figure 6.

After the solidification of PMMA, the system behaves as a composite containing the rigid fibers. Consequently, the elongational stress is greatly enhanced by the excess deformation of a matrix between the neighbor fibers. The enhancement of elongational stress for a polymer melt with rigid fibers was revealed from both theoretical [16] and experimental approaches [14,15,19].

This technique would be applicable for actual processing. Because cooling is performed at a significantly fast rate in most conventional processing operations, the technique will be applicable in industries.

REFERENCES

1. J. Meissner, T. Raible, S. E. Stephenson, *J. Rheology,* **25**, 1 (1981).
2. M. Yamaguchi, M. Takahashi, *Polymer*, **42**, 8663 (2001).
3. K. Koyama, *J. Soc. Rheol. Jpn,* **34**, 267 (2006).
4. M. Yamaguchi, C. G. Gogos, *Adv. Polym. Technol.*, **20**, 261 (2001).
5. M. Yamaguchi, M. H. Wagner, *Polymer*, **47**, 3629 (2006).
6. M. Siriprumpoonthum, S. Nobukawa, Y. Satoh, H. Sasaki, M. Yamaguchi, *J. Rheology*, **58**, 449 (2014).
7. M. Yamaguchi, H. Miyata, *Polym. J.*, **32,** 164 (2000).
8. M. Yamaguchi, *J. Polym. Sci. Polym. Phys. Ed.*, **39**, 228 (2001).
9. K. Arakawa, T. Yokohara, M. Yamaguchi, *J. Appl. Polym. Sci.*, **107**, 1320 (2007).
10. B. M. A, Mohd Amran, S. Nobukawa, M. Yamaguchi, *Pure Appl. Chem.*, 83, 1817 (2011).
11. T. Kurose, T. Takahashi, K. Koyama, *J. Soc. Rheol. Jpn,* **33**, 173 (2005).
12. B. M. A, Mohd Amran, K. Okamoto, M. Yamaguchi, A. Koshirai, T. Kasai, *J. Polym. Sci. Polym. Phys. Ed.*, **47**, 2008 (2009).
13. M. H. Wagner, A. Bernnat, V. Schulze, *J. Rheology*, **42**, 917 (1998).
14. J. Mewis, A. B. Metzner, *J. Fluid Mech.*, **62**, 593 (1974).
15. M. H. Laun, *Colloid Polym. Sci.*, **262**, 257 (1984).
16. G. K. Batchelor, *J. Fluid. Mech.,* **46**, 813 (1971).
17. T. Yokohara, S. Nobukawa, M. Yamaguchi, *J. Rheology*, **55**, 1205 (2011).
18. M. Yamaguchi, K. Fukuda, T. Yokohara, B. M. A. Mohd Amran, S. Nobukawa, *Macromol. Mater. Eng.*, **297**, 654 (2012).
19. M. Yamaguchi, T. Yokohara, B. M. A. Mohd Amran, *J. Soc. Rheol. Jpn*, **41**, 129 (2013).
20. J. Seemork, M. Siriprumpoonthum, Y. Lee, S. Nobukawa, M. Yamaguchi, *Adv. Polym. Technol.*, **34**, 21477 (2015).
21. J. Seemork, T. Itoh, S. Nobukawa, M. Yamaguchi, *J. Soc. Rheol. Jpn*, **44**, 23 (2016).
22. R. I. Tanner, Engineering Rheology, Oxford Science Publications, New York, 1985.
23. M. Tenma, N. Mieda, S. Takamatsu, M. Yamaguchi, *J. Polym. Sci. Polym. Phys. Ed.*, **46**, 41 (2008).

Rheological Behavior of Hyperbranched Copolymer Based Nano-SiO_2 Functional Monomer

Nanjun Lai*, Lei Tang and Zhongbin Ye

School of Chemistry and Chemical Engineering of Southwest Petroleum University, Chengdu, P.R. China 610500

Abstract. The rheological properties of hyperbranched copolymer based nano-SiO_2 functional monomer having a wide range of branched chain numbers and copolymer molecular weights are reported. In contrast to linear polymers, hyperbranched copolymers have a broad relaxation spectrum and a viscosity that reduces less with copolymer molecular weights. A comparison of five pairs of samples having the surface-modified degree of 20%, 40%, 60%, 80%, and 100% nano-SiO_2 functional monomer hyperbranched copolymer molecular weights showed that the molecular weights of the hyperbranched copolymer is approximately $8.20*10^6$ g/mol. For higher degrees of functionality, the molecular weights of copolymers gradually reduce. The impacts of degree of surface modification of nano-SiO_2 on the rheological properties under shearing of the copolymer were analyzed. Shear thinning tests shows, while not cut, with the increasing of the degree of surface modification of nano-SiO_2, the viscosity retention rate is higher, but, the viscosity is lower. Dynamic viscoelasticity test shows, before cutting, Polymer-40 exhibit elasticity at a lower frequency, Polymer-60, Polymer-80, Polymer-100 exhibit elastic descending order with the scanning frequency increase; after shearing, Polymer-20 still shows mainly viscous, the higher the degree of modification of nano-SiO_2, the copolymer viscoelastic shearing stability is stronger. Deformation resilience tests shows, before shearing, along with the increase of the degree of nano-SiO2 surface modification, the recovery rate of copolymer increases and then decreases; shearing effect have greater impact on Polymer-40 and Polymer-20. The first normal stress difference (N1) test shows, before shearing, Polymer-40 has the largest N1; Polymer-60 shows the biggest after shearing.

Keywords: Nano-SiO_2; Hyperbranched Copolymer; Rheological Behavior; Surface-modified Degree.
PACS: 82.35.Np, 83.80.Rs, 83.85.Lq, 83.85.St

INTRODUCTION

Polymers have traditionally been classified according to their chain architecture into three categories: linear, cross-linked, and branched molecules[1]. Recent developments in the synthesis of highly branched systems have led to a more general classification scheme that is called hyperbranched polymer. In 1952, Flory[2] first discussed the possibility of the preparation of the branched polymers via the condensation polymerization of monomer ABx in theory. However, there was not enough attention to the non-crystalline and non-chainwinding of the polymer. Until 1987, Kim[3] applied a hyperbranched polymer patent, and announced the result at Los Angeles American Chemical Society in 1988[4], people began to focusing the hyperbranched polymers. Thereafter, the synthesis of a new type of hyperbranched polymer and their application properties to become active. Kim[5] reported that the 7 arm star polyacrylamide has better shear stability than the ordinary linear polymer in 1974, and the conformation of the star is relatively shrinking. Xue[6] reported that 6 arm star PMMA had better shear stability in 2005.

Because of the close relationship between rheology and processing properties, characterization of the rheological flow behavior of hyperbranched polymers is important[7-9]. Such studies can provide insights into the intermolecular interactions and the effects that molecular variables such as surface-modified degree, nature of peripheral functional groups, and degree of branching have on bulk properties. Despite this importance, relatively few studies, particularly studies showing comparisons of the rheological behavior of hyperbranched systems, have been reported. While some advances have been made in the study of bulk properties of linear polymers, work with hyperbranched polymers is lagging[10-12]. Therefore, the goal of the present study is to investigate the effects that molecular variables such as degree of branching, molecular weight, nature of peripheral groups, and molecular architecture have on the bulk properties of hyperbranched materials. In this paper we present and compare the rheological behavior of the surface-modified degree of 20%, 40%, 60%, 80%, and 100% nano-SiO_2 functional monomer hyperbranched copolymer.

EXPERIMENTAL SECTION

Materials

Nano-SiO_2, 3-aminopropyl triethoxy silane (KH550), Grignard reagent (CH_3MgCl) (AR, Shanghai Aladdin reagent), maleic anhydride, toluene, methanol, cyclohexane, ether, hydrochloric acid, ammonium persulfate, sodium hydrogen sulfite, thymol blue indicator, N, N-dimethyl formamide (DMF), absolute ethyl alcohol, acrylic acid (AA), acrylamide (AM) (AR, Chengdu KeLong reagent). All supplementary materials (Kelong) were of analytical grade and used as received. Water was double deionized with a Millipore Milli-Q system to produce the 18 MM deionized water.

Experimental

The surface-modified degree of 20%, 40%, 60%, 80%, and 100% nano-SiO_2 functional monomer hyperbranched copolymer was prepared in consistent with Qin' s[10] well known divergent approach by gradually repeating free radical polymerization. Simple shear was tested by utilizing HAAKE MARS RS600 rheometer (Haake Technik Co., Germany) at 70 ℃, constant extended temperature modules using DG41Ti measurement system. Oscillatory, deformation resilience and the first normal stress difference were tested on a Anton Paar MCR 301 rheometer, constant stress rheometer using CP75-1-SN6126 test system. Oscillatory experiments were performed in the linear viscoelastic region in the frequency range 0.1-10 Hz, and deformation resilience experiments were completed the recovery of deformation over several 5s reached 400% deformation. Temperature control was achieved with a extended temperature modules to a precision of 0.1 ° C. In addition to viscosity, the normal stress response in simple shear flow was also measured.

RESULTS AND DISCUSSION

Weight-average Molecular Weight

The basic structural parameters of the synthesized polymers are listed in Table 1. Select unperturbed dimension to characterize the flexibility of molecular chain of synthesized hyperbranched copolymer. We can obtain the unperturbed dimension of copolymer by equation 1 and equation 2 according to the cyclotron radius and molecular weight, as shown in Table 1. The unperturbed dimension of polymer A is the unperturbed mean square end-to-end distance of per unit molecular weight. The smaller the A value is, the more flexible the molecular chain is. As can be seen from Table 1, among the five kinds of polymers, Polymer-100 has the best flexibility, followed by Polymer-80, Polymer-60, Polymer-40 and Polymer-20 have the poorest flexibility. Analysis shows that Polymer-100 has the highest modification degree, and carbon-silicon single bonds and carbon-oxygen single bonds occupy a larger proportion in molecular chains, molecular chains of Polymer-100 have the best flexibility. The better the flexibility of molecular chains is, the more the possible conformations are under the condition of the same obstruction, and the less the destroy on molecular chains is.

Unperturbed dimension of polymer A:

$$A=\sqrt{\langle h^2\rangle_0 / M} \quad (1)$$

$$\langle R_g^{\,2}\rangle=\frac{1}{6}\langle h^2\rangle_0 \quad (2)$$

Where, A - Unperturbed dimension of polymer, nm(mol/g)$^{1/2}$; $\langle h^2\rangle_0$ -Mean square end-to-end distance, nm^2; M-Molecular weight of polymer, g/mol; $\langle R_g^{\,2}\rangle$ - Mean square cyclotron radius, nm^2; $\langle h^2\rangle_0$ -Mean square end-to-end distance, nm^2.

TABLE 1 Cyclotron Radius and Absolute Weight-average Molecular Weight of Copolymer

Polymer	Cyclotron radius (nm)	M_w (10^6 g/mol)	Unperturbed dimension (nm • (mol/g) 1/2)
Polymer-20	194	9.98	0.15
Polymer-40	124	8.35	0.11
Polymer-60	102	7.89	0.089
Polymer-80	91	7.01	0.084
Polymer-100	72	6.28	0.070

Shear Thinning

Hyperbranched copolymer of shear thinning property with a similar trend, before shearing, relatively higher viscosity of Polymer-20 and Polymer-40, nano SiO_2 surface change degree higher, polymer shear dilution curve down more. After shearing, the polymer shear dilution curve appeared new changes, Polymer-20 curve in the bottom, show that shear action on the most severely damaged, Polymer-40, Polymer-60 and Polymer-80 are similar and analysis that the polymer branched degree of shear and the common influence the results.

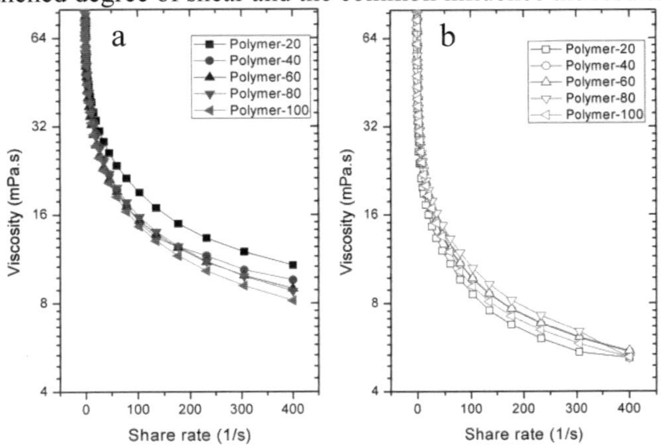

FIGURE 1. Sheer thinning of (a) hyperbranched coplymer (unsheared), (b) hyperbranched coplymer (sheared) in deionized water (polymer concentration, 2000mg/L).

Dynamic Viscoelasticity

Before shearing, when Polymer-20 scanning frequency is 0.01 ~ 2.17 Hz, Polymer-40 at 0.01 ~ 0.16 Hz, Polymer-60 at 0.01 ~ 1.01 Hz, Polymer-80 at 0.01 ~ 1.58 Hz and Polymer-100 at 0.01 ~ 1.59 Hz, The G" is greater than the G', which shows that the polymer solutions are mainly viscous in this frequency, and when the frequency is higher, the elasticity is greater than the viscosity in the test range. After shearing, when Polymer-20 scanning frequency is 0.01 ~ 6.61 Hz, Polymer-40 at 0.01 ~ 1.58 Hz, Polymer-60 at 0.01 ~ 1.59 Hz, Polymer-80 at 0.01 ~ 1.62 Hz and Polymer-100 at 0.01 ~ 3.98 Hz, The G" is greater than the G', which shows that the polymer solutions are mainly viscous in this frequency, and when the frequency is higher, the elasticity is greater than the viscosity in the test range.

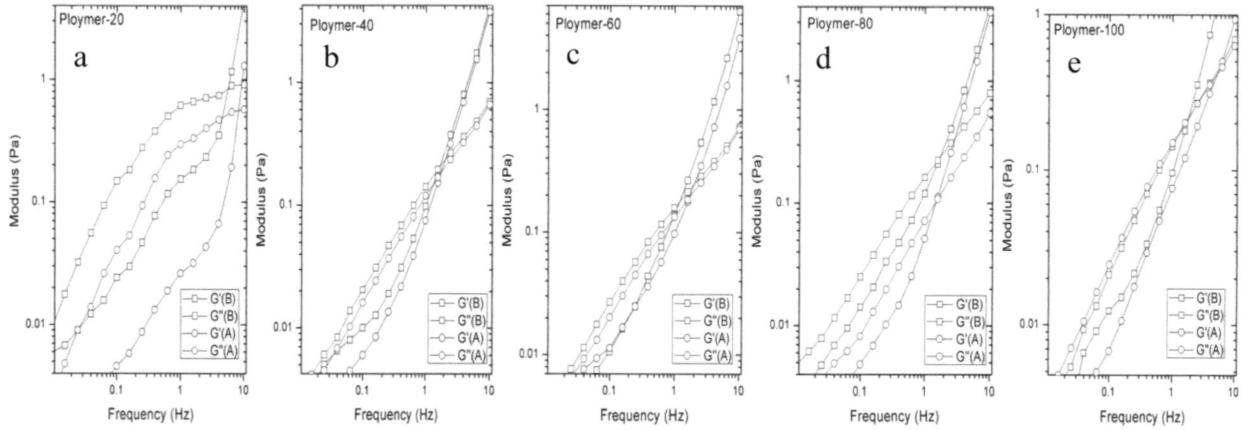

FIGURE 2. Dynamic Viscoelasticity of (a) Polymer-20 (unsheared and sheared), (b) Polymer-40 (unsheared and sheared), (c) Polymer-60 (unsheared and sheared), (d) Polymer-80(unsheared and sheared), (e) Polymer-100 (unsheared and sheared) in deionized water (polymer concentration, 2000mg/L).

Deformation Resilience

Before shearing, hyperbranched copolymers have good deformation ability to recover, The recovery rate of Polymer-20~Polymer-100 deformation follow 63.4%, 76.6%, 74.3%, 72.4% and 69.8%. To some extent, the deformation recovery ability of the polymer shows the viscoelasticity of the polymer. Hyperbranched copolymers in receives the shear action, the deformation recovery ability decreased, the deformation recovery rate follow 44.5%, 57.4%, 59.7% , 56.5% and 54.4%. Before shearing, the deformation recovery rate of Polymer-20 was the lowest. In general, with the increase of the degree of nano-SiO_2 surface modified, the deformation recovery rate of the hyperbranched copolymer increases firstly and then decreases. After shearing, the effect of shear stress on the Polymer-20 and Polymer-40 with a lower degree of modification is larger. When the degree of modification is up to 60%, the effect of shear stress on the deformation recovery rate is relatively stable. The increase of surface modification of nano-SiO_2 is beneficial to offset the negative effect caused by the shear effect, and the ability of the polymer's deformation recovery is the result of the degree of branching of the polymer and the number of branches.

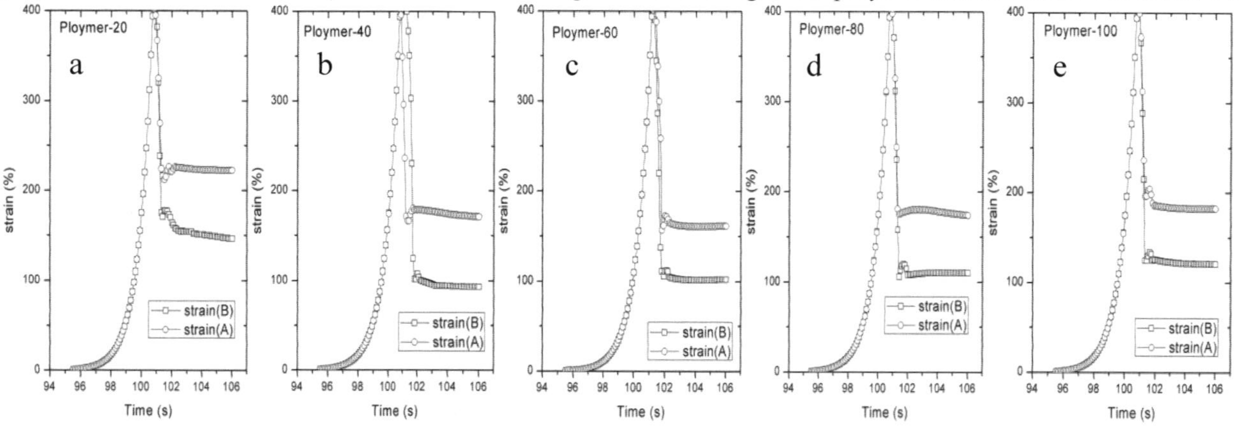

FIGURE 3. Deformation Resilience of (a) Polymer-20 (unsheared and sheared), (b) Polymer-40 (unsheared and sheared), (c) Polymer-60 (unsheared and sheared), (d) Polymer-80(unsheared and sheared), (e) Polymer-100 (unsheared and sheared) in deionized water (polymer concentration, 2000mg/L).

First Normal Stress Difference

At high shear rate, with the increase of shear rate, hyperbranched copolymers first normal stress difference increase gradually. The first normal stress difference of the hyperbranched copolymers is greatly reduced due to the shearing action. Before and after shearing, the first normal stress difference of Polymer-60 is the least. The effect of shear stress on the first normal stress difference of Polymer-60 was small.

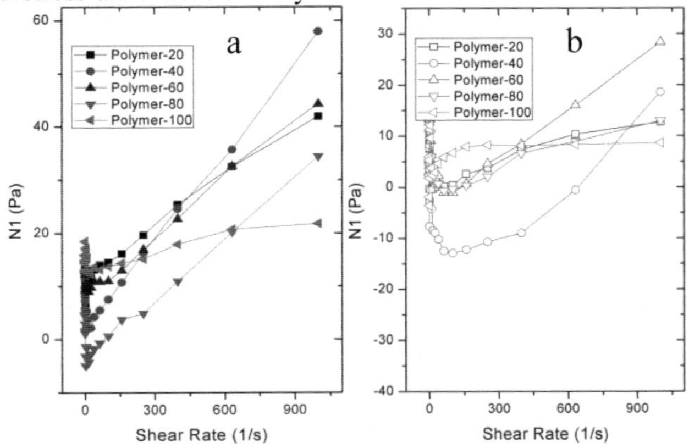

FIGURE 4. first normal stress difference of (a) hyperbranched coplymer (unsheared), (b) hyperbranched coplymer (sheared) in deionized water (polymer concentration, 2000mg/L).

CONCLUSIONS

The results of our study show the rheological behavior of hyperbranched copolymers based different degree of modification nano-SiO_2 ranging from Newtonian to viscoelastic. As expected, this behavior is affected by molecular variables, such as Polymer-20 and Polymer-100; however, the molecular architecture appears to be the most important factor determining the nature of the intermolecular interactions and consequently the rheological behavior. Five types of hyperbranched architecture have been compared in this study. The similarity of the rheological behavior of hyperbranched copolymers based different degree of modification nano-SiO_2. The hyperbranched copolymers with low degree of modification nano-SiO_2, on the other hand, are characterized by a more open structure that apparently promotes a significant degree of coil overlap and entanglement coupling. In consequence, the higher molecular weight samples exhibited viscoelastic effects that correlate with the decrease in the degree of branching. Overall, with the increase of the degree of modification nano-SiO_2, the rheological property of the hyperbranched copolymer showed the trend of increasing first and then decreasing.

ACKNOWLEDGMENTS

This work was financial supported by National Natural Science Foundation of China(51404208), The Second Session of the "Young and Middle-aged Backbone Teachers" Training Program of Southwest Petroleum University, Scientific Research Project of Southwest Petroleum University, National College Students Innovation and Entrepreneurship Training Project(201510615008), China Scholarship Council.

REFERENCES

1. Tomalia, D. A. and Esfand, R. *Chem. Ind.* **11**, 416-420 (1997).
2. Flory, P. J. *J. Am. Chem. Soc.* **74**, 2718-2723 (1952).
3. Kim, Y. H. U.S. Patent No. 4,857,630 (15 August 1989).
4. Kim, Y. H. and Webster, O. W. *Macromolecules* **25**, 5561-5572 (1992).
5. Kim, O. K. and Little, R. C. *Nature* **250**, 408-410 (1974).
6. Xue, L. and Agarwal, U. S. *Macromolecules* **38**, 8825-8832 (2005).
7. Ibrahim, S. and Anthony J. M. *Macromolecules* **33**, 590-596 (2000).
8. Lidija, S. P. and Miroslav, H. *Eur. Polym. J.* **76**, 188-195 (2016).
9. Jiang, W. Q. and Ye, F. *J. Colloid Interface Sci.* **413**, 8-16 (2014).
10. Hawker, C. J.and Farrington, P. J. *J. Am. Chem. Soc.* **15**, 4409-4410 (1995).
11. Farrington, P. J. and Hawker, C. J. *Macromolecules*, **31**, 5142-5144 (1998).
12. Uppuluri, S. and Keinath, S. E. *Macromolecules*, 31, 4498-4501 (1998).
13. Qin X. P., "The Feasible Research of Improving Sweep Efficiency Using the Modified Nano-SiO_2/AA/AM Copolymer", D. E.Thesis, Southwest Petroleum University, 2014.

Interaction between Sodium Dodecyl Benzene Sulfonate and Hyperbranched Polymer Based on Dendritic Structure

Nanjun Lai [a,*], Yan Zhang[a] and Zhongbin Ye[b]
[a] *College of Chemistry and Chemical Engineering, Southwest Petroleum University, Chengdu 610500, China*
[b] *State Key Laboratory of Oil and Gas Reservoir Geology and Exploitation, Southwest Petroleum University, Chengdu 610500, China*

Abstract. A hyperbranched polymer based on dendritic structure was compounded with sodium dodecyl benzene sulfonate (SDBS) to produce SP binary system. Meanwhile, the interaction between components was studied according to the changes of solution diameter, resonance light scattering (RLS), morphology and conductivity. And the effect of interaction on the apparent viscosity and interfacial tension (IFT) of binary system was discussed. It was found that the interaction can be divided into different characteristic ranges by SDBS concentration and the whole process was along with the increase of the conductivity. At low SDBS concentration, the electrostatic repulsion between the tail end of hyperbranched polymer branched chains and the head groups of SDBS was dominated that reduced the diameter and apparent viscosity of SP solution. With the SDBS concentration increasing, a slow reduction of diameter and viscosity together the enhancement of RLS and IFT were observed, which could be explained by the combined action of electrostatic repulsion and the micellar behavior of SDBS. The results of this letter could provide theoretical guidance for the use of such SP binary system on the basis of different actual demand.

Keywords: Composite materials; Hyperbranched polymer; Interaction; Resonance light scattering; Microstructure.
PACS: 82.70.Uv; 61.25.he;

INTRODUCTION

Considering the excellent sweep efficiency of polymer flooding and the great displacement efficiency of surfactant flooding, the polymer/surfactant (SP) combinational flooding, which promises both the viscoelasticity and lower interfacial tension, are widely used in improving oil recovery.[1] For SP systems, the interaction between components often make the conformation of polymer chains change and also affect the chemical and physical properties of surfactants, thereby generating a great influence on the system nature and the ultimate displacement effect.[2] Therefore, an intensive effort has been made to characterize the character of these interactions as well as their impact on rheological and interfacial properties.[3, 4]

Among these composite systems, linear polyelectrolyte/surfactant systems are lavishly recorded, whereas poor shear resistance limits the popular application of such widely used linear polymers and hyperbranched polymers have attracted considerable attention in oilfield.[5] To heightening the anti-shearing of polymer, our research team early synthesized a hyperbranched polymer based on dendritic structure for EOR which performed remarkable solution properties and different network structure in comparison to linear polymer.[6] It is well known that the molecular architecture plays a curial role in the SP compound system due to its impact on the interaction with surfactant.[7, 8]

Hence, in this study, the hyperbranched polymer based on dendritic structure was mixed with sodium dodecyl benzene sulfonate (SDBS) widely used in oilfield to produce SP system. The DLS, RLS, ESEM and conductivity detector were employed to gain an insight into the interaction between them. And the effect of interaction on the apparent viscosity and interfacial tension of binary system was discussed.

EXPERIMENT AND METHODS

The hyperbranched polymer used in this paper, with a mean molecular weight of 1.72×10^7 g/mol, was synthesized according the method reported elsewhere by our team.[6] The sodium dodecyl benzene sulfonate (SDBS) purchased from Chengdu KeLong Chemical Reagent Co., Ltd (China) was used after purification. Their structures were shown in FIGURE 1. The polymer and surfactant was mixed and placed for 20-40 min to form stable SP system.

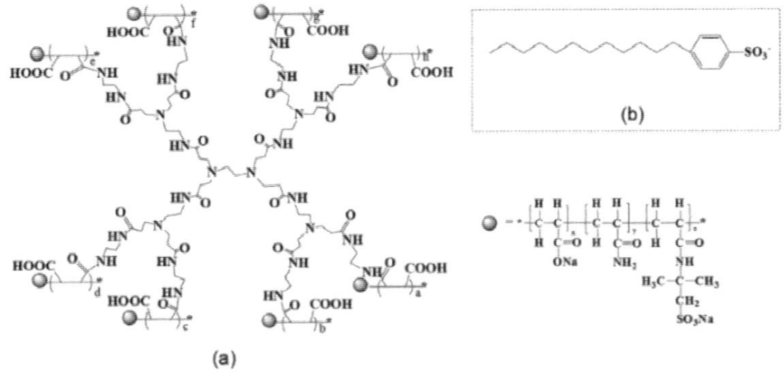

FIGURE 1. Structure of hyperbranched polymer (a) and SDBS (b).

The hydrodynamic radius (R_h) and distribution of each sample were obtained by using a BI-200SM wide angle dynamic laser light scattering (DLS) instrument. The scanning angle was 90° and the scattering wavelength was 532 nm. The static resonance light scattering (RLS) spectra were measured by scanning synchronously excitation and emission wavelengths from 200 to 800 nm on the PerkinElmer LS55 fluorescence spectrophotometer. All samples were diluted about 300 times before RLS test. The morphologies of samples were observed by Quanta 450 Environment Scanning Electron Microscope (ESEM, FEI Company, USA). The conductivity of the SP systems were tested using a DDS-11A conductivity meter (Shanghai Rex Xinjing instruments Co. Ltd, China). Finally, the Brookfield DV-III Programmable Rheometer (Brookfield Co., America) was employed to achieve the apparent viscosity of solutions at 7.34 s^{-1} and 65 °C. The interfacial tension (IFT) of systems was investigated utilizing hanging drop method by KRUSS DSA30S interfacial tensiometer (KRUSS Co., Germany). The polymer concentration was 1500mg/L throughout the whole experiments.

RESULTS

The diameter distribution of polymer, SDBS and SP solutions were presented in FIGURE 2(a). There was a single radius distribution for 300mg/L SDBS solution, and its particle size was mainly distributed at 295nm. However, the hyperbranched polymer of 1500mg/L showed a wide distribution which conformed to the characteristics of macromolecular solution. Moreover, an obvious left movement of main distribution area could be observed for SP systems in comparison to the pure polymer and SDBS solution.

The aggregation of SP system was characterized by RLS spectra as shown in FIGURE 2(b). The spectra of SDBS and hyperbranched polymer had similar shape and trend. When the 1500mg/L polymer was mixed with 300mg/L SDBS, the resulting solution cannot performed an increase scattering phenomenon. Nevertheless, a sharp improvement of RLS was obtained for SP system at the SDBS concentration of 500mg/L, on which occasion, SDBS existed in the form of micelle.

The variation of the R_h and RLS value were curved dependence on SDBS concentration (FIGURE 2(c)). With the increase of SDBS concentration, the Rh of binary system dropped at a great rate firstly and then reduced slowly. It was noted that the maximum RLS value first increased mildly, then gradually improved again after suffering a brief descent, finally decreased radically with the SDBS concentration increasing. The greater value of 6.4 was found at 500mg/L SDBS concentration.

The morphology of the hyperbranched polymer was much regular and smooth due to its hyperbranched structure (FIGURE 2(d)). From FIGURE 2(e), the thicker framework could be observed for SP system. Most significantly, the polymer skeleton and SDBS solution embedded with each other. When the SDBS concentration was 500mg/L, a clear cavity structure was discovered for SDBS solution in SP system, and it only spread on the surface of polymer's network structure (FIGURE 2(f)).

The effect of SDBS concentration on the conductivity and apparent viscosity of SP compound were exhibited in FIGURE 3(a). One observation could be found easily that the system's conductivity associated with its charge almost increased linearly with the increase of SDBS concentration. Conversely, the viscosity at 65 °C appeared to a drastic decline at earlier stage, following by a moderate change after the SDBS concentration approached 300mg/L, which was accordance with the Rh results.

The interfacial tension (IFT) of pure SDBS and SP solution based on concentration were displayed in FIGURE 3(b). Expectedly, the IFT of single SDBS first decreased and then kept stable with the concentration improving, and the CMC was 500mg/L. After mixing with the 1500mg/L hyperbranched polymer, the IFT of binary system suffered a terrible enhancement firstly and then a temperate descent could be discovered until the SDBS concentration was 100mg/L, but which was always lower than that of single one before CMC.

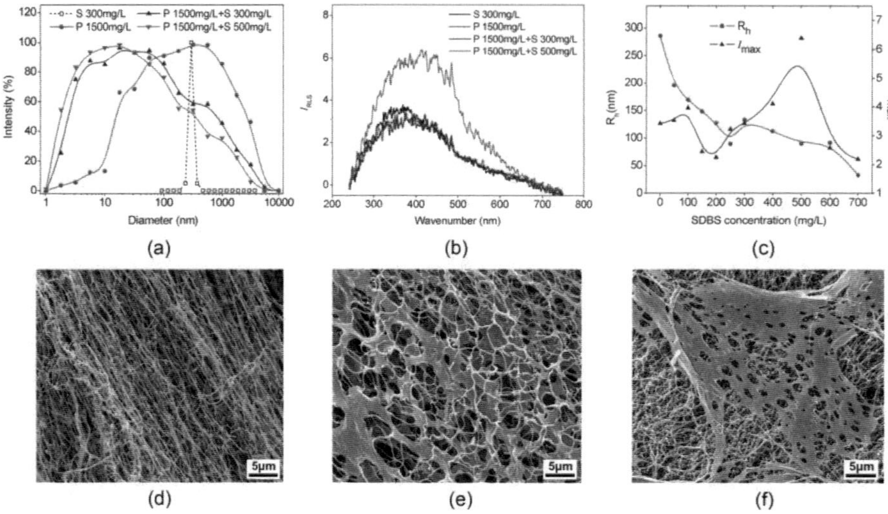

FIGURE 2. Diameter distribution (a), RLS spectra (b), Effect of SDBS concentration on Rh and Imax (c) ESEM morphologies of 1500mg/L pure polymer solution (d), 1500mg/L polymer plus 300mg/L SDBS system (e) and 1500mg/L polymer plus 500mg/L SDBS system

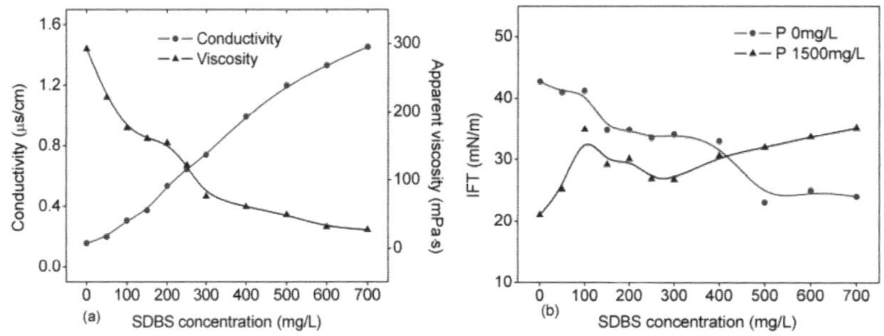

FIGURE 3. Conductivity and apparent viscosity (a) and interfacial tension (b) of SP solution.

DISCUSSION

All the experiment results indicated that the interaction of the hyperbranched polymer based on dendritic structure with SDBS can be divided into different characteristic ranges by SDBS concentration. For single polymer, dendritic structure endows its abundant branched chains which can enwind with each other to form spherical structure as shown in Range A of FIGURE 4. And the outside of structure is full of negative charge provided by the hydrolyzation of AM, AA, AMPS units. Due to the different degree of entanglement, the polymer solution generates various aggregates, finally resulting a wide radius distribution and a certain intense RLS. At a very low SDBS concentration, the hydrogen bonding interaction makes the SDBS molecular adsorb on the flexible branched chains of polymer, which causes a slight enhancement of RLS and IFT for SP system. With the SDBS concentration increasing, increasingly powerful electrostatic repulsion between the tail end of hyperbranched polymer branched chains and the head groups of SDBS causes the spherical structure to shrink (Range B), resulting the Rh, RLS and viscosity of system to reduce and the framework of network structure to get thick. When the concentration is more than 500mg/L, which on that occasion SDBS existed in the form of micelle, the electrostatic repulsion can still diminish the size of system's aggregation, but the micellar behavior of SDBS would play a role on the contrary (Range C). A combination of both finally lead that the reduction of Rh and viscosity becomes slow. At this moment, the SDBS micelles distributed in the surface of polymer spherical structure instead of the structure interior due to its large size. Meanwhile, the micelles would attract the single molecular adsorbed on the polymer chains and interface to solution, which results in a sharp drop of RLS and soft rebound of IFT.

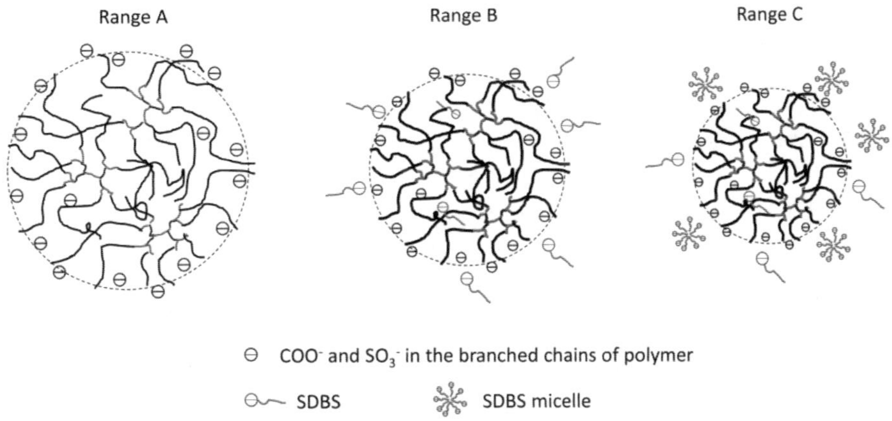

FIGURE 4. Scheme of the interaction of hyperbranched polymer based on dendritic structure with SDBS.

CONCLUSIONS

A novel polymer/surfactant binary system was formed by mixing the hyperbranched polymer based on dendritic structure and SDBS. The DLS, RLS, ESEM, conductivity, apparent viscosity and interfacial tension measurements confirmed that the interaction between components was the electrostatic repulsion between the tail end of hyperbranched polymer branched chains and the head groups of SDBS at low SDBS concentration. When the SDBS was greater than the CMC, a combination of both the electrostatic repulsion and micellar behavior of SDBS could be observed. To broaden scope of application, a further study of this hyperbranched polymer with other types of surfactant is progress.

ACKNOWLEDGMENTS

This research was supported by the National Natural Science Foundation of China (51404208), China Scholarship Council, Second Youth Backbone Teachers Project and Scientific research project of Southwest Petroleum University (2014QHZ014), Open Fund (PLN1433) of the State Key Laboratory of Oil and Gas Reservoir Geology and Exploitation (Southwest Petroleum University).

REFERENCES

1. K. Rai, R. T. Johns, M. Delshad, L. W. Lake and A. Goudarzi, *J. Petrol. Sci. Eng.* **112**, 341-350 (2013).
M. P. Brown and K. Austin, *The New Physique*, Publisher City: Publisher Name, 2005, pp. 25-30.
2. S. L. Yuan, G. Y. Xua, Y. X. Luan and C. B. Liu, *Colloid Surface A.* **256**, 43-50(2005).
3. R. Mészáros, L. Thompson, M. Bos, I. Varga and T. Gilányi, *Langmuir* **19**, 609-615 (2003).
4. C. Flood, C. A. Dreiss, V. Croce and T. Cosgrove, *Langmuir* **21**, 7646-7652 (2005).
5. W. Kuang and Y. X. Xia, *Mater. Lett.* **115,** 109-112 (2014).
6. N. J. Lai, Y. Zhang, Q. Xu, N. Zhou, H. J. Wang and Z. B. Ye, *RSC Adv.* **6**, 32586-32597 (2016).
7. W. T. Wu, L. Shi, Q. R. Zhu, Y. S. Wang and W. M. Pang, *Mater. Lett.* **62**, 2762-2765 (2008).
8. A. M. Percebom and W. Loh, *J. Colloid Interf. Sci.* **466**, 377-387 (2016).

Analysis of the viscoelastic properties of carbon nanotube reinforced polypropylene composites

Jun Huang[a,b,*], Denis Rodrigue[c], Linzhu Sun[b]

[a]*College of Civil Engineering and Architecture, Guangxi University of Science and Technology, Liuzhou City, 545006, China*

[b]*College of Civil Engineering and Architecture, Wenzhou University, Wenzhou City, 325035, China*

[c]*Department of Chemical Engineering, Laval University, Quebec City, G1V 0A6, Canada*

Abstract: Finite element method (FEM) is used to analyze the mechanical properties of carbon nanotubes (CNT) reinforced polypropylene (PP) composites. Firstly, polypropylene is assumed as a viscoelastic material and carbon nanotubes are assumed as linear elastic materials to study the effect of carbon nanotubes on the viscoelastic properties of polypropylene based composites and to compare between the mechanical behavior of nano-composites based on carbon nanotubes and the neat matrix. Secondly, the relaxation time of polypropylene was studied as a function of temperature to determine also the viscoelastic properties of the nano-composites. Thirdly, the effect of carbon nanotube volume fraction was studied by comparison of the stress distributions between the fillers and the matrix with different numerical specimens including the nano-composites and neat polypropylene. Finally, the effects of carbon nanotubes on the elastic, elastic-plastic, and viscoelastic properties of CNT-PP nano-composites are compared.

Keywords: Carbon nanotube; Polypropylene; Viscoelastic properties

1 Introduction

Polypropylene (PP) is used in a wide variety of applications including packaging, textiles, automotive components, etc. However, PP has a very narrow thermoforming window, so the mechanical properties of the material rapidly decrease with increasing temperature [1]. Related studies were discussed by many researchers [2-5]. Lesan-Khosh et al. [6] measured the local and global mechanical properties at various temperature and strain rates and reported some relations between both mechanical properties. Peng et al. [7] proposed a novel phenomenological model consisting of springs, stress-locks and sliders to describe the nonlinear viscoelastic behavior.

In order to improve the mechanical properties of composites, fillers with specific advantages are often added into the matrix. Nano-clay with different content was blended with PP and Ali et al. [8] built a 3-element Maxwell model to describe with good agreement the viscoelastic properties of these nano-composites. Peng et al. [9] proposed a phenomenological thermo-mechanical viscoelastic constitutive model for PP wood composites. Karaduman et al. [10] investigated the viscoelastic properties of jute/PP nonwoven reinforced composites. Shim et al. [11] studied the viscoelastic behavior of glycidyl methacrylate-multi wall carbon nanotubes/PP nano-composites.

Carbon nanotubes (CNT) have been extensively studied due to their unique combination of mechanical, thermal, and chemical properties [12, 13]. For this reason, they have high potential to modify polymer matrices [14]. In previous works [15, 16], single wall carbon nanotubes (SWCNT) were randomly distributed into PP, and numerically studied via finite element analysis. In this work, some experimental results from reference [6] are adopted. Finite element method (FEM)

[*] Corresponding author. Tel.: +86-13587607056, E-mail: huangjun@scut.edu.cn (J. Huang)

will be used to investigate the viscoelastic properties of SWCNT reinforced PP composites at four temperatures (20, 40, 60 and 80°C).

2 Numerical models of composites

A representative volume element (RVE) is used to evaluate the viscoelastic properties of PP based composites. For studying the mechanical behavior of RVE via finite element method, the effect of SWCNT content on the stress-time curves of SWCNT/PP composites under different temperatures are investigated. The numerical specimen is assumed to be a cylinder for which the length and end diameter are constant at 600 nm and 200 nm, respectively. SWCNT is randomly distributed in the cylinder (PP matrix) along the deformation direction. The diameter of SWCNT is taken as 1.4 nm and the length is 400 nm [16]. As a first step, SWCNT volume fraction is fixed at 0.5%. Furthermore, the SWCNT content changes from 0.5% to 1.5%. All specimens will be created with MATLAB and tested for uniaxial tension. Fig. 1(a) shows the mechanical model of the nano-composites. The node displacements of the bottom along x, y, and z directions are constrained. A displacement taken as 0.1 nm for tensile numerical simulations is applied to the nodes of the top surface of the nano-composites. The reaction forces are calculated on the surface where the displacement conditions are given.

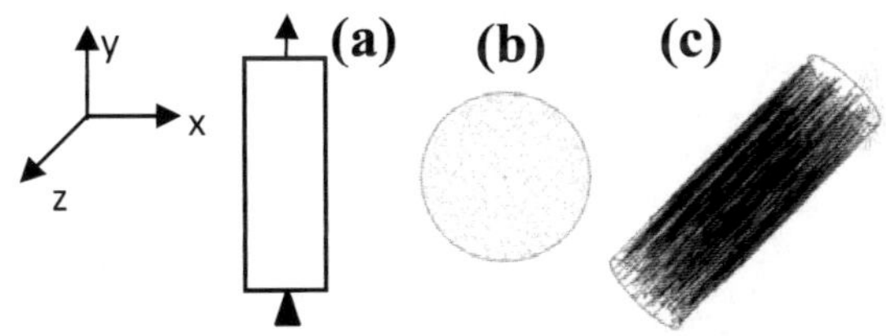

Figure 1. Numerical model for: (a) mechanics model; (b) top view of the model; (c) SWCNT/PP.

The commercial finite software MSC.MARC is used because of its wide range of element library [17]. Two types of finite elements are used here. Element 7 is used to simulate PP, while element 98 is used to model the SWCNT.

3 Mechanical properties of components

MSC.MARC has two models to represent viscoelastic materials. Here, the stress relaxation form is used. In this work, PP is assumed as a viscoelastic material for the temperature range studied: 20 to 80°C. When the temperature is 20°C, the relaxation modulus (G) is taken as 0. For higher temperatures, the relaxation modulus is taken from ref. [6]. The bulk moduli are not considered here. SWCNT is globally isotropic and is characterized by a linear elastic behavior. In this case, the following elastic constants are needed: Young modulus E = 1030 GPa and Poisson's ratio ν = 0.063 [15].

Table 1. Mechanical parameters of the components

Mechanical properties	SWCNT	Polypropylene (matrix)
Young's modulus (GPa)	1030	1.88
Tensile strength (GPa)	30	0.0348
Poisson ratio	0.063	0.4

Finally, the SWCNT/PP composite is approximated as a two-phase material. As a first step, the interface between SWCNT and PP is assumed to have perfect adhesion. The mechanical properties of each component are listed in Table 1.

4 Results

In order to characterize the viscoelasticity, the relaxation modulus is used to depict the time dependence of the nano-composites. On the other hand, SWCNT content will play an important role on the mechanical properties of the nano-composites. In these numerical simulations, the effect of SWCNT volume fraction is investigated.

4.1 The effect of temperature

In this work, the global mechanical properties of PP are selected [6]. The elastic modulus of PP is taken as 1880 MPa which is the value at 20°C. When the temperature increases to 40°C, 60°C, and 80°C, the elastic modulus is taken as 1536 MPa, 946 MPa, and 586 MPa, respectively.

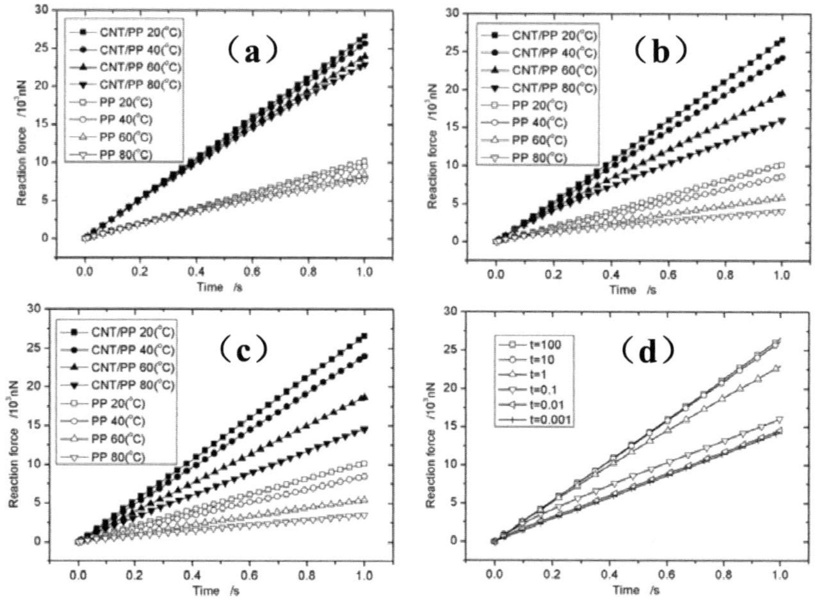

Figure 2. Tensile reaction force – time curves of the numerical specimens: (a) relaxation time t=1 s; (b) relaxation time t=0.1 s; (c) relaxation time t=0.01 s; (d) for SWCNT/PP composites at 80°C.

In order to investigate the viscoelastic properties of PP and SWCNT/PP composites, the loss modulus can be transformed to the time dependent relaxation modulus which is 122.8 MPa, 333.5 MPa or 462.1 MPa in terms of the relationship between the elastic modulus and the shear modulus. Here, the relaxation time is assumed to change to simulate the effect of SWCNT on the viscoelastic properties of the composites at various temperatures. The numerical results are plotted in Figure 2. In all numerical tests, the loading displacement is fixed at 0.1 nm, so the reaction forces which are the response to the displacement are obtained. Figure 2 (a) shows the reaction force - loading time curves of PP and SWCNT/PP composites at a constant relaxation time of 1 s. Similarly, when the relaxation time is taken as 0.1 or 0.01 s, the related numerical results are plotted in Figure 2 (b) and Figure 2 (c). Compared with Figure 2 (a), the relaxation time also has a direct effect on the viscoelastic properties of PP and SWCNT/PP composites. Figure 2 (d) shows the effect of the relaxation time on the viscoelastic properties of SWCNT/PP composites at 80°C.

In order to determine the effect of SWCNT on the mechanical properties of the nano-composites, one fiber (SWCNT) was isolated from the matrix; i.e. with the isolated SWCNT, stresses, strains and other mechanical properties of the SWCNT can be separately determined. Figure 3 (a) shows the stresses-displacements curves of a specified node of one isolated fiber (SWCNT) from the matrix. Figure 3 (b) and (c) show the stresses distributions at different nodes for one SWCNT isolated and non-isolated elements. To better understand stresses distributions on the nodes, stresses at different loading steps are presented in Figure 3 (d).

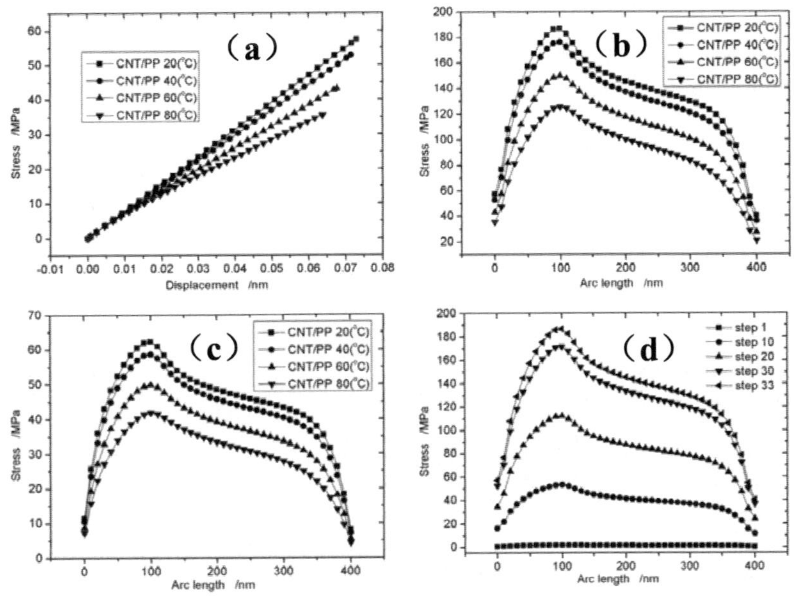

Figure 3. Results for one SWCNT in a nano-composites at a relaxation time of t=0.1 s: (a) stress-displacement curves for a specified node at four temperatures; (b) stress-arc length curves for isolated SWCNT; (c) stress-arc length curves for non-isolated SWCNT; (d) stress-arc length curves at different loading step.

4.2 The effect of SWCNT content

Assuming a relaxation time of 0.01 s, three volume fractions of SWCNT (0.5%, 1.0%, and 1.5%) are selected to simulate the viscoelastic properties of SWCNT/PP composites. The numerical results for 20 and 80°C are plotted in Figure 4 (a) and (b).

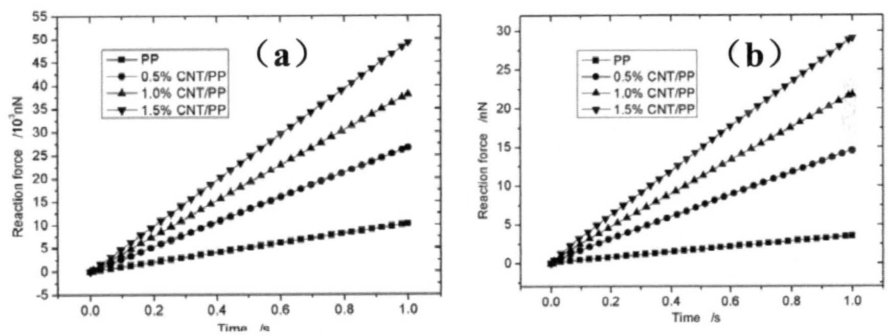

Figure 4. Reaction force - time curves of SWCNT/PP composites for different SWCNT content with a relaxation time of 0.01 s at different temperature: (a) 20°C and (b) 80°C.

5 Conclusions

In this work, the relaxation time and relaxation modulus were used to depict the viscoelastic properties of PP. Considering the effects of temperature and SWCNT content, numerical simulations on the mechanical properties of PP and SWCNT/PP composites were carried out. The main results are listed as follows:

1. When the temperature increased, the relaxation modulus of PP increased, but stiffness and strength of PP and SWCNT/PP composites decreased.

2. For higher elastic modulus, SWCNT can efficiently improve the mechanical properties of the composites, even at low content. With increasing SWCNT volume fraction, SWCNT/PP nano-composites have higher stiffness and strength.

3. By isolating the SWCNT from the matrix, it can be seen that SWCNT can undertake more stress than PP in SWCNT/PP composites.

Acknowledgement

This study was supported by National Natural Science Foundation of China (Grant Number: 51568009; 51378398; 51178356) and Education Department of Guangxi Government (Grant Number: 200103YB103).

References

[1] He CX, Costeux S, Wood-Adams P, etc. Molecular structure of high melt strength polypropylene and its application to polymer design. Polymer, 2003;44(23):7181-7188.
[2] Kobayashi Y, Otsuki Y, Kanai T. Viscoelastic flow analysis of surface morphology on injection-molded polypropylene. Polymer Engineering & Science. 2010;50:2182-2189.
[3] O'Connor CPJ, Martin PJ, Menary G. Viscoelastic material models of polypropylene for thermoforming applications. International Journal of Material Forming. 2010;3:599-602.
[4] Drozdov AD. Effect of temperature on the viscoelastic and viscoplastic behavior of polypropylene. Mechanics of Time-Dependent Materials. 2010;14:411-434.
[5] Tscharnuter D, Jerabek M, Major Z, Pinter G. Uniaxial nonlinear viscoelastic viscoplastic modeling of polypropylene. Mechanics of Time-Dependent Materials. 2012;16:275-286.
[6] Lesan-Khosh R, Bagheri R, Asgari S. Nanoindentation of isotactic polypropylene: Correlations between hardness, yield stress, and modulus on the local and global scales. Journal of Applied Polymer Science. 2011;121:930-938.
[7] Peng G, Ma Y, Feng Y, Huan Y, Qin C, Zhang T. Nanoindentation creep of nonlinear viscoelastic polypropylene. Polymer Testing. 2015;43:38-43.
[8] Ali MA-h, Elleithy RH. Viscoelastic properties of polypropylene/organo-clay nano-composites prepared using miniature lab mixing extruder from masterbatch. Journal of Applied Polymer Science. 2011;121:27-36.
[9] Peng X, Yin H, Chen J, Liu X. A Phenomenological thermal-mechanical viscoelastic constitutive modeling for polypropylene wood composites. Advances in Materials Science and Engineering. 2012;2012:1-7.
[10] Karaduman Y, Sayeed MMA, Onal L, Rawal A. Viscoelastic properties of surface modified jute fiber/polypropylene nonwoven composites. Composites Part B: Engineering. 2014;67:111-118.
[11] Shim Y-S, Min B-G, Park S-J. Effects of functional grafting on viscoelastic and toughness behaviors of multi-walled carbon nanotubes-reinforced polypropylene nano-composites. Macromolecular Research. 2012;20:540-543.
[12] Tung VC, Chen LM, Allen MJ, Wassei JK, Nelson K, Kaner RB, et al. Low temperature solution processing of graphene-carbon nanotube hybrid materials for high-performance transparent conductors. Nano Letters. 2009;9:1949-1955.
[13] Choi W, Hong S, Abrahamson JT, Han JH, Song C, Nair N, et al. Chemically driven carbon-nanotube-guided thermopower waves. Nature Materials. 2010;9:423-429.
[14] Yang K, Gu MY, Guo YP, Pan XF, Mu GH. Effects of carbon nanotube functionalization on the mechanical and thermal properties of epoxy composites. Carbon. 2009;47:1723-1737.
[15] Huang J, Rodrigue D. The effect of carbon nanotube orientation and content on the mechanical properties of polypropylene based composites. Materials and Design. 2014;55:653-663.
[16] Huang J, Rodrigue D. Analysis of multi-axial properties of carbon nanotubes/polypropylene and nanocrystalline cellulose/polypropylene composites. Polymer Composites. 2016;37(4):1180-1189.
[17] MSC.Marc version 2010. Volume B: Element library. MSC Software Corporation, USA (www.mscsoftware.com). 2010.

Effect of Fibre Orientation and Concentration on the Viscosity of Polymer Composites

Deverpiran Vishak Perumal[a]*, Sati N Bhattacharya[a], Rahul K Gupta[a] and Franco S Costa[b]

[a]*Rheology and Materials Processing Group, School of Engineering, RMIT University, Melbourne, VIC, Australia*

[b]*Autodesk Australia Pty Ltd, 259-261 Colchester Road, Kilsyth, VIC, Australia*

Abstract: Rheological behaviour of polymer composites containing short glass fibres and the influence of fibre orientation on rheology are presented in this paper. Shear flow experiments are performed on a rotational rheometer using short glass fibre/Polyamide-6 compression moulded samples. The samples are subjected to clockwise shearing for different durations and cooled down rapidly to preserve the fibre orientation. X-ray Computed Tomography scanning is performed on 6mm wide rectangular strips cut from sheared samples to obtain 3-dimensional representations of fibre orientation. Additional thermal characterisation and compositional analysis were also performed on the samples. Development of a constitutive model containing empirical parameters will be attempted to relate fibre orientation and composite melt viscosity.

Keywords: Fibre orientation, Transient viscosity, Computed tomography, Polymer composites

PACS: 83

INTRODUCTION

Composites made using thermoplastic polymers that are reinforced with glass fibres are increasingly being used in industrial applications due to the enhanced thermal and mechanical properties they provide in comparison to the neat polymer. The rheological behaviour of molten composites is complex due to the anisotropy introduced by the fibres and the additional interactions between the fibres and matrix. In industrial manufacturing processes such as injection moulding, fibre orientation distribution throughout a geometry varies based on the type of deformation (shear or extensional) experienced by the fibre. Initial research work found that the shear viscosity of molten composites increases when fibres are oriented in the through-thickness direction which is transverse to the shearing direction [1]. Rheological effects of fibre orientation are better observed at lower shear rates as the fibre orientation evolution is slower [2]. Rheological characterization of filled systems also found deviations from the Cox-Merz rule as differences between the steady state viscosity and complex viscosity were observed [3]. This has been explained by the different fibre orientation distributions present during steady and oscillatory testing. With steady state shearing, the fibres are increasingly orienting in the shearing direction while in oscillatory shearing the fibre orientation distribution retains a more random orientation state.

As for constitutive modelling, some researchers have attempted to extend rheological models to fit experimental results obtained from stress growth experiments carried out on fibre filled systems [4]. The Lipscomb model coupled with the Folgar-Tucker orientation model is an example which has been used to characterize steady state and transient rheological behaviour [5, 6]. Even with detailed modelling, predictions of transient viscosity and normal stress overshoot have not been accurate based on their findings [7]. Recent works have found that the transient response of filled systems was not solely due to fibre orientation but also due to possible additional interactions between the fibres and matrix such as fibre contact [7, 8]. Researchers have attempted to incorporate the effects of fibre contact into constitutive models, but experimental validation was only carried out for fibres in a Newtonian medium [9]. This shows that accurate quantitative rheological modelling of complex filled systems is still a work in progress.

Techniques for measuring fibre microstructure are also expanding rapidly with recent advancements allowing for 3-dimensional measurement with minimal sample damage [10]. X-ray Computed Tomography (X-CT) offers the possibility of measuring fibre orientation distribution, fibre length distribution and fibre volume fraction which makes it a promising option [11]. This technique provides an avenue for characterising fibre filled polymer systems containing different orientation and length distributions. The aim of this work is to characterise the transient rheological behaviour of filled systems at low shear rates as a function of fibre orientation and also determine the changes in fibre orientation that take place during shearing.

MATERIALS AND METHODS

In this research, samples were prepared using pellets obtained from extrusion containing short glass fibres (SGF) in a Polyamide-6 (PA-6) matrix. The pellets were compression moulded at 250°C with a pressure of 90kN for 5 minutes to obtain 25mm diameter discs suitable for rheological testing.

In order to establish the thermal behaviour of the samples and confirm the mass fraction of glass fibres, a Modulated Differential Scanning Calorimetry (MDSC) instrument from TA Instruments and a hyphenated Differential Scanning Calorimetry (DSC)-Thermogravimetric analysis (TGA) instrument from Perkin Elmer were used respectively. Thermal stability of the composite samples was tested by holding them at 270°C for 1 hour in a Nitrogen environment. Heating and cooling cycles were maintained at 2°C/min. Mass fraction analysis was carried out by heating the sample to 600°C in a Nitrogen environment and then heating it up further to 900°C and holding it for 20 minutes in an air environment to provide sufficient time for the polymer matrix to burn off.

Rheological measurements were performed using an ARES G2 rotational rheometer with a 25mm diameter parallel plate configuration. A rheometer gap of 2.0mm was fixed for all experimental runs to allow for reproducibility and testing was carried out at 250°C. In order to obtain information on the effects of fibre orientation on shear viscosity, rheological experiments were carried out in transient mode to observe the changes in viscosity before steady state conditions are reached. Shear rates ranging from $0.1s^{-1}$ to $0.5s^{-1}$ were applied on fresh samples in the clockwise direction. The total strain units applied on each sample was kept at a constant value of 60 for all shear rates by varying the shearing duration. This allowed for a more effective comparison on the effect of shear rate. Upon cessation of shearing, the sample was cooled rapidly using a fan blower and removed from the rheometer for morphological analysis.

A GE Vtomexs 240 X-CT machine was used to capture the 3-dimensional fibre morphology of the filled system. The resolution of the scans was set at 2µm to allow for a clear distinction between the glass fibres and polymer matrix. In order to obtain the required resolution, the circular 25mm diameter samples were cut into 6mm wide strips using an electric saw blade machine. Parameters such as voltage, current and power were fine tuned to achieve the best contrast between the fibre and matrix. Table 1 provides the settings that were used for the X-CT scans.

TABLE 1: Parameter used for X-CT scanning of fibre filled sample

Parameters	Value
Voltage	75kV
Current	27µA
Power	2W
Voxel size	2µm

RESULTS AND DISCUSSION

Rheological Results

The results in Figure 1 show the transient rheological response of the molten composite at different shear rates. A viscosity overshoot is observed which is due to the additional stress caused by the evolution of fibres before a steady state viscosity value is reached corresponding to a fully evolved fibre orientation state. As the shear rate is increased, it is observed that the viscosity overshoot value decreases and the steady state viscosity values are also lower. This reduction in steady shear viscosity is due to the shear thinning behaviour of the polymer system and also due to the increased alignment of fibres.

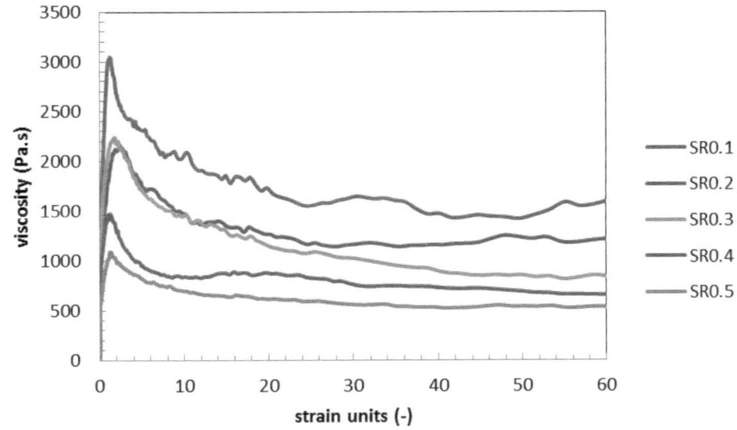

Figure 1: Transient viscosity at start-up of shear flow for 30% weight SGF PA-6 samples for shear rates $0.1s^{-1}$, $0.2s^{-1}$, $0.3s^{-1}$, $0.4s^{-1}$ & $0.5s^{-1}$

Morphological Study

Based on the results from Figure 1, the different fibre orientation distributions for each steady state condition are to be quantified. Figures 2a & 2b show the 2-dimensional image of fibre distribution obtained from X-CT scanning in samples subjected to a shear rate of $0.1s^{-1}$, $0.2s^{-1}$ and $0.5s^{-1}$ respectively. The 3-dimensional representation developed using multiple slices of 2-dimensional data will be used to compute fibre orientation tensors in the planar direction and fibre volume fraction. With an increasing shearing rate, an increased alignment factor in the shearing direction is expected.

2a 2b 2c

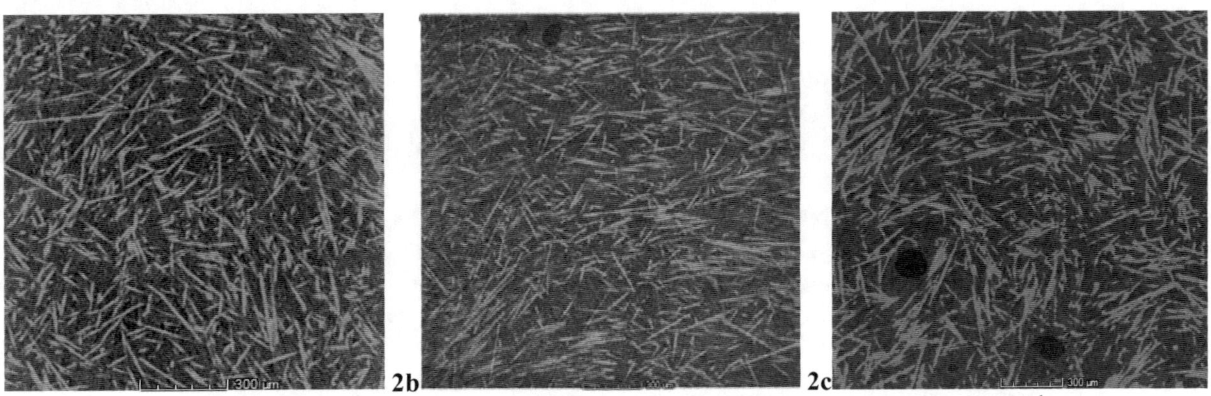

Figure 2A: X-CT images of a SGF PA-6 sample subjected to shear rate of $0.1s^{-1}$
Figure 2B: X-CT images of a SGF PA-6 sample subjected to shear rate of $0.2s^{-1}$
Figure 2B: X-CT images of a SGF PA-6 sample subjected to shear rate of $0.5s^{-1}$

Thermal and Compositional Analysis

Thermal analysis of the composite samples containing glass fibres were carried out with the aim of establishing if prolonged heat exposure causes degradation to the sample. Results from the MDSC heating cycles are presented in Figures 3a & 3b which show the thermal behaviour of the sample before and after being held at 270°C for 1 hour in a Nitrogen environment respectively. The results show that the sample did not undergo significant thermal degradation or damage as the thermal curves and melting points are similar.

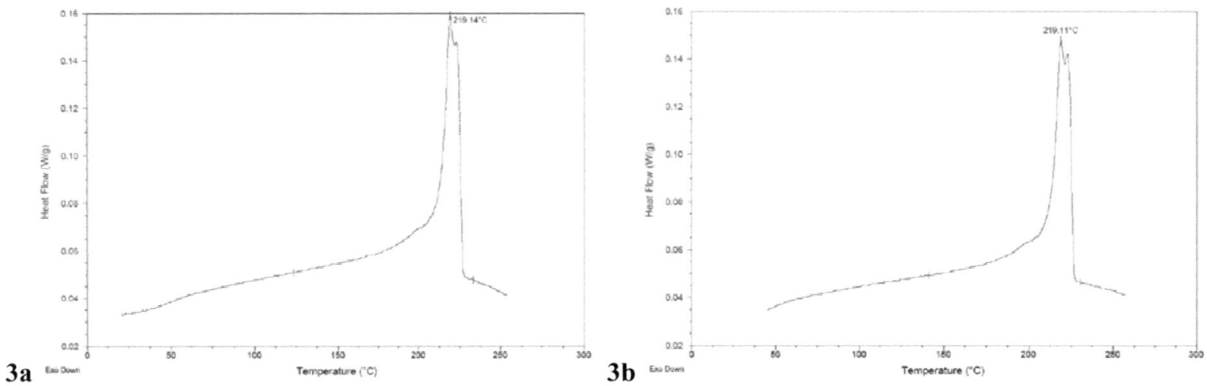

Figure 3A & 3B: Thermal behaviour of SGF PA-6 sample in a Nitrogen environment (3a: First heating cycle, 3b: Second heating cycle after holding sample at 270°C for 1 hour)

Using the hyphenated DSC-TGA instrument, compositional analysis on the SGF PA-6 samples was performed. The results in Figure 4 show the mass changes taking place in the composite sample as a function of temperature. It can be seen that thermal degradation of the polymer takes place between 440°C-540°C and the glass fibres are left as residue. Using this method, the mass fraction of glass fibres was determined to be 32%.

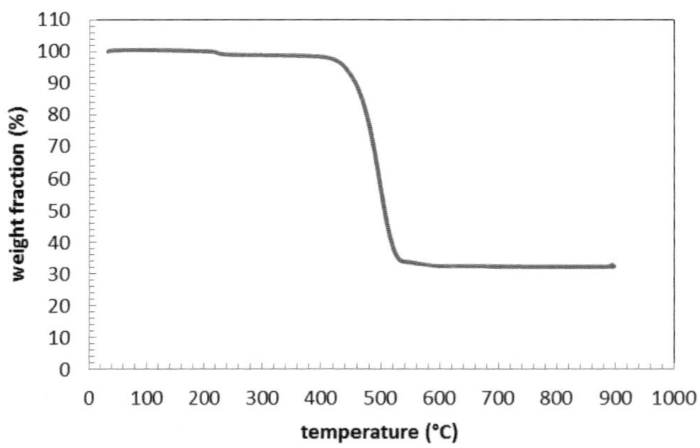

Figure 4: Mass fraction analysis of SGF PA-6 sample from

CONCLUSION AND FUTURE WORK

The aim of this work was to relate the fibre orientation distribution to the rheology of glass fibre filled polymer systems. Clockwise transient stress growth experiments were carried out between shear rates $0.1s^{-1}$ and $0.5s^{-1}$ to determine the viscosity overshoot and the steady state viscosity values. It was observed that the viscosity overshoot and steady state viscosity values decreased with increasing shear rate. In order to relate the rheological measurements to specific fibre orientation distribution, sheared samples were scanned using an X-CT machine to calculate the fibre orientation distribution. Future work will be performed to relate the extent of fibre orientation to shear viscosity and develop a constitutive model that is applicable for a wide range of processing conditions observed in industrial processes such as injection moulding.

ACKNOWLEDGEMENTS

This project is financially supported by Autodesk Australia Pty Ltd as part of a research collaboration with RMIT University, Australia and their contribution is deeply appreciated. Use of instruments and facilities at the Rheology and Materials Processing Centre, RMIT University and School of Aerospace, Mechanical and Manufacturing Engineering, RMIT University, Australia is also gratefully acknowledged.

REFERENCES

1 Laun, H.M.: 'Orientation effects and rheology of short glass fiber-reinforced thermoplastics', Colloid & Polymer Sci, 1984, 262, (4), pp. 257-269

2 Djalili-Moghaddam, M., and Toll, S.: 'Fibre suspension rheology: effect of concentration, aspect ratio and fibre size', Rheol Acta, 2006, 45, (3), pp. 315-320

3 Guo, R., Azaiez, J., and Bellehumeur, C.: 'Rheology of fiber filled polymer melts: Role of fiber-fiber interactions and polymer-fiber coupling', Polymer Engineering & Science, 2005, 45, (3), pp. 385-399

4 Eberle, A.P., Baird, D.G., and Wapperom, P.: 'Modeling the Transient Rheology of a Polypropylene Melt Reinforced with Long and Short Glass Fibers'. Proc. ANTEC2006 pp. Pages

5 Eberle, A.P.R., Baird, D.G., Wapperom, P., and Vélez-García, G.M.: 'Obtaining reliable transient rheological data on concentrated short fiber suspensions using a rotational rheometer', Journal of Rheology (1978-present), 2009, 53, (5), pp. 1049-1068

6 Eberle, A., Baird, D., and Wapperom, P.: 'Modeling the Rheology and Orientation Distribution of Short Glass Fibers Suspended in Polymeric Fluids: Simple Shear Flow', in Editor (Ed.)^(Eds.): 'Book Modeling the Rheology and Orientation Distribution of Short Glass Fibers Suspended in Polymeric Fluids: Simple Shear Flow' (2007, edn.), pp. 2806

7 Eberle, A.P.R., Baird, D.G., Wapperom, P., and Vélez-García, G.M.: 'Using transient shear rheology to determine material parameters in fiber suspension theory', Journal of Rheology, 2009, 53, (3), pp. 685-705

8 Cieslinski, M.J., Wapperom, P., and Baird, D.G.: 'Influence of fiber concentration on the startup of shear flow behavior of long fiber suspensions', Journal of Non-Newtonian Fluid Mechanics, 2014, (0)

9 Férec, J., Ausias, G., Heuzey, M.C., and Carreau, P.J.: 'Modeling fiber interactions in semiconcentrated fiber suspensions', Journal of Rheology (1978-present), 2009, 53, (1), pp. 49-72

10 Shen, H., Nutt, S., and Hull, D.: 'Direct observation and measurement of fiber architecture in short fiber-polymer composite foam through micro-CT imaging', Composites Science and Technology, 2004, 64, (13–14), pp. 2113-2120

11 Köpplmayr, T., Milosavljevic, I., Aigner, M., Hasslacher, R., Plank, B., Salaberger, D., and Miethlinger, J.: 'Influence of fiber orientation and length distribution on the rheological characterization of glass-fiber-filled polypropylene', Polymer Testing, 2013, 32, (3), pp. 535-544

Effect of Extrusion Temperature on the Elongational Properties of Polyphenylene Sulfide using Melt Spinning Technique

Changbin Tan [a,*], Yan Yang[b], Youwei Zhang[b], Hengbo He[b], Jiaqiang Ning[b]

[a] *Analytical and Testing Center, Sichuan University of Science and Engineering, Zigong 643000, China*
[b] *College of Materials Science and Engineering, Sichuan University of Science and Engineering, Zigong 643000, China*

Abstract. The elongational properties of polyphenylene sulfide (PPS) melt were measured using a melt spinning technique. The relationship between extrusion temperature and melt strength (*MS*) as well as between elongational viscosity and drawability were investigated with respect to the effects of extrusion temperature and extensional strain rate on the melt extensional stress and elongational viscosity. The results showed that the stretching force for the PPS melt decreased with a rise of extrusion temperature while increased roughly with an increase of extensional rate. The *MS* decreased with an increase of temperature, and the ln *MS* was a linear function of $1/T$ when the extrusion velocity was constant. Both the melt extensional stress and elongational viscosity decreased with the increase of the extrusion temperature. With increase of the extensional strain rate, the extensional stress increased while the melt elongational viscosity first decreases and then increases gradually. A low melt elongational viscosity might be beneficial to improve the melt drawability.

Keywords: polyphenylene sulfide, extensional rheology, elongational viscosity, melt strength, extrusion temperature, melt spinning technique

INTRODUCTION

In polymer processing, melt extensional flow is a common flow pattern, such as entrance converging flow in die extrusion or runner injection of polymer melts from an extruder barrel[1-5], blow molding, blowing film and melt spinning. In general, the flow status of polymeric materials will significantly affect the processing technology and the final properties of products. Therefore, the melt extensional flow behavior of polymeric materials and its mechanisms have been paid attention by rheological researchers and polymer processing engineers[6].

The extensional rheological properties of polymeric materials are usually measured using special extensional rheological instruments[7-9]. Handge and Schmidheiny[7], Hachmann and Meissner[8] as well as Wagner et al.[9] introduced the testing instruments and methods for elongation properties of polymer melts. However, these special rheometers are expensive. Melt spinning measurement process is similar to isothermal uniaxial tension, and the extensional rheology parameters such as elongational viscosity may be obtained. Hence, one may comprehensively study the extensional rheological properties of polymer melts. Muke et al.[10] studied the extensional rheology of two polypropylene melts from uniaxial stretching after extrusion (Rheotens test), and found that the viscosity determined from the Rheotens test can be used to estimate transient elongational viscosity. Baldi et al.[11] compared different experimental techniques and researched the determination of the elongational viscosity of two high-density polyethylene melts by melt spinning experiments. The results showed that for high strain and strain rate levels, the melt spinning experiments provide elongational viscosity data quite close to the transient elongational viscosity values obtained from the melt tensile tests.

Polyphenylene sulfide (PPS) fiber is a kind of engineering thermoplastic fiber, it has an excellent high-temperature thermal stability, and has a very good performance of denial in dry and wet environments. Its chemicals resistance capacity is also excellent, it can maintain a stable acid-resistance, alkali, organic solvent resistance. So it is very suitable as high temperature filter material, especially the pocket filter material[12, 13]. The melt flow status of PPS has a significant impact on the physical properties of PPS fiber. However, the study of the elongational behavior of PPS melt is rarely involved. The objectives of this article are to measure the extensional rheological properties of a PPS using a melt spinning technique, to investigate the effects of extrusion temperature and extensional strain rate on the melt extensional stress and elongational viscosity, as well as to study the relationship between the melt elongational viscosity, the melt strength and the melt drawability.

EXPERIMENTAL

Materials

The resin used in this study was a polyphenylene sulfide (PPS, DHPPS-YD-I) with density 1.32 g/cm^3 and melt flow index (MFI) 165 g/10min. The PPS was supplied by Dunhuang Western Region Special New Material Co., Ltd., Jiuquan city, China.

Instruments and Methodology

The melt elongation rheological measurements were carried out by means of a capillary rheometer (CEAST SR50) supplied by Instron Corporation, which was equipped with a specific stretching unit.

The polymer melt was stretched in uniaxial extension according to the melt spinning principle to measure the rheological properties. For each measurement, the resin was heated to a given temperature and then the melt was extruded at a constant flow rate through a capillary die. Subsequently, the melt was drawn by a set of pinching rotating wheels positioning at some distance downstream. At the beginning, the take-up speed of the wheels was adjusted to the extrusion velocity of the polymer melt. Then, the speed was increased with a constant acceleration rate to stretch the melt until it broke. During the test, the draw-down force measured was plotted as a function of velocity of the rotating wheels. For each temperature, the tests were repeated at least 3 times, the one with the highest draw ratio at break being used to calculate the elongational viscosity. For all the measurements, the length of the spinline was 600 mm, the diameter of extrusion die was 1 mm, the ratio of length to diameter was 30:1, the entrance angle was 90°, the extrusion velocity was 16.875 mm/s and the acceleration rate was 5 mm/s^2. The melt temperatures of the PPS were from 300 to 320 °C.

According to the expression derived by Wagner et al.[9], the extensional strain rate, extensional stress and melt elongational viscosity may be determined from the extension flow curves, and the calculation equation is given by

$$\eta_{el} = \frac{\sigma}{\dot{\varepsilon}} = \frac{F \cdot L}{Q \cdot \ln(\lambda)} \quad (1)$$

where λ is the extension ratio, $\dot{\varepsilon}$ is the extensional strain rate, in s^{-1}; σ is the extensional stress, in Pa; L is the distance between the die exit and the traction wheel, in m; F is the stretching force, in N; Q is the volume flow rate, in m^3/s.

RESULTS AND DISCUSSION

Extension flow curves

Fig. 1 shows the relationship between the stretching force (F) and the extension rates for the PPS melt under test conditions of extrusion temperature range from 300 to 320 °C and extrusion velocity of 16.875 mm/s. It can be seen that F increases roughly with the increase of extension rate (v), and the tendency becomes more obvious with a rise of extrusion temperature. Furthermore, the macromolecular chain of PPS is a rigid chain because of the benzene ring. In general, a melt with rigid chain needs more extension force to form macromolecular chain orientation and produce elongational flow than a melt with flexible chain such as HDPE and LLDPE at a constant temperature. Consequently, the extension force of the PPS is much higher than that of the LLDPE and HDPE[14]. It may also be observed that F decreases with a rise of extrusion temperature because the movement ability of the macromolecular chain of PPS will be enhanced with the increase of temperature, and the force needed for its orientation reduces. Hence, the stretching force of the PPS melt decreased with a rise of temperature during elongational flow.

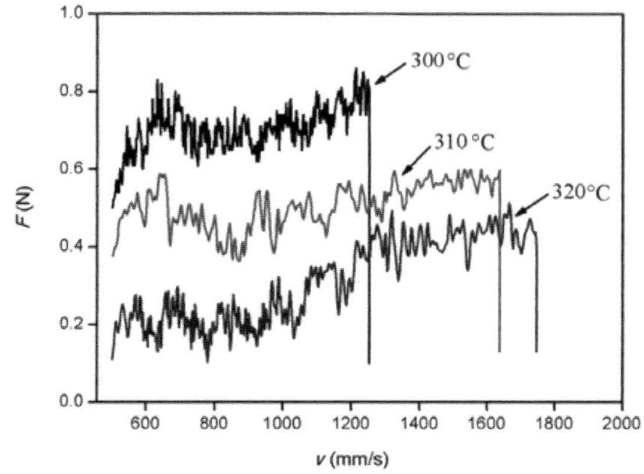

FIGURE 1. Melt spinning curves at extrusion rate of 16.875 mm/s.

Dependence of Melt Strength on Extrusion Temperature

In general, the cross section of the melt thread is difficult to determine under test conditions, hence the melt stretching force at break is usually defined as the melt strength (*MS*)[6].Fig. 2 displays the dependence of the *MS* of the PPS melt on the extrusion temperature. It can be seen that the *MS* of the PPS melt decreases with a rise of temperature, and the slope of the curve of the melt strength versus temperature of the PPS melt is much greater than that of the polyethylene melts, as shown in Liang's work[14]. This indicates that the dependence of the *MS* of the PPS resin on extrusion temperature is significant.

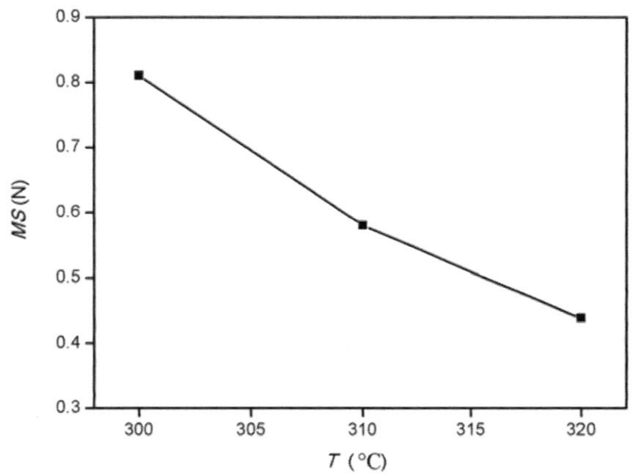

FIGURE 2. Relationship between melt strength and extrusion temperature.

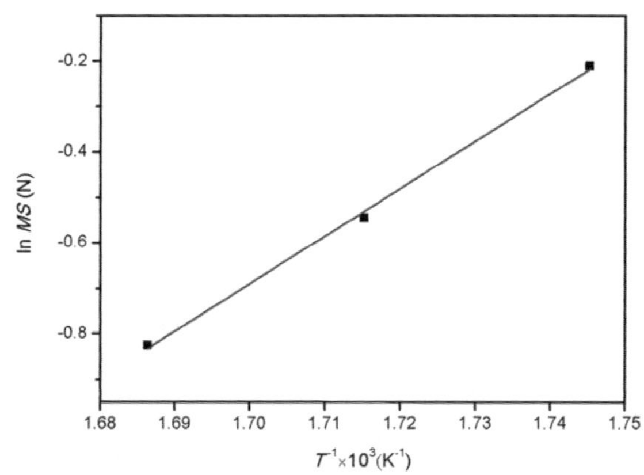

FIGURE 3. Dependence of melt strength on extrusion temperature at extrusion rate of 16.875 mm/s.

To further investigate the influence of extrusion temperature on the *MS* of the PPS resin, we plot the curve of ln *MS* against the reciprocal of temperature (T^{-1}) when the extrusion velocity is 16.875 mm/s. The result is shown in Fig. 3. It may be observed that the relationship between the ln *MS* and T^{-1} is a linear function. This indicates that the dependence of the *MS* for the PPS on the extrusion temperature obeys the Arrehnius expression, that is

$$MS = A \exp\left(\frac{E_{MS}}{RT} \right) \tag{2}$$

where A is a constant, R is the Molar gas constant, 8.314 J/(mol·K). T is the absolute temperature in K, E_{MS} is the melt strength activation energy, in kJ/mol.

The parameter E_{MS} reflects the sensitivity of the melt strength to temperature. The value of E_{MS} of polymer melts can be calculated from the slope of the curves of the melt strength versus temperature. On the basis of *Eq*. 2 and Fig. 3, the E_{MS} of PPS is 86.88 kJ/mol, which is much larger than that of the polyethylene melts (20 kJ/mol for LLDPE, 19.1 kJ/mol for HDPE)[14]. As stated above, the macromolecular chain of PPS is a rigid chain with benzene ring. It is generally believed that the motion ability of the molecular chains of polymer melt will be enhanced with a rise of temperature. In this case, the rigid chain of the PPS melt will be oriented easily and the melt strength will be reduced obviously during elongational flow with a rise of temperature. For flexible chain structure resin, such as HDPE and LLDPE, the factor affecting the molecular chain orientation is mainly strain rate, while the influence of temperature is relatively weak. Therefore, the sensitivity of the *MS* of the PPS to temperature is more significant than that of the polyethylene.

Effects of extensional strain rate and extrusion temperature on melt extensional stress and elongational viscosity

Melt extensional stress is also an important characterization of melt strength of polymeric materials. Fig. 4 illustrates the dependence of the melt extensional stress (σ_e) of the PPS on extensional strain rate ($\dot{\varepsilon}$) under different extrusion temperatures when the extrusion velocity is 16.875 mm/s. When the extrusion temperature is fixed, $\ln\sigma_e$ increases nonlinearly with the increase of $\ln\dot{\varepsilon}$, while $\ln\sigma_e$ decreases with a rise of temperature with $\dot{\varepsilon}$ constant. In general, the storage and dissipation of deformation energy are coexistent during polymer melt flow. The stored energy usually increases with increasing extensional strain rate, while it decreases with a rise of extrusion temperature during melt flow. Consequently, the melt extensional stress increases with an increase of extensional strain rate, whereas it reduces with a rise of temperature during the elongational flow of the PPS melt[6].

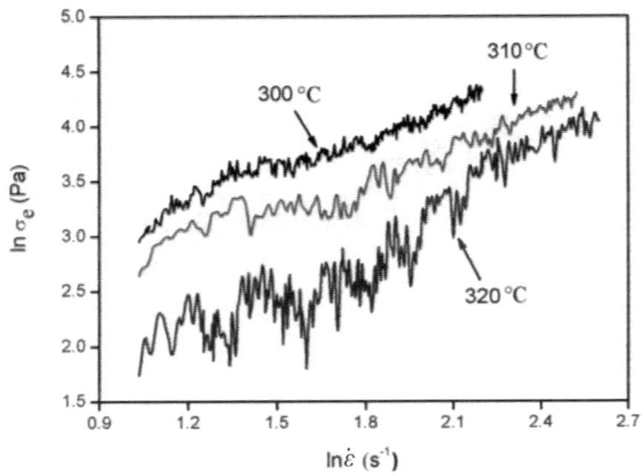

FIGURE 4. Extensional stress versus strain rate at extrusion velocity of 16.875 mm/s.

Melt elongational viscosity is a basic parameter for presenting the elongational flow property of polymeric materials, it is very important for melt spinning, foaming and blowing technology. Fig. 5 displays the relationship between the elongational viscosity (η_{el}) and extensional strain rate ($\dot{\varepsilon}$) of the PPS melt at different extrusion temperatures when the extrusion velocity is also 16.875 mm/s. It can be seen that the $\ln\eta_{el}$ of the PPS melt first decreases and then increases gradually with increasing extensional strain rate. The reason for this can be illustrated as follows: the increase of the extensional strain rate might be beneficial to the disentanglement of the macromolecular chains of the PPS melt, resulting in decreased melt elongational viscosity. On the other hand, the orientation degree of the macromolecular chains along the extension direction will increase, leading to the increase of the melt elongational viscosity[14]. Therefore, the change of the elongational viscosity with the extensional strain rate depends on the competition between the disentanglement and the orientation degree of the macromolecular chains of the PPS melt. In the beginning stage, the disentanglement of the macromolecular chains plays a dominant role, and the melt elongational viscosity decreases slightly. When the extensional strain rate reaches a certain level, the orientation effect of the macromolecular chain is enhanced gradually and plays a major role, so the melt elongation viscosity begins to increase. As mentioned previously, the macromolecular chain of PPS is a rigid chain with benzene ring, and the melt extension hardening effect is significant during elongational flow. It may also be observed that the elongational viscosity decreases with a rise of temperature, resulting in increased fluctuations in the elongational viscosity. It is known comparing Figs. 1 and 5 that the melt strength is low as the melt elongational viscosity is low. However, the melt drawability is good, because a low melt elongational viscosity may be beneficial in this respect.

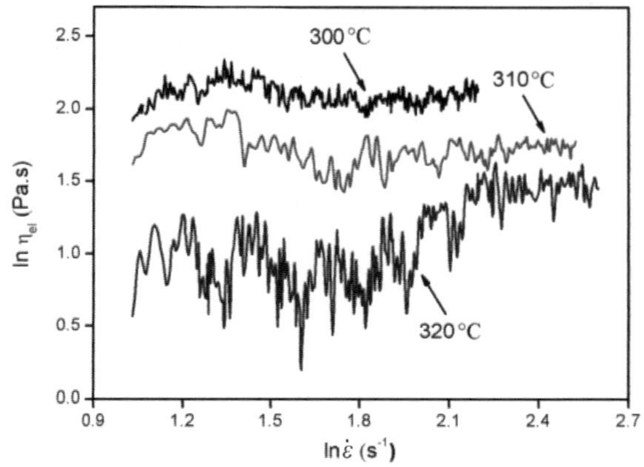

FIGURE 5. Elongational viscosity versus strain rate at extrusion velocity of 16.875 mm/s.

CONCLUSIONS

The effects of extrusion temperature and extensional strain rate on the melt extensional stress and elongational viscosity of PPS resin were significant. The stretching force for the PPS melt decreased with a rise of extrusion temperature while increased roughly with increase of the extensional rate. The melt strength decreased with an increase of extrusion temperature, and the ln *MS* was a linear function of $1/T$ when the extrusion velocity was constant. The melt strength activation energy could be determined based on the curves of the ln *MS* versus $1/T$.

With increase of the extensional strain rate, the extensional stress increased while the melt elongational viscosity first decreased and then increased gradually under the action of both the disentanglement and the flow-induced orientation of the macromolecular chains of the PPS melt. Both the melt extensional stress and elongational viscosity dropped with a rise of extrusion temperature. A low melt elongational viscosity might be beneficial to improve the melt drawability.

ACKNOWLEDGMENTS

This work was financially supported by the Talent Introduction Funds of Sichuan University of Science and Engineering (No. 2015RC33), General Science Project of Education Department of Sichuan Province (No. 16ZB0249), Undergraduate Training Programs of Sichuan Province for Innovation and Entrepreneurship (201510622079, 201510622083).

REFERENCES

1. F. Baldi, A. Franceschini, F. Bignotti, et al. *Rheologica Acta* **48**, 73-88 (2009).
2. J.-Z. Liang, *Chemical engineering science* **53**, 3185-3187 (1998).
3. J.-Z. Liang, *Journal of Applied Polymer Science* **80**, 1150-1154 (2001).
4. J.-Z. Liang, *Polymer-Plastics Technology and Engineering* **46**, 475-480 (2007).
5. J. Z. Liang, *Plastics rubber and composites processing and applications* **25**, 495-498 (1996).
6. J.-Z. Liang, L. Zhong, *Polymer Testing* **29**, 972-976 (2010).
7. U. A. Handge, W. Schmidheiny, *Rheol Acta* **46**, 913-919 (2007).
8. P. Hachmann, J. Meissner, *Journal of Rheology* **47**, 989-1010 (2003).
9. M. H. Wagner, A. Bernnat, V. Schulze, *Journal of Rheology* **42**, 917-928 (1998).
10. S. Muke, I. Ivanov, N. Kao, S. N. Bhattacharya, *Journal of Non-Newtonian Fluid Mechanics* **101**, 77-93 (2001).
11. F. Baldi, A. Franceschini, T. Riccò, *Rheol Acta* **46**, 965-978 (2007).
12. O. Yemul, T. Imae, *Biomacromolecules* **6**, 2809-2814 (2005).
13. J. Yang, M. Zhao, *Advanced Materials Research* **557-559**, 1668-1671 (2012).
14. J.-Z. Liang, L. Zhong, *Polymer Engineering & Science* **51**, 2490-2494 (2011).

Preparation and Characterization of PPO/PA66/Carbon black/Sisal Fibers Electromagnetic Shielding Composite

Hezhi He*, Shuwen Cheng, Guangjian He and Yue Xing

The Key Laboratory of Polymer Processing Engineering, Ministry of Education, National Engineering Research Center of Novel Equipment for Polymer Processing, South China University of Technology, Guangzhou 510640, China

Correspondence to: Hezhi He; E-mail: pmhzhe@scut.edu.cn

Abstract. Electromagnetic shielding composites consisting of hybrid fillers and polyphenylene oxide(PPO)/polyamide 66(PA66) immiscible polymer blends were prepared by melt blending. Carbon black (CB) was used as the foremost conducitve filler and steam exploded sisal fibers (SF) were used as the assistant filler. The cooperation of hybrid fillers and it's effect on electrical conductivity and electromagnetic interference shielding effectiveness of composites have been investigated. It was observed that SF greatly contributed to effectively decrease the percolation threshold of CB in PPO/PA66 matrix and adsorb carbon particles and PA66 phase of enriched CB particles to form a conductive network. Furthermore, when the mass ratio of CB/SF was 1/1, the electromagnetic shielding effectiveness of PPO/PA66/CB-SF achieved the most excellent performance.

Keywords: carbon black; sisal fibers; electromagnetic interference shielding; composite
PACS: 81.20.Hy

INTRODUCTION

The rapid increment of electrical and electronics devices emit electromagnetic energy, which has inevitably caused electromagnetic interference (EMI) problems. It becomes essential to avoid and shield all sources of unwanted electromagnetic waves to protect the environment and workplace. Various materials have been selected for shielding against EMI depending on their shielding effectiveness (SE) over different frequency ranges [1-4].

For an EMI shielding materials it is needed to have excellent electrical conductivity. Traditionally polymers are electrically insulating and transparent to electromagnetic waves. However, when insulative polymers incorporate with conductive fillers, there is a dramatic improvement in both electrical conductivity and shielding properties of composites. A number of works have endeavored to prepare conductive polymer composites to contain carbon black (CB), carbon fibers (CF), carbon nanotubes (CNTs), carbon nanofibers (CNFs) and nickel-coated carbon fibers (NCCF) [5-10]. CB has the lowest cost advantage over other carbon-based fillers and especially double percolation phenomena will occur for CB-filled binary immiscible polymer blend composites, in which the percoaltion concertration of CB particles reduce [11-13].

Natural fibers-filled polymer composites appear to have attracted more and more research interests because of their environmental friendly nature, low cost, abundant availability, moderae tensile and flexural properties. Sisal fibers (SF) as the assistant fillers are commonly used in polymer composites [14-17]. Conductive fillers coorporating with SF produce a positive synergy effect on the electrical conductivity and other properties of composites.

In this study, an attempt has been made to expole the incorporation of CB particles and sisal fibers treated by steam-explsion and their appropriate mass ratio. In particular, the effect of new macro-conductive networks on electrical conductivity and EMI SE of composites due to the addition of assistant fillers was systematically investigated. Futhermore, compared with PPO/PA66/CB ternary composites, we studied the internal mechanism of performance improvements for PPO/PA66/CB-SF composties.

EXPERIMENTAL

Materials and Preparation of Samples

PPO (1000H) and PA66 (1300S) were supplied from the Asahi Kasei Group (Japan). SEBS-g-MAH was obtained from Nanjing Sutai Polymer Technology Co. (Nan Jing, China). Conductive CB (VXC72) was purchased from Cabot Co. Sisal fibers were purchased from Guangdong Dong Fang Sisal Co., Ltd. (Zhanjiang, China). SF bundles were cut into 5mm short fibers and then blasted into cotton wool-shapes by steam explosion technology. Polymer matrix (PPO and PA66), hybrid fillers (CB and CB-SF) and SEBS-g-MAH were compounded together in proportion by internal mixer and then granulated. Then the PPO/PA66/CB-SF composites were hot pressed into to quadrate plates (3×280×280 mm^3) by compression modeling.

Characterizations

Morphology of cross-sectional surfaces of the composites were observed with a JEOL JSM-5900LV scanning electron microscope at accelerating voltage of 20kV.

Electrical condutivity (EC) was measured by four probe method. The electromagnetic intereference shielding effectiveness (SE) was measured by the screened room method [18]. The set-up consisted of an anechoic shielded room, a signal generator (Agilent E8257D-532), a spectrum analyzer (Agilent 4408B), transmitting and receiving antennas. The electromagnetic frequency range was from 400 MHz to 18 GHz.

RESULTS AND DISCUSSION

Morphology

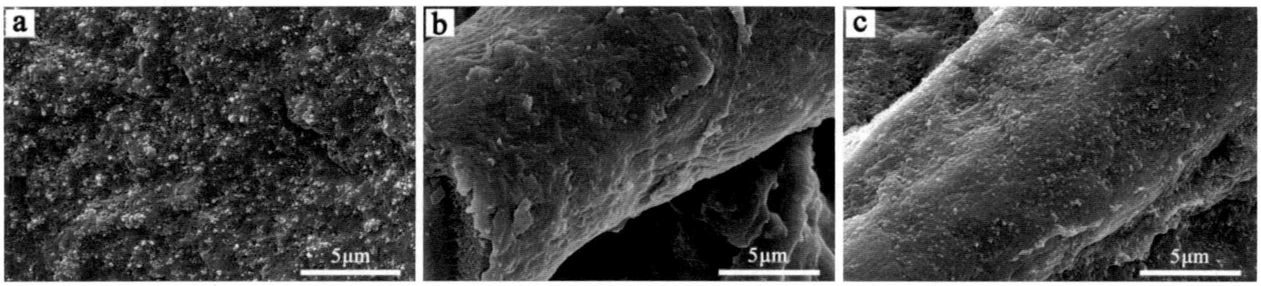

FIGURE 1. SEM micrograph of PPO/PA66 blends filled with (a) 20wt% CB, (b) 10wt% CB-SF, (c) 40wt% CB-SF (PPO/PA66 mass ratio =1/3 and CB/SF mass ratio =1/1)

Figure 1 depicts the SEM micrograph of PPO/PA66 blends filled with CB and CB-SF. For the PPO/PA66/CB ternary composites, the CB preferentially distributed in the PA66 phase and a double-percolated network occurred in Figure1a. For the PPO/PA66/CB-SF quaternary composites, the surface of steam-explosion treament sisal fibers were coated with PA66/CB system to form three-dimensional conductive networks. Compared with 10wt% CB-SF in PPO/PA66 matrix (Figure 1b), much more conducting phase coated on the SF (Figure 1c), which resulted in high electrical conductivity and shielding performance.

Electrical Conductivity

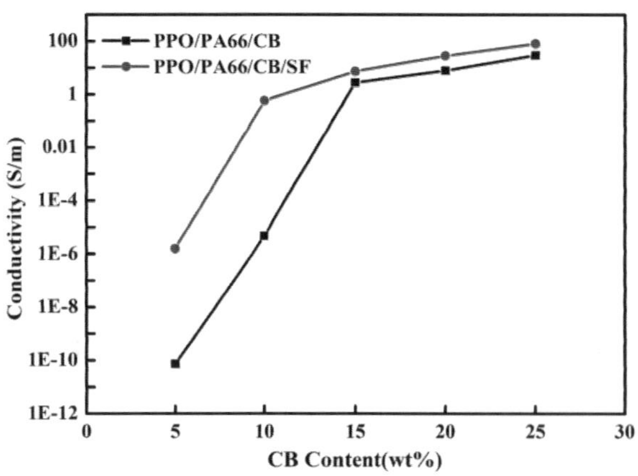

FIGURE 2. Electrical conductivity of the PPO/PA66/CB and PPO/PA66/CB-SF composites as a function of CB content (PPO/PA66 mass ratio =1/3 and CB/SF mass ratio =1/1)

TABLE 1. Electrical conductivity of the PPO/PA66/CB-SF composites

Component	Electrical Conductivity (S/m)
20%CB, 0%SF, PPO/PA66 = 1/3	7.9
20%CB, CB/SF =1/2, PPO/PA66 = 1/3	9.5
20%CB, CB/SF =1/1, PPO/PA66 = 1/3	28.5
20%CB, CB/SF =2/1, PPO/PA66 = 1/3	10.9

Figure 2 shows the effect of CB content on the Electrical conductivity of PPO/PA66/CB and PPO/PA66/CB-SF composites when the mass ratio of CB/SF was 1/1 and PPO/PA66 was 1/3. From Figure 2, It is observed that the EC of PPO/PA66/CB and PPO/PA66/CB-SF composites increased with an incerese in the CB content. For the ternary composites, when the CB content was 10～15wt%, the EC increased dramatically(approximately seven orders of magnitude). However, for quarternary composites, the EC increased progressively at 5~10wt% CB loading. That is, for the PPO/PA66 based composties mixed with CB and SF, the percolation threshold of CB in PPO/PA66/CB and PPO/PA66/CB-SF composites was 10～15wt%, 5~10wt%, respectively. This finding indicates that the percolation threshold of CB loading can be decrease effectively in PPO/PA66 matrix with the synergy of SF assistant filler.

Table 1 shows the electrical conductivity of PPO/PA66/CB-SF composites as a function of mass ration of CB/SF. It can be observed that the EC of PPO/PA66/CB-SF composites achieves a maximum (28.5S/m) as the mass ratio of SF/CB was 1/1. The results presented in Table I indicate that CB-SF can exhibit great synergy effect with the appropriate proprotition and could be an effective assistant filler.

Electromagnetic Shielding Effectiveness Analysis

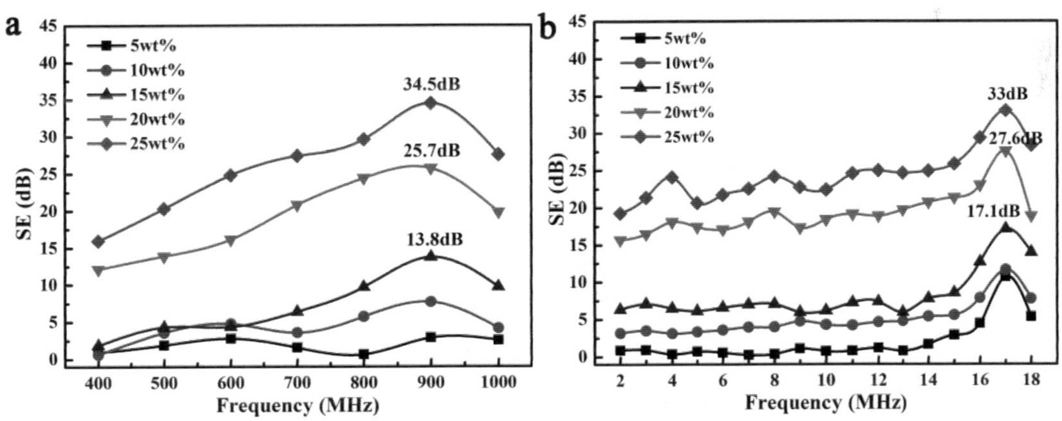

FIGURE 3. EMI SE of the PPO/PA66/CB composites as a function of CB content in the (a) low-frequency range, (b) high-frequency range (PPO/PA66 mass ratio =1/3)

Figure 3 shows the EMI SE of PPO/PA66/CB composites as a function of CB content when the mass ratio of PPO/PA66 was 1/3, over low- and high- frequency ranges, respectively. The SE values of PPO/PA66/CB composites increased with the CB content. At the same time, when the CB loading was 5-15wt%, the SE rise was relatively slow. However, the SE values exhibited a dramatic rise at 15-20wt% CB loading. Therefore, for the PPO/PA66/CB composites, the percolation concentration about EMI SE was 15wt% CB content. Finally, the highest SE value was 34.5dB (900MHz) in low-frequency and 33dB (17GHz) in high-frequency with 25wt% CB loading.

According to the double percolation theory of Sumita, the carbon black is selectively distributed in the PA66 phase and the PPO/PA66 phase interface[17]. When the CB content reached the percolation threshold, those microconductive shains would form a conductive network in PA66 phase. In addition, conductive paths were formed in the PPO/PA66 immiscible blends due to its co-continuous phase structure. In other words, double percolation phenomena had occurred. In a wide frequency range of 400 MHz to18GHz, the SE of PPO/PA66/CB composites was greater than 20dB and the shielding performance was stable when the CB content was higher than 25wt%.

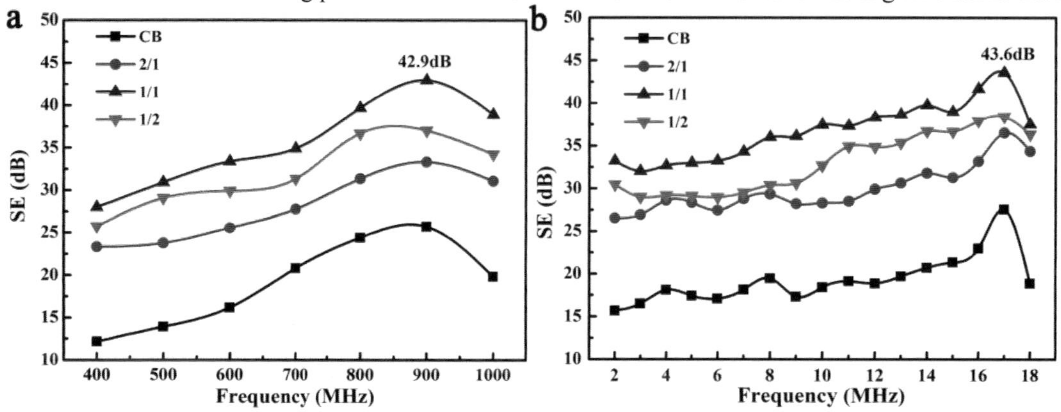

FIGURE 4. EMI SE of the PPO/PA66/CB-SF composites with different CB/SF mass ratio in the (a) low-frequency range, (b) high-frequency range (20wt% CB and PPO/PA66 mass ratio =1/3)

Figure 4 show the EMI SE of PPO/PA66/CB-SF composites with different CB/SF mass ratio when the CB loading was fixed at 20wt% and the mass ratio of PPO/PA66 was 1/3, over low- and high- frequency ranges, respectively. It is observed that the SE of the PPO/PA66/CB-SF composites improved progressively with the addition of steam-explosion treated sisal fibers over the measured frequency ranges. Owing to the steam-explosion treament, the aspect ratio and surface activity of sisal fibers increased. At the same time, the fibrillar structure of cotton wool microfibers was appeared. In spite of SF itself being insulating, the CB concentration raised in the PPO/PA66 matrix by the addition of SF due to the Volume Exclusion Effect, and then the distance between CB particles reduced, which helped a lot in forming the micro-conductive chain. Furthermore, PA66 phase enriched with CB was coated on SF by the reaction of between a mass of free hydroxyl groups on the surface of steam-explosion treated sisal fibers and amine groups of PA66 matrix, which resulted in the formation of conductive sisal fibers.

As shown in Figure 4, the best performance of shielding in PPO/PA66/CB-SF composites accomplished as the mass ratio of CB/SF was 1/1. It exhibited shielding level of 42.9dB (900MHz) in low-frequecny and 43.6dB (17GHz) in high-frequency range. However, when the steam-explosion treated sisal fibers content were too mcuh or less, for instance, the mass ratio of CB/SF was 1/2 or 2/1, the combination of Volume Exclusion Effect and the mount of conductive sisal fibers perform less coordinated at the same CB loading. This indicated that conductive fillers (CB particles) and assistant filler (steam-explosion treated sisal fibers) produced a effective synergy effect on the formation of three-dimensional conducitve network as the mass ratio of CB/SF was appropriate.

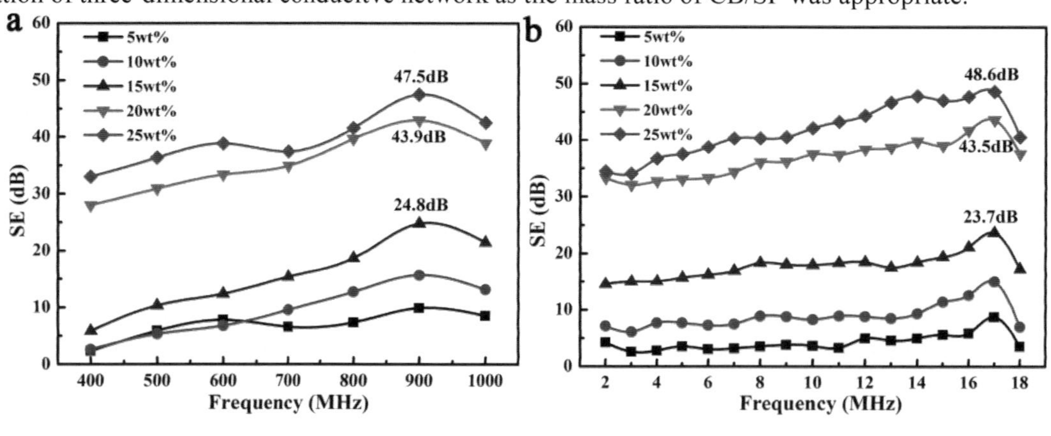

FIGURE 5. EMI SE of the PPO/PA66/CB-SF composites as a function of CB content in the (a) low-frequency range, (b) high-frequency range (PPO/PA66 mass ratio =1/3 and CB/SF mass ratio =1/1)

Figure 5 show the EMI SE variation of PPO/PA66/CB-SF composites containing different CB content when the mass ratio of PPO/PA66 was 1/3 and that of CB/SF was 1/1, over low- and high- frequency ranges, respectively. It was apparent that on increased CB loading, shielding level increased, but the rate was much faster at 15-20 wt% loading. Thus, for PPO/PA66 blends filled with CB-SF, 15-20 wt% CB loading was as the shielding percolation region. Compared with PPO/PA66/CB composties in Figure 3 and Figure 4, the incorporation of CB-SF increased the conductivity performance and reduced the percolation threshold of quarternary composites. In this case, the maximum SE value of PPO/PA66/CB-SF composites was about 47.5 dB (900 MHz) and 48.6 dB (17 GHz), over low- and high- frequency ranges, respectively. Morever, with CB loading being above 20wt%, the SE shows greater than 30dB over the measured frequency ranges, which can meet the civil application microwave shielding requirement[17].

CONCLUSION

PPO/PA66 immiscible polymer blends filled with CB and CB-SF were fabricated for EMI shielding application and the morphology and electrical characteristics were investigated. SEM micrographs showed that CB particles were selectively distributed in the PA66 phase and PA66/CB systems were coated on the surface of steam-explosion treated sisal fibers which form the three-dimensional conductive networks. Compared with PPO/PA66/CB ternary composites, the percolation threshold of PPO/PA66/CB-SF quarternary composites was much lower and the shielding performance was enhanced at the same CB loading due to the addition of SF. An appropriate mass ratio of CB/SF (1/1) helped a lot in achieving the best shielding performance; the highest SE value of quarternary composites was 47.5-48.6 dB over the measured frquency ranges. All those results implied that PPO/PA66 polymer-blends filled with CB-SF was a simple and low cost method to prepare the new promising shielding material for commercial aplication.

Acknowledgments

This work was supported by the Natural Science Foundation of China (No.51273068).

REFERENCES

1. D. D. L. Chung, Carbon 39, 279-285 (2001).
2. C. Y. Huang and C. C. Wu, *Eur Polym J* **36**, 2729-2737 (2000).
3. J. S. Im, J. G. Kim and Y. S. Lee, *Carbon* **47**, 2640-2647 (2009).
4. J. C. Huang, *Adv Polym Technol* **21**, 299-313 (2002).
5. J. Hwang, J. Muth and T. Ghosh, *J Appl Polym Sci* **104**, 2410-2417 (2007).
6. M. Sumita, K. Sakata, S. Asai, K. Miyasaka and H. Nakagawa, *Polym Bull* **25**, 265-271 (1991).
7. M. Sumita, K. Sakata, Y. Hayakawa, S. Asai and K. Milyasaka, *Colloid Polym Sci* **95**, 197-202 (1978).
8. N. Sgriccia, M. C. Hawley and M. Misra, *Compos Part A* **39**, 1632-1637 (2008).
9. S. J. Kim, J. B. Moon, J. H. Kim and C. S. Ha, *Polym Test* **27**, 801-806 (2008).
10. H. Ku, H. Wang, N. Pattarachaiyakoop and M. Trada, *Compos Part B-Eng* **42**, 856-873 (2011).
11. J. C. K. D. Verney, M. F. S. Lima and D. M. Lenz, *Mater Res* **11**, 447-451 (2008).
12. M. E. Malunka, A. S. Luyt and H. Krump, *J Appl Polym Sci* **100**, 1607-1617 (2006).
13. J. M. L. Reis, *Constr Build Mater* **37**, 177-180 (2012).
14. M. P. Brown and K. Austin, *Appl. Phys. Letters* **85**, 2503-2504 (2004).
15. B. Wessling, *Polym Eng and Sci* **31**, 1200-1206 (1991).
16. J. Li and J. K. Kim, *Compos Sci Technol* **67**, 2114-2120 (2007).
17. F. Dalmas, R. Dendievel, L. Chazeau, J. Y. Cavaille and C. Gauthier, *Acta Mater* **54**, 2923-2931 (2006).
18. Y. Li, C. X. Chen, S. Zhang, Y. W. Ni and J. Huang, *Appl Sur Sci* **254**, 5766-5711 (2008).

The Study on the Stress-Photooxidation Aging of Polyethylene Cable Sheath under Different Processing Conditions

Ren Xiancheng [a,*]

[a] College of Polymer Science and Engineering, Sichuan University, Chengdu, 610065, China.

Abstract. Polyethylene sheath materials were under stress, ultraviolet light, heat and other comprehensive factors in the process of actual use. Different molding process had influence on the state of aggregation structure and internal stress of polyethylene sheath materials, which lead to the difference of product aging resistant performance. The long-term stress and photo-oxidative aging behavior of highdensity polyethylene (HDPE) under different processing conditions was studied using a stress-aging apparatus. The aging behavior was investigated through the methods of the surface morphology observation, fourier transform infrared spectroscopy, differential scanning calorimetry. It was found that the aging resistance of sheath had been associated with sheath surface layer crystallinity. Sample in the lower extrusion temperature system and higher extrusion speed had lower surface layer crystallinity, which the aging resistance was the best. In addition, the stress and photo-oxidative aging mechanism had been explored, there were two stages in the stress and photo-oxidative aging, reflected in the surface layer crystallinity increased at first and then decreased. It was found that the service life of sheath material was effected by the extrusion technology and materials largely. This study may be useful for the rational selection of HDPE for the sheath material of bridge cable.

Keywords: polyethylene, stress and photo-oxidative aging, cracking, molding technology
PACS: 81.05.Lg

INTRODUCTION

High-density polyethylene (HDPE) has been widely used as the sheath material of bridge cable due to its excellent and comprehensive properties, such as chemical resistance, corrosion resistance, UV aging and environmental stress cracking resistance, etc[1]. .Over the past decades, abundant researches have been accumulated in the photo-oxidative aging of HDPE relating to the changes of chemical structure, mechanical properties and other aspects of the process [2-6].

However, the HDPE sheath gets earlier cracking in the service period on the effects of sun exposure, rain, air environment and stress (stress generates during processing and installation), the cable has to be replaced or repaired far less than the actual design life about 50 years. The process of molding has influence on the state of aggregation structure and internal stress of polyethylene sheath materials, which lead to the difference of product aging resistant performance. The long-term stress and photo-oxidative aging behavior of high-density polyethylene (HDPE) under different processing conditions was studied using a stress-aging apparatus.

In this work, the cracking behavior of HDPE under different processing conditions during stress and photo-oxidation was studied through the methods of the surface morphology observation, structure analysis. This study may be useful for the rational selection of HDPE for the sheath material of bridge cable deployment under different processing.

EXPERIMENTAL

Materials and Sample Preparation

HDPE HE3490-LS was supplied as pellets by Borstar. The material data of HDPE used was showed in TABLE 1.

HE3490-LS sheath was extruded in SJ150X25B single screw extruder at two different processing temperature and two different processing speed. The processing temperature and speed were showed in TABLE 2.

Afterwards the extruding sheath samples were detached from the cable and measured the shrinking percentage. Then cut into standard dumbbell-shaped specimens. (length 150 mm, width 25 mm).

Stress and Photo-oxidative Aging

Stress and photo-oxidative aging experiments of HDPE specimens were conducted in a self-made aging apparatus

(Chengdu yibang analysis instrument Co., Ltd) at 50 °C ± 1 °C with air circulation and relative humidity 30% ± 1%.The initial stress applied was 6.3~7.8 MPa, which was calculated by the multiplication of tensile modulus and shrinking percentage . The UV light was provided by a 1000 W Ga-In source lamp with a spectrum range of 300~405 nm, and with a maximum intensity at 365 nm. The irradiation intensity was 15.5 Wm^{-2} tested by a UV irradiance meter model UV-A Tester (Photoelectric Instrument Factory of Beijing Normal University, Beijing, China). The samples were taken out to at interval time and used for characterization.

TABLE (1). Material data of HDPE used

Sample	MFR (g/10 min)	Density (g/cm^3)	Yield strength (MPa)	Tensile Strength (MPa)	Tensile modulus (MPa)
HE3490-LS	0.25	0.950	25	32	500

TABLE (2). The numbers of sample

Sample numbers	Extrusion temperature	Extrusion speed
1#	140~190 °C	4 m/min
2#	140~190 °C	6.5 m/min
3#	175~220 °C	4 m/min
4#	175~220 °C	6.5 m/min

Characterization

Metallurgical Microscopy Observations

Surface morphologies of stress and photo-oxidative aged samples were observed using a UM200i metallurgical microscope (Chongqing UOP Photoelectric Technology Corp. Ltd, China)

Fourier Transform Infrared Spectroscopy

Infrared spectra of aged samples were recorded by using a Nicolet 6700 instrument equipped with a diamond ATR accessory. Carbonyl index (CI) value was defined as the ratio of the carbonyl stretching peak area range 1670 to 1800 cm^{-1} and CH_3 stretching peak area at 2917 cm^{-1}.

$$CI = \frac{A_{C=O}}{A_{2917}} \quad (1)$$

where $A_{C=O}$, A_{2917} are respectively carbonyl group absorption peak areas and internal reference peak areas.

Differential Scanning Calorimetry

A TA instrument DSC Q200 was used for the thermal analysis of different aging time specimens. Approximately 5mg powder samples scraped from the surface of specimens were heated from 40 °C to 210 °C at rate of 10 °C min^{-1}. The heat of fusion (ΔH) of experimental melting peak area was derived to calculate crystallinity (Xc):

$$X_C = \frac{\Delta H}{\Delta H_{100}} \times 100\% \quad (2)$$

Here ΔH_{100} is defined as the heat of fusion of 100% crystalline HDPE (293 J/g [7-8])

In addition, oxidation induction time (OIT) was measured on the same instrument. 5mg samples were heated to 210 °C at the same rate, with 50 ml min^{-1} N_2 flow. After thermal equilibration for 5 min, the purge gas was switched to O_2. OIT was the time interval between O_2 insertion and oxidation onset point measured within ± 0.1 min.

RESULTS AND DISCUSSION

Metallurgical Microscopy Observations

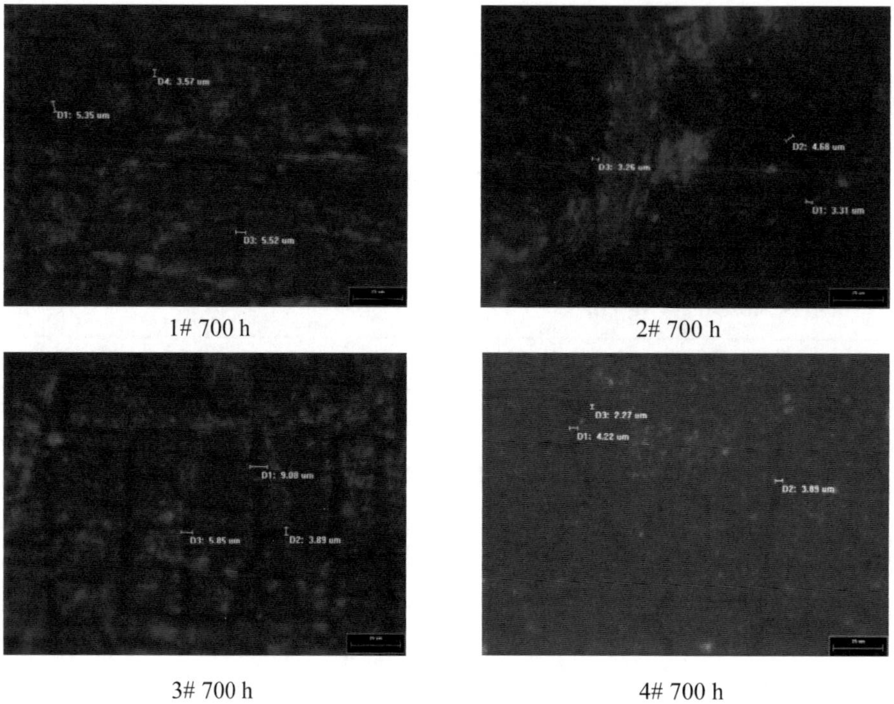

FIGURE 1. Micrographs of samples surface after 700 h stress-photooxidation aging（×200）

To investigate the microstructure variation of HDPE surface under different processing conditions during stress and photo-oxidation, the surfaces of samples were observed with metallurgical microscopy, the micrographs were shown in FIGURE 1. The stress direction was horizontal. The surface of 3# sample appeared cracks after 100 h, then appeared the cracks that both perpendicular to the stress direction and parallel to the stress direction, formed a T-shaped cracks. The surface of 1# sample appeared cracks after 300 h. The surface of 2# and 4# sample appeared cracks after 500 h, and the width of cracks of 2# sample was narrower than 4#.

This phenomenon indicated that the sheath material was effected by the extrusion technology largely.

Fourier Transform Infrared Spectroscopy

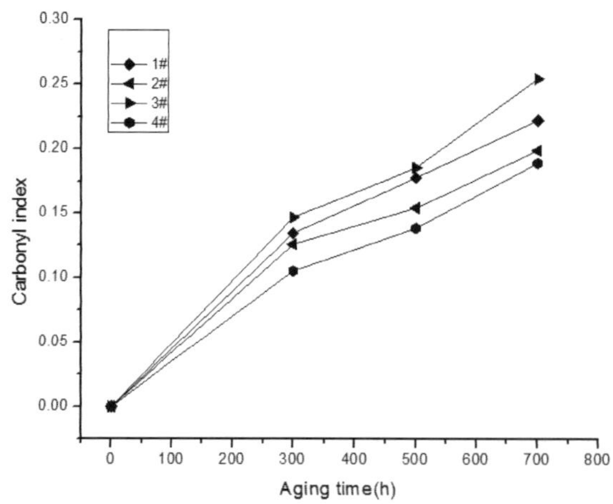

FIGURE 2. Carbonyl index of four materials during stress and photo-oxidative aging

IR spectroscopy was used to characterize the oxidation products that were formed during polyethylene samples degradation reactions [9-11]. In this work, ATR-FTIR was characterized the chemical changes on the surface of samples after stress photo-oxidative aging. It was clear that carbonyl groups were formed after aging from the absorption band at 1715 cm-1 which was the C=O stretching vibration.

The degree of oxidation for HDPE samples during aging was characterized by the carbonyl index. The carbonyl index of four materials during stress and photo-oxidative aging were shown in FIGURE 2. It could be seen that CI increased significantly with increasing aging time, compared the CI under different processing conditions, it can be found that at 700 h the size of carbonyl index was 3#>1#>4#>2#, which was the same as the development speed of surface crack.

Differential Scanning Calorimetry

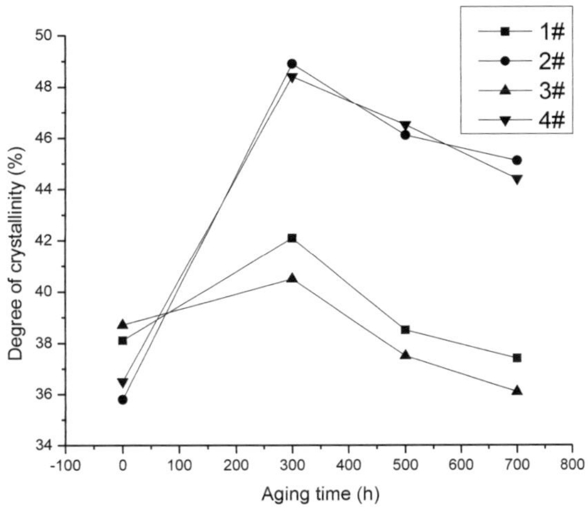

FIGURE 3. Crystallinity changes of four samples during stress and photo-oxidative aging

The crystallinity changes of four samples during stress and photo-oxidative aging were shown in FIGURE 3. It could be seen that there were two stages in the stress and photo-oxidative aging, reflected in the surface layer crystallinity increased at first and then decreased. Because recrystallization occurred at first, then under the effects of stress intergranular amorphous area tangled chain rupture, uproot from crystal zone, crystal damage made the reduce of crystallinity. In addition it was found that the aging resistance of sheath had been associated with sheath surface layer crystallinity. Sample in the lower extrusion temperature system and higher extrusion speed had lower surface layer crystallinity, which the aging resistance was the best.

CONCLUSION

Through the stress-ultraviolet aging test, it was found that the aging resistance of sheath had been associated with sheath surface layer crystallinity. Sample in the lower extrusion temperature system and higher extrusion speed had lower surface layer crystallinity, which the aging resistance was the best. In addition, the stress and photo-oxidative aging mechanism had been explored, there were two stages in the stress and photo-oxidative aging, reflected in the surface layer crystallinity increased at first and then decreased. The sheath life was effected by the extrusion technology largely. This study may be useful for the rational selection of extrusion technology for the sheath material of bridge cable.

ACKNOWLEDGEMENTS

The author would like to thank the National Natural Science Foundation of China (grant 51273130) for financial support of this work.

REFERENCES

1. S.S. Wu, J. Zhang, and X. Xu, *Polym. Int.* **52**, 1527-1530 (2003).
2. X.M. Shi, J.D. Wang, S. Stapf, C. Mattea, W. Li, and Y.R. Yang, *Polym. Eng. Sci.* **51**, 2171-2177 (2011).

3. F. Carrasco, P. Pages, S. Pascual, and X. Colom, *Polym. J.* **37**, 1457-1464 (2001).
4. S. A. Jabrain and A. L. Elizabeth, *J. Appl. Ploym. Sci.* **53**, 411-423 (1994).
5. R. Yang, Y. Liu, J. Yu, and D. Q. Zhang, *Polym. Eng. Sci.* **48**, 2270-2276 (2008).
6. J. Pabiot and J. Verdu, *Polym. Eng. Sci.* **21**, 32-38 (1981).
7. Wunderlich, B. and Dole, M, *J. Polym. Sci.* **24**, 201-213(1957).
8. Wunderlich, B. and Czornyj, G, *Macromol.* **10**, 906-913(1977).
9. G.R. Rideal and J.C. Padget, *Polym. Degrad. Stab.***57**, 1-15 (1976).
10. F. Gugumus, *Polym. Degrad. Stab.* **67**, 35-47 (2000).
11. T. Zanasi, E. Fabbri, and F. Pilati, *Polym. Test.* **28**, 96-102 (2009).

Influences of Resin Components and Injection Rates on Generation of Tiger-striped Flow Marks in Injection Molding

Shigeru OWADA and Hidetoshi YOKOI

Institute of Industrial Science, The University of Tokyo
4-6-1, Komaba, Meguro-ku, Tokyo, 153-8904, JAPAN

Abstract. In this study, the effects of materials and molding conditions on the generation of tiger-striped flow marks in the injection molding of polypropylene (PP)/rubber/talc blends were investigated. We conducted investigations on the relationship between flow mark occurrence and viscosity characteristics of resins having different component ratios as well as that and the velocity of flow front using a bar-flow cavity with 650 mm flow length. In previous studies, we reported the phenomenon whereby flow-marks occur due to the oscillation of the flow front, which is the swinging movement of melt spurting from both sides of the cavity wall during the flow process. Furthermore, we proposed a model on the flow mark generation based on the results of our visualization analysis. Applying this model, we further expanded and investigated the range of influential factors in detail, and obtained the following new interesting results;

By evaluating the effects of the resin component ratio of PP/rubber/talc, we found that the flow marks occurred in conjunction with the oscillation of the flow front for the base PP containing talc. In addition, the flow marks and the oscillation movements became more distinct as the rubber content increased. These resins in which flow marks are generated showed such phenomena as sudden drop in the extensional viscosity curve at a certain range of strain. This suggests that melt fractures due to the shape factor of the filler components could occur through the melt extension process. In addition, we also confirmed that no flow marks were observed for any component ratio in high speed injection molding in which flow front velocities were measured over 2800 mm/s. It clearly suggests the possibility that there may be a flow front velocity threshold restricting the oscillating or swinging movement of the flow front, irrespective of rising melt temperature and melt pressure under high injection rates.

Keywords: Polypropylene, Rubber, Talc, Tiger-striped flow mark, Injection molding, Fountain flow
PACS: 81.05.Lg, 81.20.Hy, 83.50.Uv, 83.60.Wc, 83.80.Tc

INTRODUCTION

Tiger-striped flow marks are typical external defects appearing on injection molded product surfaces. These flow marks consist of glossy and cloudy bands appearing alternately on the surface of molded products. One major feature is that their phases are reversed on the front and back sides of the molded products. In the molding of car bumpers using PP compounds containing rubber or talc, usually tiger-striped flow marks tend to occur mainly in areas where the melt flow ends. As these flow marks affect the aesthetic design of molded products, their causes need to be clarified urgently.

The authors have discovered that these flow marks occur due to the asymmetrical fountain flow phenomenon in the flow process [1]-[3]. Following this, the conditions for the numerical analysis of these causes and the conditions of the rheological characteristics of materials were studied by various researchers [4]-[9], and it was further reported that the signs of unstable flow remaining in the molded products as flow marks become conspicuous due to aging degradation [10]. However, how asymmetrical fountain flow occurs and its generation mechanism have yet to be determined.

The authors have been conducting dynamic visualization analysis of the flow mark generation process and investigations on the flow history using molded products of two colors. Based on the results obtained, we proposed a model on asymmetric fountain flow which temporarily occurs to relax the extensional front surface at the moment the melt fractures near one side of cavity wall. It results in periodic differential transcription between the melt spurting side and the opposite side [11]. In this study, we further investigated the factors affecting the two stages of the flow mark generation process of this model, which are the occurrence of oscillating movements of the flow front and the mold transcription process, under a broad range of conditions.

Measures to prevent such defects are being studied, such as giving the melt a higher elasticity [7, 9], adjusting the viscosity of rubber component [12], reducing the filling amount of talc, hiding the flow marks by applying grain processing to the cavity surface, etc. But, the problem of flow marks has not been completely resolved.

Against this background, we studied the correlation between the oscillating movement of the flow front and viscosity characteristics of the resin for different component ratios. We also investigated the influence of the injection rate as a way to control the state of mold transcription by using a bar flow mold with a flow length of 650mm, which is a length simulating the flow length of actual molded products.

TABLE 1. Experimental materials

Resin material No.	Component ratio (wt %) Block-PP/ Rubber/ Filler	MFR (g/10min)
M-0	100/ 0/ 0	12
M-1	76/ 4 / 20(Talc)	12
M-2	80/ 0 / 20(Talc)	--
M-3	80/ 20/ 0	--
M-4	60/ 20/ 20(Talc)	--
M-5	80/ 0 / 20(Spherical silica)	--

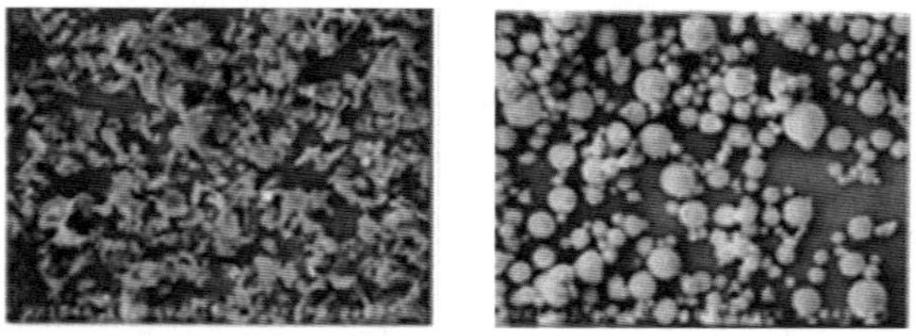

(a) Talc (b) Spherical silica

FIGURE 1. SEM images of fillers which are components of resin

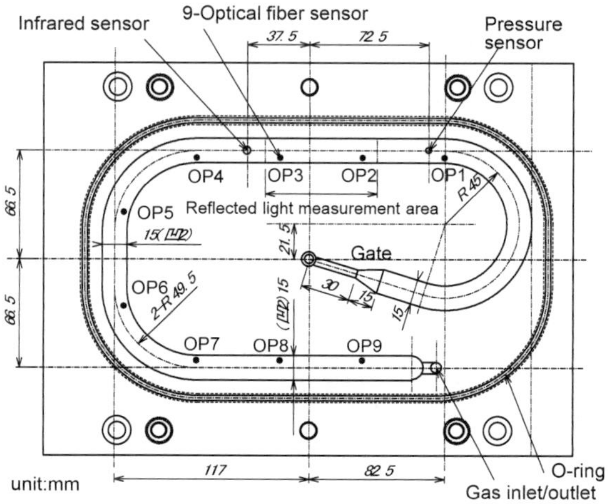

FIGURE 2. Cavity shape and measuring positions of melt temperature, cavity pressure and melt passing time [Movable side view of bar-flow mold]

EXPERIMENTAL METHOD

Resin Components and Flow Mark Generation Test

The component ratios of the resins used are shown in Table 1. Resin material M-0 contains only PP copolymer, and M-1~5 contain PP copolymer, rubber, and filler which is either talc or spherical silica particles. Carbon black (CB) was used to color the resins. Here, the Mooney viscosity of the rubber contained in the resins was 60(125°C, ML1+4). Figure 1 shows the photographs of the fillers used. The average particle sizes of the talc and spherical silica were 6μm and 5μm respectively. A rectangular cavity measuring 180mmL×40mmW ×2mmT in size was used. The cavity surface had a mirror finish with calculated average surface roughness (Ra) of less than 0.1μm. The size of the side-gate was 1.0mmL×2.0mmW×1.0mmT. The molding conditions in the tests were: cylinder temperature of 220°C, mold temperature of 30°C and holding pressure of 22.1MPa for 5s. For the injection molding machine, the Sumitomo Heavy Industry ULTRA75-C250 was used. During the injection molding, the flow front behavior was observed with a visualization system used for videoing the path of the flow front during the cavity filling process. Details are omitted as they have been discussed in a previous report [13].

Measurement of Melt Viscosity

The share viscosity characteristics were measured for each composite resin by capillary rheometer RH7D (Rosand Corp.) at the temperature of 220°C. Extensional viscosity characteristics were also measured using a Meissner-type rheometer RME at the temperature of 170°C and the extensional strain rate of 0.01, 0.1, 1.0/s.

Flow Mark Generation Tests using Bar-flow Mold

Figure 2 shows the cavity shape of bar-flow mold. The cavity surface has a machined finish. The mold has a cavity thickness of 2mm and a flow length of 650mm. A pressure sensor 6001B (piezoelectric sensor type, PRIAMUS SYSTEM TECHNOLOGIES AG) was mounted at the location of 192mm from the gate and the infrared thermometer RPH-OA013MVA (response time to 95% of less than 10μs, Tokyo Seiko) mounted at the location of 302mm from the gate. In addition, optical fiber sensors OP1~OP9 were mounted at the locations shown in Figure 1 on the surface of the stationary cavity block. The actual flow front velocity was calculated from the time at which the flow front passes over the optical sensors and the positions of the optical sensors. The gas vent was set up at the end of the cavity. For the injection molding machine, the Nissei Plastic Industrial FN4000-9HTN (maximum screw speed of 1000mm/s, translated maximum injection rate of 804cm^3/s) was used. To measure the flow mark generation area, the intensity of the reflected light was measured.

RESULTS AND DISCUSSION

Flow Front Behavior during Filling Process and Flow Mark Generation Status

Table 2 shows the state of flow mark generation in the visual observation of molded products for each component of the resin. This table also shows the results of observing the behavior of the flow front during the

TABLE 2. Evaluation of results in flow mark generation test under various components of resin

Material No.	Injection rate Q (cm^3/s)	Maximum shear rate γmax (s^{-1})	Flow mark occurrence	Behavior of melt front
M-0	18.1	679	×	×
M-1	18.1	679	○	○
M-2	18.1	679	○	○
M-3	9.0 18.1 27.1 45.2	338 679 1016 1695	× ○○ ○ ×	– ○ – –
M-4	18.1	679	○○	○
M-5	18.1	679	×	×

·Cavity shape is rectangular: 40.0Wx180.0Lx2.0T(mm)
·Maximum shear rate is calculated by $\gamma max=6Q/BH^3$, B=40 mm (cavity width), H=2 mm (cavity thickness)
·Flow mark occurrence is classified as very visible (○○), visible (○) and nothing (×).
·Behavior of melt front is classified as oscillating flow (○) and normal fountain flow (×).

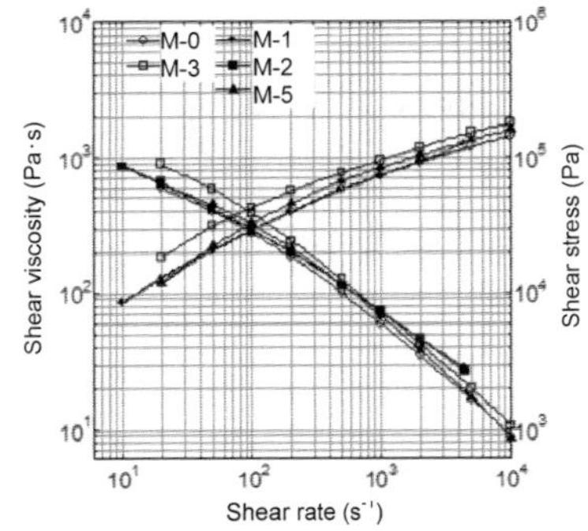

FIGURE 3. Share viscosity measured for various composite resins by capillary rheometer at 220°C

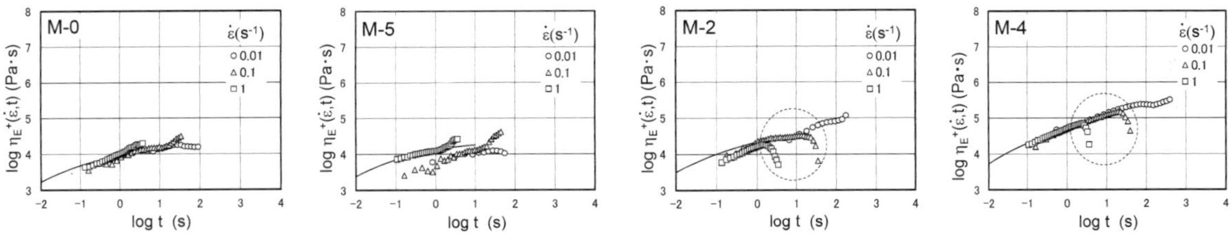

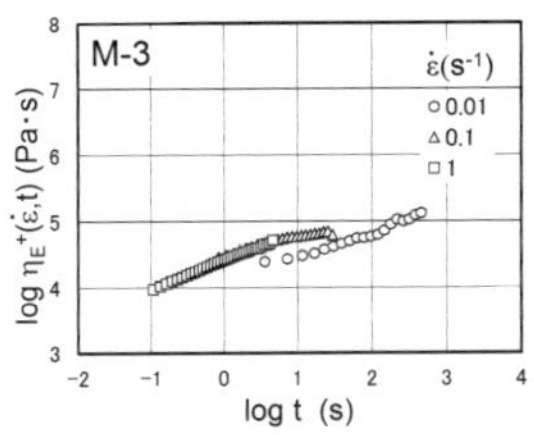

FIGURE 4. Extensional viscosity measured for various composite resins by Meissner-type rheometer at 170°C

filling process. Since the flow marks for each resin were observed in the injection rate range from 9.0 to 45.2cm^3/s, it can be seen that the 18.1cm^3/s is the representative injection rate except for the resin material M-3. In addition, flow marks were not found to occur in the range of the examined injection rate for M-0 which consists only of PP copolymer and M-5 which contains spherical silica particle as a filler. The oscillating movement of the melt front was not also observed with these resins. On the other hand, flow marks occurred when only talc was contained as the filler (M-2) and when rubber was added (M-3), and they became more distinct as the rubber content increased for the resin containing talc (M-1, M-4). Resins with which flow marks occur also showed oscillating movements of the melt at the surface of the flow front during the filling process. It is thought that the main reason for the oscillating movement of the flow front is the effect of flake-shaped talc on the orientation in the flow direction, because this phenomenon was not observed with M-5 which contains spherical silica particle as a filler.

Melt Viscosity Characteristics

Figure 3 and Figure 4 show the measurement results of shear viscosity and extensional viscosity, respectively. In Figure 3, all resins demonstrate more or less similar trends for shear viscosity characteristics, and no distinct differences between each resin can be observed regardless of whether flow marks generated or not. In Figure 4, M-0 and M-5 with which flow marks did not occur showed that extensional viscosity increases with time. The influence of the spherical filler was also insignificant within the investigated range. On the other hand, M-2 and M-4 with which the flow marks are generated showed phenomena such as sudden drop in the extensional viscosity curve at a certain range of strain. This suggests that adhesive failure occurs at the interface between the matrix resin PP and flake-shaped talc which orient in the flow direction, causing melt fractures through the melt extension process. Though the resin temperature differs from that of injection molding, this suggests that the melt may have fractured near one side of cavity wall subject to maximum stress due to accumulating extensional stress up to a certain level on the surface of the flow front. It is thought that the decentering of the spurting melt due to the melt fracture triggers the generation of asymmetric fountain flow. However, M-3 which contains only rubber showed neither strain hardening nor drop in the extensional viscosity for the examined range. This is considered to be related to the phenomenon that no flow mark is observed at low

TABLE 3. Evaluation results in flow-mark occurrence test by using bar-flow mold

Screw speed setup Vs (mm/s)	Theoretical injection rate Q (cm^3/s)	Theoretical flow-front velocity Vf (mm/s)	Mold temp. 40°C			Mold temp. 85°C	
			Measured average flow-front velocity Vf1 (mm/s)	Flow-front contact temperature on the infrared sensor (°C)	Flow-mark occurrence	Measured average flow-front velocity Vf2 (mm/s)	Flow-mark occurrence
10	8.0	267	185	204	○	–	–
30	24.0	800	589	208	○	596	○
60	48.2	1607	–	–	–	1164	○
80	64.3	2143	1539	215	○	1547	Δ
100	80.4	2680	1917	216	Δ	1932	×
150	120.6	4020	2873	219	×	2892	×
200	160.8	5360	3778	222	×	3784	×
500	402.0	13400	8687	229	×	–	–

·The resin used is M-1.
·Molding conditions are 220°C (cylinder temperature), 40°C or 85°C (mold temperature).
·Theoretical flow-front velocity is calculated by Vf=Q/(BH) : B=15 mm (cavity width), H=2 mm (cavity thickness)
·Flow mark occurrence is classified as visible (○), difficult to see (Δ) and nothing (×).

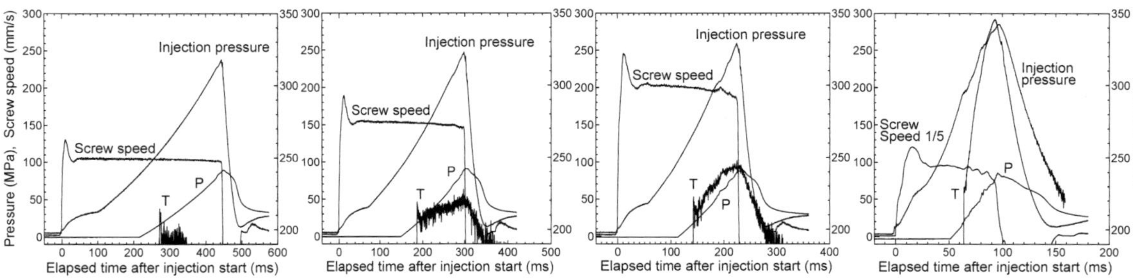

FIGURE 5. Change curves of indicated properties with time during one shot cycle of injection molding under different injection rates [The resin used is M-1. Molding conditions are 220°C (cylinder temperature) and 40°C (mold temperature). T is temperature data measured by infrared sensor and P is pressure data measured by pressure sensor]

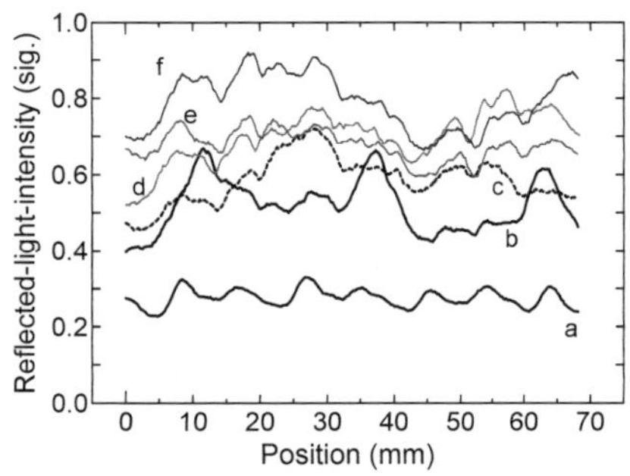

FIGURE 6. Measurement results of reflected light intensity distribution on surface of flow mark generation area shown in FIGURE 2 [The resin used is M-1. Screw speeds are 10 mm/s (a), 30mm/s (b), 80mm/s (c), 100mm/s (d), 150mm/s (e) and 200mm/s (f). The other molding conditions are 220°C (cylinder temperature) and 40°C (mold temperature)]

injection rates as shown in Table 2. Melt fracture is also thought to occur due to the deformation behavior of rubber particles at higher injection rates where extensional strain rates are higher.

Flow Mark Generation Status under the Condition of High Injection Rate

Table 3 shows the evaluation results of the flow-mark generation test using a bar-flow mold at each injection rate for M-1.This table also shows the average value of the actual flow front velocity detected by the optical sensors. As shown in the table, under the mold temperature of 40°C, the generation of the flow marks is suppressed at high flow front speeds of 2800mm/s or more. Furthermore, we also confirmed that no flow marks even observed for all resin materials M-1~ 4 during high speed injection molding.

Figure 5 shows the change curves for the various properties with time during one shot injection molding cycle at different injection rates for M-1. As shown in this figure, the melt contact pressure P at each injection rate linearly increases with the increase in the injection pressure from the moment when the melt is in contact with the sensor. The pressure reached after the injection process is similar value at about 90MPa at each injection rate, and no difference was observed between the different injection rates. On the other hand, when the screw speed Vs is less than 100mm/s at which flow marks are generated, the melt contact temperature T shows a maximum value at the moment when the melt is in contact with the sensor, followed by a decrease. However, when Vs is more than 150mm/s at which flow marks are suppressed, T shows a small peak value at the moment when the melt contacts the infrared sensor, and it continues to rise thereafter until the flow stops. The temperature reaches up to about 350°C at Vs=500mm/s.

In our view, the flow marks are formed at the moment both edges of the flow front contact a low temperature mold surface accompanied by the oscillating movement of the flow front. For this reason, it is not likely that that this rise in temperature at the internal layer near the surface due to the shear heating causes the phenomena of the re-melting and re-transfer of resins after the skin layer forms. Consequently, the rising temperature in the internal layer at high injection rates is most probably not the factor which suppressed the flow mark. Table 3 also shows the flow-front temperature, which peaks at the moment the melt comes into contact with the infrared sensor at each Vs. The temperature at Vs=30mm/s at which flow marks were found to occur and that at 150mm/s at which flow marks did not occur were compared and the difference was about 10°C. It clearly suggests the possibility that the oscillating or swinging movement of the flow front may be restricted by the flow front velocity, irrespective of rising melt temperature and melt pressure under high injection rates.

The results of measuring the intensity of the reflected light for the molded samples at each screw speed are shown in Figure 6. As shown in the figure, peaks of the periodic changes in the surface gloss can no longer be seen at Vs=100mm/s, and similar curve patterns corresponding to the improvement of the transcription of the finishing patterns to the mold surface start to appear. This result suggests that the periodic changes in the gloss of the surface of molded products are unlikely to occur because of uniform and high transfer due to the high stretching rate of the melt regardless of whether irregular fountain flow occurs during high speed injection molding.

On the other hand, mold temperature serves also as a factor changing the transcription property. From Table 3, it can be seen that the mold temperature of 85°C tends to slightly shift the boundary of the flow mark generation towards the low injection rate side Vs=80mm/s. But it was clarified that this has no significant effects within this mold temperature range.

CONCLUSIONS

We conducted investigations on the relationship between flow mark occurrence and viscosity characteristics of resins with different component ratios as well as between that and the velocity of the flow front using a bar-flow cavity with flow length of 650 mm. In our view, flow-marks occurs due to the periodic difference in the transcription state due to the oscillating movement of the flow front, and the results obtained in this study show the possibility that flow marks can be suppressed by controlling mold replication and restricting flow front movement. Specifically, the following results were obtained.

(1) Flow marks occur in conjunction with the oscillating movement of the flow front during the filling process. This behavior occurs in the base PP containing talc, and becomes more distinct as the rubber content increases.

(2) Resins with which flow marks occurred showed such phenomena as a sudden drop in the extensional viscosity curve in a certain strain range. This suggests that melt fractures occur due to the shape factor of the filler components.

(3) We confirmed that flow marks were not observed for any component ratio during high speed injection molding at flow front velocities over 2800mm/s. It suggests the possibility that the oscillating or swinging movement of the flow front may be restricted.

REFERENCES

1. Yokoi, H., Niki, Y. and Seki, T., *JSPP* Annual Meeting 1993, pp.239-242
2. Yokoi, H. and Narita, J., *JSPP* Autumnal Meeting 1998, pp.277-280
3. Yokoi, H. and Narita, J., *JSPP* Autumnal Meeting 1999, pp.125-128
4. Grillet, A.M., Bogaerds, A.C., Peters, G.W., Baaijens, F.P. and Bulters, M., *J. Rheol.*, **46**(3), 651-669 (2002)
5. Bogaerds, A.C., Hulsen, M.A., Peters, G.W. and Baaijens, F.P., *J. Rheol.*, **48**(4), pp.765-785 (2004)
6. Patham, B., Papworth, P., Jayaraman, K., Shu, C. and Wolkowicz, M.D., *J. Appl. Polym. Sci.*, **96**(2), pp.423-434 (2005)
7. Maeda, S., Fukunaga, K. and Kamei, E., *J. the Society of Rheology, Japan*, **35**(5), pp.293-299 (2007)
8. Baltussen, M., Hulsen, M. and Peters, G., *J. Non-Newton Fluid Mech.*, **165**(11), pp.631-640 (2010)
9. Iannuzzi, G. and Rigdahl, M., *J. Appl. Polym. Sci.*, **118**(1), pp.604-610 (2010)
10. Hirano, K., Tamura, S. and Kanai, T., *J. Appl. Polym. Sci.*, **105**(4), pp.2416-2426 (2007)
11. Owada, S. and Yokoi, H., PPS Annual Meeting, Jeju, Korea, pp.350-354 (2015)
12. Hirano, K., Suetsugu, Y. and Kanai, T., *J. Appl. Polym. Sci.*, **104**(1), pp.192-199 (2007)
13. Yokoi, H., Masuda, N. and Mitsuhata, H., *J. Mater. Process. Technol.*, **130**, pp.328-333 (2002)

The Function of Detecting Backflow Phenomenon on Injection Molding Machine and the Evaluation

Satoshi Takatsugi
FANUC Corporation

The injection molding process consists of metering process which reserves resin with melting at the front of the screw and injection process which injects the metered resin at the front of screw into the mold. At the in-line screw type injection molding machine, resin movement at the above two processes is controlled by a check ring. When metering process on the injection molding machine is ended, the check ring is positioned at the front end of the movable stroke, and keeps state of resin flow path opened. At the following injection process, the check ring moves to the rear end of the stroke and the flow path is sealed by the pressure building up generated at the front of the screw. It is known resin backflow occurs during the time. The backflow generation causes injected resin amount error and then results in molding defect such as short shot. Therefore, making backflow amount constant every cycle is mandatory to make molding stable. Conventionally, as for methods to detect the backflow, minimum cushion (the most advanced screw position in injection and pack process) and peak pressure (maximum injection pressure in injection and pack process) are known. However, they are not able to detect backflow phenomenon accurately, since the detected value relies on mold or material condition. This study mentions the method of detecting backflow phenomenon which is developed as a new function of injection molding machine.

Keywords: Check ring, Backflow, wear

INTRODUCTION

On the injection molding machine with inline screw, the check ring installed at head of the screw performs close motion of resin flow path at the beginning of injection. During the closing motion, small amount of resin causes backflow toward to rear side of the screw, then the backflow stops as completion of the close. Therefore, stable injected resin amount is led by repeating same timing of close motion every cycle. In the meantime, on a long-term point of view, unstable backflow occurring in injection due to wearing or abnormality of check ring is possible to cause error in specified injected resin amount. Therefore it is important to have ability to decide the stability on a point of long and short term view with detecting check ring motion. When it is decided that the check ring motion is unstable based on the decision, taking proper countermeasure such as adjusting molding parameters or replacing the check ring is very useful in mass production molding factory.

As for methods of detecting check ring motion, a method of estimating check ring motion with pressure sensors installed at front and rear side of the screw researched by Oka[1)] and others, a method of directly observing check ring motion via visualized cylinder researched by Oka and others as well, and a method of using position sensor (magnetostrictive sensor) which utilizes magnetostriction phenomenon caused by check ring motion with magnet installed on the check ring researched by Ono[2)] are known.

However, the method of estimating check ring motion with pressure sensors installed at front and rear side of screw had an issue of lowering the cylinder strength since the resin pressure sensor had to be installed to the cylinder. In the same manner, the method of using visualized cylinder had difficulty to use it under high injection pressure condition due to the strength issue. Furthermore, the method of using magnetostrictive sensor had an issue of the strength since the check ring body had to be non-magnetic material because of the installation of magnetic material. As they are described above, the conventional methods of detecting check ring motion are not practicable since the usage in mass production molding factory is restricted in the use conditions. Therefore, the author established another method of detecting check ring motion and backflow phenomenon in mass molding production and installed the function into all electric injection machine. Hereafter, the author introduces the function. Additionally, usage of the function as an index of maintenance of check ring is introduced as well.

Principle of detecting check ring motion

FIGURE 1 indicates check ring close motion at the beginning of injection process. Until starting injection, the check ring is positioned at front of the screw (left hand side in the figure). When the screw starts to advance as injection process starts, the resin pressure at front of the screw increases, therefore the check ring starts close motion by the action of force pushing back the check ring to rear side (right direction in the figure). Until it completes close motion, the resin causes backflow toward to rear side of the check ring, though it is small amount, the author calls it initial backflow. It was achieved to understand the backflow state and the backflow motion by detecting load given to the screw resulted by the back flowing resin. FIGURE 2 indicates backflow screen which graphically displays the detected value. The horizontal axis indicates elapsed time after injection start, and the vertical axis shows the detected value. It can be decided that the check ring closes at the same timing, and the injected resin amount into the mold is stable when the same waveform is duplicated every cycle. Thus, it has been enabled to understand the backflow phenomenon visually.

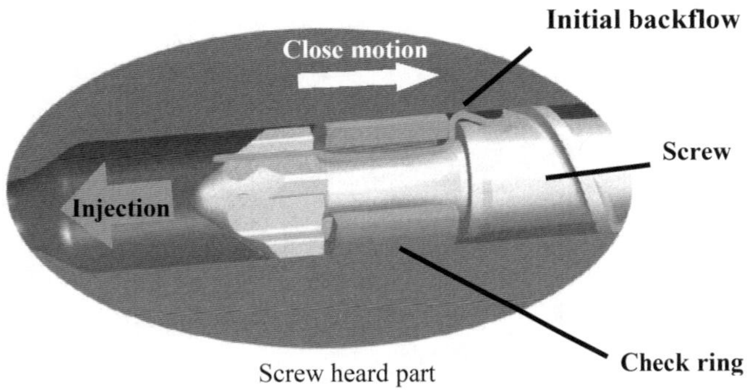

FIGURE 1 Initial backflow at the beginning of injection process

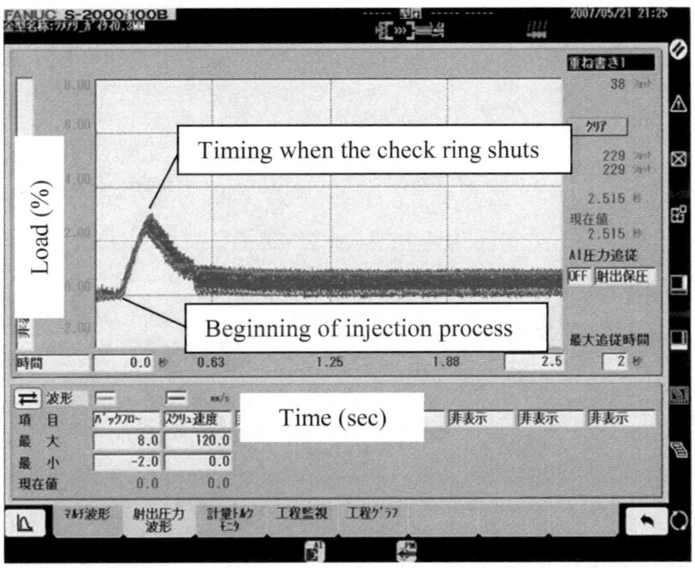

FIGURE 2 Screen of displaying backflow phenomenon (Backflow monitor)

EXPERIMENTAL METHOD

Comparison between methods of detecting backflow

The author compares the above described backflow monitor and conventionally known method of measuring resin pressure at front and rear of the check ring. FIGURE 3 shows major dimensions of the cylinder and the screw of the injection molding machine, and the resin pressure sensor installation points. On the experiment, consecutive molding cycles are performed under two different resins with different flow rate and three different injection speeds. The injection machine specification for the experiment is shown in TABLE 1 and experimental condition is shown in TABLE 2.

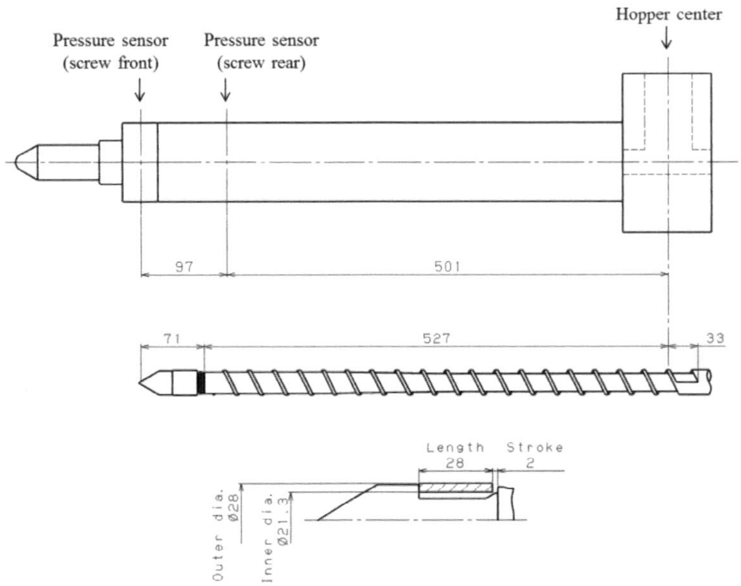

FIGURE 3 Pressure sensor location and major dimensions of screw and check ring

TABLE 1 Tested injection molding machine specifications

Injection machine specification	Machine type	FANUC ROBOSHOT Alpha-S50iA (All electric)
	Clamping force	500(kN)
	Screw diameter	28(mm)
Pressure sensor	Sensor	Dynisco PT465XL-30M
	Amplifier (Screw front side)	Dynisco SJ600E-04
	Amplifier (Screw Rear side)	Dynisco N600C1-60
Molding parts	Connector (2 cavities)	

TABLE 2 Tested injection molding machine specifications

Resin	PBT	POM
Manufacturer	Mitsubishi Engineering- Plastics Corporation	Polyplastics corporation
Grade	NOVADURAN 5010R5	DURACON M90-44
MFR	-	9.0(g/10min)
MVR	25(cm3/10min)	-
Density	1.31(g/cm3)	1.41(g/cm3)
Tm	224(℃)	165(℃)
Temperature	245-250-250-240-70(℃) (Nozzle, Cylinder1,2,3, Hopper)	200-200-200-190-60(℃) (Nozzle, Cylinder1,2,3, Hopper)
Injection speed	10, 100, 200(mm/s)	

Correlation between detected backflow and check ring wearing

At the next, the author compares the correlation between wearing of the check ring and the motion by the backflow monitor. The part where wearing occurs depending on type of the check ring either the ring turns along with screw so called "castle ring" or the ring freely turns against the screw, so called "non-castle ring". Therefore, the author prepared four kinds of check rings with different wearing per check ring type. The specifications of the injection machine used for the experiment is shown in TABLE 3.The resin is PBT shown in TABLE 2 and the injection speed is 50mm/s.

TABLE 3. Tested injection molding machine specifications

Injection molding machine specification	Machine type	FANUC ROBOSHOT S-2000i100B (All electric)
	Clamping force	1000(kN)
Molding parts	Connector (2 cavities)	

RESULT AND DISCUSSION

Deviations between methods of detecting check ring motion

FIGURE 3 indicates result of the comparison between method of detecting backflow phenomenon and conventional method of detecting resin pressure at front and rear of the check ring. It is known that the timing of rear side resin pressure of check ring becomes maximum is the timing of the check ring closes. According to the result, it is confirmed that the timing of rear side resin pressure of check ring becomes maximum well fits to timing of the backflow monitor value becomes maximum regardless of the resins kinds and the injection speeds.

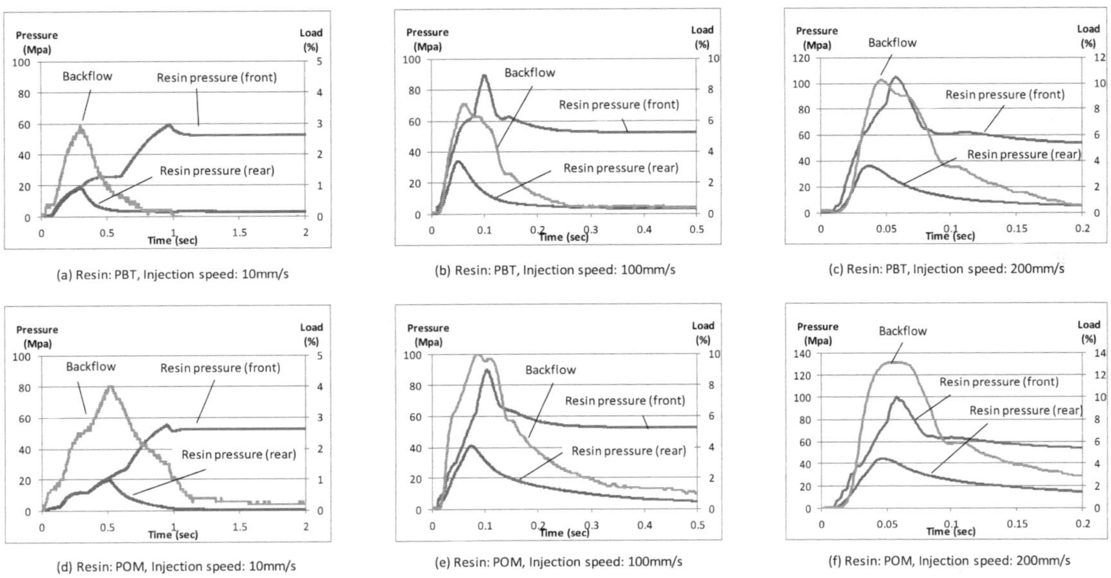

FIGURE 4 Resin pressure and backflow

Result with non-castle type check ring

During metering process, edge face of the check ring wears in case of non-castle type check ring according to the screw rotation. As the wearing develops, motion stroke of the check ring becomes longer, therefore amount of the backflow increases accordingly. FIGURE 5 indicates results of comparison with 30 consecutive molding cycles between four kinds of check rings with different motion stroke (stroke:2mm (reference dimension), 4mm, 6mm, 8mm).

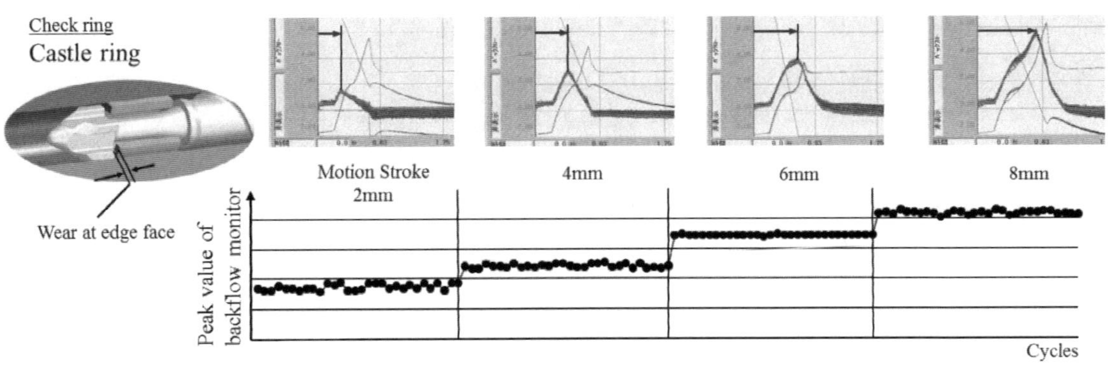

FIGURE 5 Backflow and non-castle type check ring stroke

From the waveform in FIGURE 4, it is indicated that timing of the check ring close delays according to increase of check ring motion stroke. Furthermore, it is confirmed that the time based integral value from the start of injection to the close increases according to the delay, and it indicates the time based integral value is an index of showing amount of backflow whether it is large or small.

Result with castle type check ring

Meanwhile, in case of castle type check ring, wearing develops at outer wall of the check ring during metering process in injection molding. When the outer wall wears, gap between the outer wall part and the inner wall surface of the cylinder becomes wider, therefore it causes backflow due to resin transfer from the screw front to the rear through the gap. FIGURE 6 indicates results of comparison with 30 consecutive molding cycles between four kinds of check rings with different diameter (diameter: ϕ 28mm (reference dimension), ϕ 27.8mm, ϕ 27.7mm, ϕ 27.6mm)

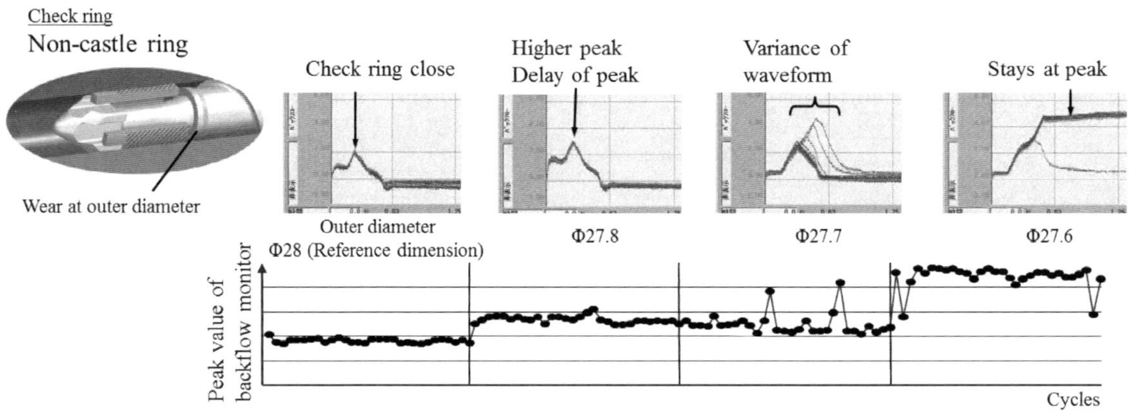

FIGURE 6 Backflow and castle type check ring wearing

FIGURE 6 indicates trend that amount of backflow increases according to decrease of the outer diameter. Additionally, although the waveforms are stable and duplicates every cycle for both diameter ϕ 28 and ϕ 27.8, the waveform variance is started to be observed and an accidental large backflow is confirmed when the diameter decreased to ϕ 27.8. Furthermore, when the outer diameter reached to ϕ 27.6, most of the cycles show waveform which stays at the peak value, it indicates once the check ring closes however, backflow continues at later injection process. Therefore, the check ring with less 0.3mm of outer diameter is possible to cause defective molding parts by accidental backflow, and it is indicated that the check ring reaches to its replacement time.

CONCLUSION

On this paper, the backflow monitor function which enables to understand backflow state and check ring motion in injection is introduced. The conclusions led by the research of the function and the experiment are shown below.

(1) The backflow monitor function which enables to display backflow state graphically is developed.
(2) It is confirmed that the check ring close timing is detected regardless of resin kinds and injection speeds.
(3) The check ring state and backflow state during injection are compared with four kinds of check rings with different motion stroke and another four kinds of check rings with different outer diameter. At the result, variance of close timing and accidental increase of backflow amount according to decrease of the outer diameter are confirmed.
(4) It is confirmed the backflow monitor is capable to use for an index of replace timing of check ring.

REFERENCES

1. Oka and other three JSPP'91 Tech. Papes, 19(1991)
2. Yokoi, Ono: JSPP '99 Tech. Papers, 287(1999)

Application Progress of In-mold Decoration

Yang Wang [a,*], Jiansheng Chen[a]
[a] *ABB Corporate Research, 10 Jiuxianqiao Road, Beijing 100015, China*

Abstract. Differentiation design and rising requirement in product appearance cause new challenge to traditional thermoplastic processing technology. Various promising surface decoration technology is emerging. In-mold decoration (IMD) can finish injection molding and surface decoration in one production cycle. Automatic level of production line, productivity, aesthetical appearance can be improved by IMD. Technical branches and corresponding mechanism of IMD was reviewed. Technical requirement of thermoplastic material, ink, film, and apparatus was introduced. And current material and equipment suppliers were summarized. The limitation and future trend of IMD was outlooked.

Keywords: In-mold decoration, IMD, thermoplastic material, injection molding

INTRODUCTION

In-mold decoration (IMD) is a new technology for injection molding process. The technical principal of IMD is injection molding in the presence of a decorated film, and the product forming and labeling can be finished in one-shot. Thermoplastic product with different appearance, such as pure color, wooden texture, metal or shinning effect can be produced with the appropriated combination of graphic film, printed graphic (ink), thermoplastic resin, and special designed mold.

In traditional technology, product is made from injection molding process and subsequent printing process such as screen printing or pad printing; the printing or labeling process is combined with injection molding in IMD process. The most promising technical benefit of IMD is shorten and simplify the production process, reduce the production stage and numbers of component, and thus reduce the production time and cost; together with that, the product quality, flexibility and durability can also be improved [1-4].

The IMD technology can be categorized as IML (in-mold label) and IMR (in-mold roller). The main difference between IML and IMR is whether there is a transparent protection layer above graphic ink (Figure 1). The main technical feature of IML is ink layer exist between thermoplastic resin and transparent protection layer. The protection layer is effective to provide anti-scratch, wear-resistance property, and the durability of color. The ink-printed protection layer could be flat or 3D in geometry, and a separate production stage is required to produce the ink-printed protection layer.

The main technical feature of IMR is graphic ink pre-printed on continuous film roll; the graphic ink will be heat transferred onto thermoplastic resin surface during injection molding process. The final product of IMR is without the transparent protection layer as IML process [5]. As the graphic is printed on a continuous film roll, IMR is suitable for high automatic level production. Without the protection layer, IMR suffer from lower wear resistance, and is limited to gentle application such as mug cup or office device.

FIGURE 1. Scheme of IML and IMR

Material property and forming process are the fundamental factors to achieve satisfied decorative effect, with the increasing demand of functional decorated product, it becomes essential to understand the material and process theory in IMD.

REQUIREMENTS ON MATERIAL

Graphic Film

Graphic film is one of the most important material factors in IMD, and the substrate material related with the product quality and yield rate. Generally, the substrate material need to meet following requirements:

(1) Good thermal resistance. The thermoplastic resin melt flow has high temperature, the substrate material need to have good thermal resistance to reduce the shrinkage during injection process;

(2) Adhesion to ink. Low adhesion to ink will result in graphic disturbance during hot resin melt injection

(3) Good thermal forming property. The good forming property will facilitate for film or sheet forming process of graphic layer.

The most commonly adopted film material include PC, PET, PC/PBT, PC/PET, etc [6]. PET film is a colorless, transparent, high gloss material, and can provide good toughness, elasticity, thermal resistance and dimensional stability, it is an ideal material for graphic substrate in IMD and received widely application in industry. PC is alternative material to PET with lower material cost, but the wear resistance is lower than PET, can be used to produce IMD with balance cost and quality [7].

Wrinkling, rupture, warpage, graphic dislocation, whitening is the common defect in IMD production. Huang [8] found the graphic clearness, surface quality is rely on thermal forming capability of IMD film. The thermal forming capability is related to the material type, process parameter, mold design, etc. The thermal forming capability of IMD film is key factors in IMD process, and is a hot topic in theoretical and experimental research. Philips [9] study the performance of pure PC and PC/polyester blend film by theoretical analysis and experiment study. It was found that the draw depth of blend film is higher than PC film at high temperature, the blend film also has higher maximum elongation. In conclusion, blend film require lower forming pressure than PC film and is suitable for low temperature forming process. Chen [10] studied the effect of film material and mold temperature on the IMD process. It was found that for PC film, the retardation of heat transfer will cause temperature drop of mold cavity, the temperature difference could be 17.7°C compared with injection process without IMD film. While for PET film, the temperature difference is 13 °C. And higher film thickness will result in higher temperature difference. Xue [11] studied the heat transfer behavior of injection molding in the presence of film by ANSYS tool. The numerical model based on 0.1 mm to 0.5 mm film was built to simulate the thermal field. PET film will cause 10°C to 20°C temperature drop on mold cavity temperature, and the temperature difference will becomes smaller along with the process time.

Leong [12] investigated the injection speed and annealing process on the molecular orientation and orientation-induced crystallization behavior of film and substrate. High molecular orientation from high injection speed will increase crystallization rate, and is considered to provide high adhesion property of film-substrate interface. The strong interface bonding is beneficial to final product appearance.

The thermoforming capability, thermal resistance, thermal conductivity is bottle neck of film material. Investigation on new polymer structure, polymer blend and process condition are trends to achieve balanced performance and provide industry applicable solutions.

IMD Ink

Due to the nature of IMD process has significant difference with traditional printing process, the ink need to meet specific requirement for IMD process. The key requirement of IMD ink is compatibility with injection process, therefore, the basic requirement for IMD ink include:

(1) Thermal resistance. Ink material should be durable to temperature under the time scale of injection process, and keep the decorated graphic intact under the impact of melt flow.

(2) IMD ink should have good adhesion to graphic film.

(3) Fast surface drying capability.

Kessel [13] studied the IMD ink by incorporating pigment in urethane acrylate oligomer that has a polycarbonate backbone. The ink is photo curable by introduction of UV or heat initiator. The developed ink has good adhesion and high elasticity, and the solvent content is also reduced by its process. But the low drying speed limits the industry application. Fujimaki [14] investigated IMD ink with different composition, including polyester, polyimide with silicone and fluorocarbon resin. Results showed that silicone and fluorocarbon is effective to increase the thermal resistance of ink; however, due to the high material cost and difficulty in processing of polyimide, the industry application is limited.

Polyurethane (PU) is widely accepted as a material with anti-scratch, wear resistance, solvent resistance, good adhesion and high flexibility properties [15]. PU can be customized to various requirements by fine-tuning monomer formulation. However, traditional PU suffers from its weak thermal stability; thermal degradation is anticipated above 200°C process temperature [16]; and the poor thermal stability hampers the application of PU as IMD ink. By introducing isocyanurate, oxazolidone, imide into PU's backbone, thermal resistance of PU is expected to be improved. Wang [17] studied the effect of isocyanurate in PU, result showed the initial thermal degradation temperature of modified PU is 23.7°C higher than pristine PU. Liu [18] prepared epoxy modified PU, and the resulting polymer contain oxazolidone functional group. The modified PU has thermal resistance up to 300°C. Hamid Yeganeh [19] prepared new generation of PU, it was found that the incorporation of polyimide can increase 10% thermal degradation temperature by 100~120°C

Chattopadhyay [20] reviewed the thermal degradation mechanism of PU. An effective way to improve thermal resistance is to increase content of aromatic group in soft and hard segment of PU. The effect of incorporate of filler such as metal oxide, POSS, CNT were also reviewed. Lin [21] studied the effect of nanosilica and reduction modified graphene oxide (RMGEO) on epoxy modified PU, results showed the nanofiller and graphene oxide are effective to improve the impact resistance, flexibility and thermal stability. The T_{10} can be increased by 8.2 to 17°C by introduction of nanosilica. The RMGEO modified PU can increase the T_{10} by 11.2 to 25.8°C depends on composition level.

IMD ink need to meet various challenge apart from thermal stability, such as provide good adhesion to substrate, accurate pattern on 3D surface, competitive cost compared to traditional printing technology. Single composition is hard to meet all these requirement, and the material study to provide satisfied and balance performance is promising and necessary in the future.

Thermoplastic Resin

The thermoplastic resin in IMD usually include raw material with easy processing, high mechanical strength and low cost. Commonly candidates include PC, PET, transparent ABS, PMMA, PS, etc. In the above materials, PC and PET can provide high performance with high cost; PS has lowest cost but the material is brittle in application; transparent ABS can provide balanced performance and cost. In IMD process, resin selection has more limitations than traditional injection molding, such as graphic film shrinkage, the compatibility of ink and resin.

Wang [22] prepared the blend of PC/PET, the polymer blend can provide high adhesion under damp heat condition. The transparency, impact resistance and shrinkage performance is also better than ABS. There are many technical progress in PC alloy recently, some PC alloy can keep the high mechanical strength of PC and also reduce melt temperature thus improve processability. ABS/PC blend is kind of promising raw material, combine the benefit of ABS and PC, with good flowability and also thermal resistance [23-24].

COMMERCIAL MATERIAL AND EQUIPMENT COMPANY

IMD process has been adopted by industry field for years, there are companies to provide specialized resin, ink and equipment for IMD.

Macdermid autotype provide series of hard coated PC, PET, PMMA film, which can provide promising scratch, solvent and chemical resistance performance. The hard coated material can improve the texture and gloss appearance on finish product surface [25]. Bayer, Sabic, Kumho Chemicals Inc also provide injection grade PC, ABS and PET that is suitable for IMD.

Close cooperating with Convestro AG, Proell provide various IMD ink and adhesion promoters [26]. The graphic film can be printed on various materials, including polyester, PMMA, polyamide. The Proell ink can be printed on plane or 3D geometry with good appearance.

TABLE 1. Product line from Proell

Grade	Substrate	Feature
NORIPHAN® HTR N	PC	Thermal resistance
NORIPHAN® XWR	PC, PET	Washout resistance, high cohesion
NORIPHAN® N2K	PC	No curling, suitable for thin film (50 micron)
NORIPHAN® PCI N	PC	Halftone painting, thermal resistance
NoriPET®	Polyester	High flexible, thermal resistance
NoriCryl®	PMMA	Solution for all acrylic IMD
NoriAmid	PA	Washout resistance, high cohesion, long screen opening time, formable
NoriAmid APM	NoriCryl, NoriAmid, NORIPHAN XWR	Adhesion promoter
NoriPress® SMK	PA, PC, PMMA, PVC, ABS	Adhesion promoter
AquaPress®	Unspecified	Screen printable hotmelt adhesive

Taikoku Printing Inks offer molding ink covered several kinds of graphic films, such as PC type, PET type and TPU type; and provide features such as stretch capability, flexibility, wash out resistance, depend on grade. Ink binder is also provided to promote the adhesion force at the ink/resin interface [27].

TABLE 2. Molding inks from Teikoku Printing Inks

Grade	Substrate	Feature
ISX-HF	PC	high stretch capability, no pot life
IPX-HF	PET, PC	excellent wash out and heat resistance
INQ-HF	TPU	Enhanced flexibility, printed surface resistance
IMB-HF006/009	Unspecified	Ink binder

There are also equipment companies that working on solutions for IMD. Arburg can provide automatic systems including injection molding machine and robotic system, which provide high accuracy of film pay-off device (1/100mm) and gentle part removal device [28]. OECHSLER can provide vertical integration service from mold design, mold manufacture, injection molding, machining and assembling, can provide high standard high cleanliness solution for customer at high, middle and low cost [29].

OUTLOOK OF IMD TECHNOLOGY

IML and IMR are to two main technical branches of IMD process, even though they share lots of common technical aspect, but they still has own advantages and limitations.

The most significant technical feature of IML is decorated film and injection molded part form a unibody component, which can make the color more durable and provide wear-resistance to graphic. The limitation for IML is the product development need longer time, the surface protection layer has a risk of deformation and shedding.

IMR technology is intrinsically suitable for automatic production with high productivity, however, the limitation come from the absence of protection layer on graphic surface. Without the protection layer, the decorated graphic is vulnerable to oxidation and wear. To provide high adhesion to plastic resin and weather resistance, the technical requirement for IMR ink is higher than IML ink. Therefore, the nature of high automatic level and high ink cost make IMR only cost effective under large volume mass production.

With various advantage such as simplified production process, higher automatic level, enhanced and durable label color and versatile decorated texture, IMD becomes an attractive injection molding solutions; and IMD need involve more and more cross area studies, such as graphic film, IMD ink, injection molded resin, accurate localization system, ink printing system, and automatic mold system. The following field may become hot topic in the future:

(1) Material with high thermal resistance and adhesion performance;

(2) Injection molding with large size, thick/thin wall, complex geometry;

(3) Close integration with other process, such as foam molding, gas assisted injection molding, thermoforming;
(4) High accuracy and high automatic level production.

REFERENCES

1. Y. P. Chang and S. C. Chen, *Annual Technical Conference - Society of Plastics Engineers* **67**, 1228-1232 (2009).
2. Y.C. Wang, "Study on the mold surface temperature control and analysis using thin film for heat hesitation", Master thesis, Chung Yuan Christian University, Chung-Li, Taiwan, 2009.
3. T. A. Osswald and G. Menges, *Materials Science of Polymers for Engineers,* Hanser (1996).
4. A. C-Y. Wong and K. Z. Liang, *Journal of Materials Processing Technology* **63**, 510–513 (1997).
5. J. Rothe, *Kunststoffe-Plast Europe* **87**, 1564-1574(1997).
6. C. O. Phillips and E. H. Jewell, *Measurement Science & Technology* **19**, 28-28(2008).
7. Y. K. Li, "Experimental Study on Thermoformability of IMD Film", Master thesis, College of Mechanical Engineering Zhejiang University of Technology, 2015.
8. Z.D. Huang, "Study of injection molding characteristics for PC/TPU in-mold decoration", Master Thesis, Chung Yuan University, Taiwan, 2003.
9. C. O. Philips and D. C. Bould, *Materials and Design* **30**, 537-539(2009).
10. S. C. Chen and H. M. Li, *International Communications in Heat & Mass Transfer* **37**, 501-505(2010).
11. Dan Xue, "Influnce of Film on Cooling Process in IML Injection Molding", Master Thesis, Huazhong University of Science & Technology, 2009.
12. Y. W. Leong and M. Kotaki, *Journal of Applied Polymer Science* **104**, 2100-2107(2007).
13. S. Kessel and C. L. Lawrence, US: 20060052477A1 (9 Mar 2006).
14. S. Fujimaki, US: 5376433 (27 Dec 1994).
15. V. D. Athawale and M. A. Kulkarni, *Pigment & Resin Technology* **40**, 49-57(2011).
16. E. Dominguez-Rosado and J. J. Liggat, *Polymer Degradation and Stability* **78**, 1–5(2002).
17. J. J. Wang and T. J. Shi, *China Synthetic Rubber Industry* **24**, 347-349 (2001).
18. X. H. Liu and M. Q. Kang, *China Synthetic Rubber Industry* **20**, 333-379(1997).
19. H Yeganeh and MA Shamekhi, *Polymer* **45**, 359-365(2004).
20. D. K. Chattopadhyay and D. C. Webster, *Progress in Polymer Science* **34**, 1068-1133 (2009).
21. Jing Lin, "Synthesis of high-temperature resistant polyurethane composites and its application in IMD ink", Ph.D. Thesis, South China University of Technology, 2012.
22. Y. Y. Wang and Y. M. Dong, CN101619157 (6 Jan 2010).
23. J. Zhou and C. X. Wu, *CIESC Journal* **62**, 3588-3594(2011).
24. M. L. Ge, *Guangzhou Chemistry* **33**, 13-15(2005).
25. http://autotype.macdermid.com/categories/7-film_insert_moulding
26. http://www.proell.de
27. http://www.teikokuink.com/en/index.html
28. https://www.arburg.com
29. http://www.oechsler.com

Improvement of Molding Quality of Light Guide Plate Using Multi-Coating Surface Treatment for Injection Molding Screw

Ranran Jian [a], Weimin Yang[a, b], Zhiwei Jiao[a], Yumei Ding[a] and Pengcheng Xie[a,*]

[a] *College of Mechanical and Electrical Engineering, Beijing University of Chemical Technology, Beijing, China*

[b] *State Key Laboratory of Organic-Inorganic Composites, Beijing University of Chemical Technology, Beijing, China*

Abstract. Precision-injection-molding as a replication method is one of the key technologies for precision-manufacture, such as light guide plate, microfluidic chip. An important stage in precision-injection-molding which can affect the accuracy and mechanical properties of the produced component is part plasticizing. During this stage, the failure of surface treatment of screw can cause a variety of defects to light guide plates, including black discoloration and yellow streaks. Therefore, in this paper, the effects of different surface treatments on the screw are investigated. In particular, the molding quality of a representative precision component, light guide plate, was studied as a function of different coatings of screw, in combination with three coating features such as stripping performance, surface microstructure and adhesion strength, employing the design of experiment approach. In addition, the results obtained using different combinations of coating features were analyzed to identify the best surface treatment methods in regard to large number of industry tests of polycarbonate light guide plate when utilizing these coatings. The results indicate that the light guide plate parts using the screw with the TiAlN/Cr multi-coating had a better performance in view of the fact that the defective percentage had been decreased to 0.11%, which was much lower than the 19.21% of the general screw with hard chrome.

Keywords: multi-coating, screw surface treatment, injection molding quality, light guide plate
PACS: 68.35.Ct; 81.05.Bx; 81.65.-b; 82.35.Gh

INTRODUCTION

With widely use of personal computers, smart phones, digital cameras, smart learning machines , etc., light guide plate (LGP) parts are facing a huge potential market demand as one of the primary parts of Back Light. And injection molding is the common available, efficient molding approach for mass production of LGP [1, 2]. Mostly, the part thickness is nearly under 0.5mm and the microstructure array exists on the surface, so that injection molding processing for LGP is very different from conventional parts [3-6]. And owing to the intense abrasion and corrosion on the screw surface, a surfaces failure is easy to occur in case that screw and screw tip had an undesired surface treatment, which may create gaseous decomposition in the products [7] and its macro presents visible black discoloration and yellow streaks on the surface. Unfortunately, such defects significantly influence the transparency and uniformity of parts. Therefore, injection molding screw surface treatment for LGP should be taken into consideration seriously to enhance its wear and corrosion resistance. Our team has conducted systemic research around the effects of different surface treatments on the screw. Nano-multilayered coatings technology was first proposed to apply in the screw surface treatment of precision injection molding for optical plastic parts, which has been extensively studied in the area of super-alloy, hardened tool steels, etc.[8-11]. And our study results demonstrate it works. The LGP of less defective percentage have been obtained using the screw with the TiAlN/Cr multi-coating.

EXPERIMENTAL AND TESTING EQUIPMENT

Three coating features such as stripping performance between polymer and coatings film, surface microstructure of coatings film and adhesion strength between metal substrate and coatings film were analyzed to identify the best surface treatment methods in regard to large number of industry tests of polycarbonate light guide plate when utilizing these coatings.

In order to compare stripping stress between polymer and various coatings film, we built up a stripping performance testing platform as shown in FIGURE 1. Atomic force microscope (AFM: dimension V SPM, Veeco, America) was used to investigate the surface profile of various coating film. Meanwhile, images of 3D surface microstructure could be created. The adhesion strength between metal substrate and coatings film was measured by material surface properties tester (MFT-4000, Zhongke Kaiteke Industrial and Trading Co., Ltd, Lanzhou, China). What's more, all-electric injection molding machine (TIANRUI VE900, Ningbo ZHAFIR Plastics Machinery CO.,

LTD, China) equipped with various coated screws including conventional plating hard chrome, PVD TiN, PVD TiAlN and PVD TiAlN/Cr coating was employed to produce the LGP used material of polycarbonate (LC1500, NATURAL ED 76366)(shown as FIGURE 2).

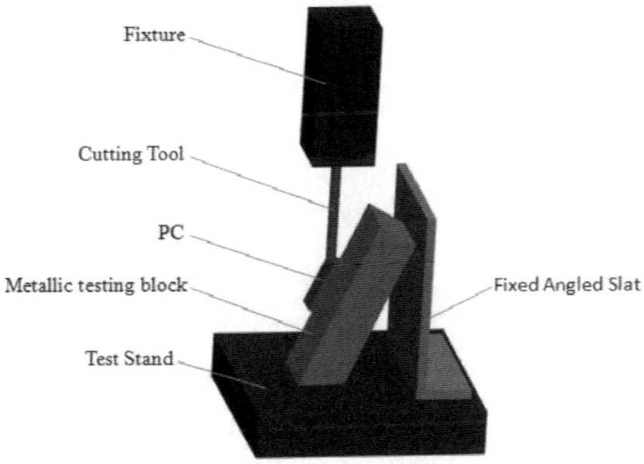

FIGURE 1. The stripping performance testing platform

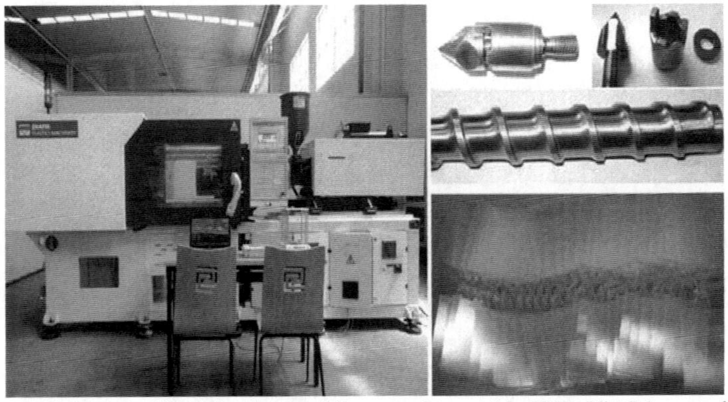

FIGURE 2. The injection molding machine with PVD TiAlN coated screw and molded LGP parts

RESULTS AND DISCUSSIONS

Stripping Performance between Polycarbonate and Coatings Film

The stripping performance between polycarbonate and various coatings film is shown in TABLE (1).From the results, chromium hemi trioxide by thermal spraying treatment process (sample number 3) got the lowest average stress value, that is to say, it presented the best stripping performance. But undesirably, the coating film together with PC specimen was peeled off from the metallic block during the experiment, so did sample number 2. However, Samples number 1, 5 and 6 could greatly improve the stripping performance compared with uncoated metallic block (sample number 7) and those coating surfaces remained integrated and smooth after testing.

TABLE (1). The stripping performance between polycarbonate and various coatings film

Sample Number	Testing Material	Treatment Process	Stripping Force(N)	Area (mm^2)	Average Stripping Stress(N/mm^2)
1	Nickel	Brushing electroplating	507	900	0.563
2	Tungsten carbide	Thermal spraying	548	900	0.609
3	Chromium hemi trioxide	Thermal spraying	161	900	0.179
4	Nanphase material	Brushing electroplating	701	900	0.779
5	Nickel tungsten	Brushing electroplating	573	900	0.637
6	TiAlN	Physical vapor deposition(PVD)	430	900	0.478
7	SKD61	No coating	1155	900	1.283

Also, we choose another material of polycarbonate (Bayer 2805) with higher viscosity to find out if materials properties have an influence on the stripping stress. As shown in FIGURE 3, both PVD TiN and brushing electroplating Nickel Tungsten alloy had excellent stripping performance compared with conventional hard chromium coating and no coating SKD61 block, just agreed with the results of TABLE (1). The only difference was that the stress values were bigger than that ones with lower viscosity polycarbonate. And experimental studies indicated that brushing electroplating approach was much more difficult to apply in complex screw surface. Therefore, TiN and TiAlN coating by physical vapor deposition would be the recommended methods for the special screw for light guide plate injection molding.

Analysis of Coating Properties

The various surface treatment methods are applied on the metal substrate that possess different properties in hardness, surface microstructure etc., which is one of the critical factors dominating stripping performance of coatings film because of the microscopic surface texture including crystal growth, particle size, the arrangement of crystals, and the existence of large grooves or not. It also affects the adhesion strength between substrate and coatings film. Therefore, the influence of the surface roughness of substrate, the thickness of coatings film on the surface quality and adhesion strength of coatings were investigated.

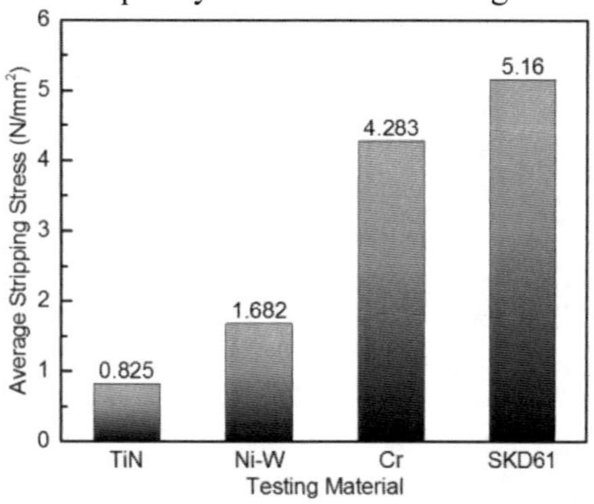

FIGURE 3. Average stripping stress of different testing material

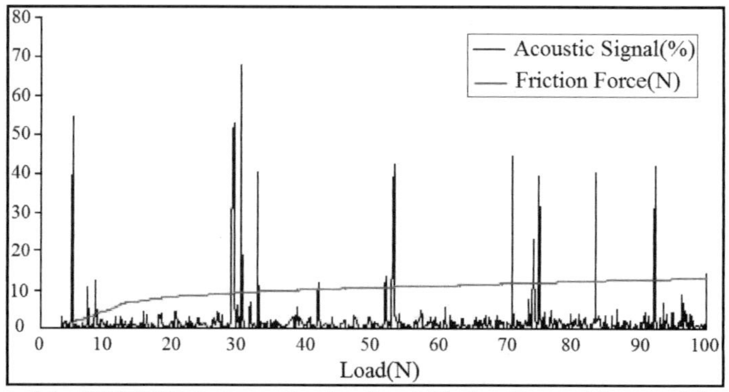

FIGURE 4. Acoustic signal and friction force with material surface properties tester (substrates roughness=1.6)

Four cylindrical substrate samples with different surface roughness were prepared and coated with PVD TiAlN, the coatings film thickness was 1μm. The coated surface quality and their adhesion strength with different roughness substrates are shown in TABLE (2). Along with the increase of substrates surface roughness, the values of R_a, R_q and R_{max} of coatings film increased significantly, and the critical adhesion strength that was measured by material surface properties tester decreased constantly (FIGURE 4). Therefore, the surface roughness of screw substrates should be as low as possible to develop coatings with improved comprehensive performance.

TABLE (2). Surface qualities of coatings film with different roughness substrates

Substrates Roughness	**0.4**	**1.6**	**3.2**	**6.3**
R_a(nm)	9.35	21.5	51	89.9
R_q(nm)	17	30.3	74.5	117
R_{max}(nm)	295	330	650	1078
Critical adhesion strength(N)	30.71	30.17	17.26	14.65

Where R_a is the arithmetic average of the absolute values of the surface height deviations measured from the mean plane; R_q means the root mean square average of height deviation taken from the mean image data plane; R_{max} indicates the maximum vertical distance between the highest and lowest data points in the image following the plane.

Also, different thickness of TiAlN coatings with the same substrate surface roughness, 0.4, was prepared. The measured results are shown in TABLE (3). From results, surface quality could be improved to some extent along with coating thickness increase. But the thickness shouldn't become too small as adhesion strength was the worst among all samples. And thickness was not the bigger the better. The adhesion strength began to decline when thickness reached more than 5μm. A possible reason is that residual stress of coating simultaneously becomes greater as the coating thickness increases, which lead to cracking and delamination between substrate and coatings film. Then lower film adhesion strength appears. Considering both, surface quality and adhesion strength, coating thickness of about 1-5μm for PVD TiAlN adopted in this study would be preferable.

TABLE (3). Surface qualities of coatings film with different thickness of coatings

Thickness of Coatings(μm)	**0.1**	**1**	**5**	**10**
R_a(nm)	4.31	9.35	5.29	6.05
R_q(nm)	6.02	17	10.7	8.11
R_{max}(nm)	74.8	295	154	65.2
Critical adhesion strength(N)	16.06	30.71	33.32	31.26

Multi-Coating Surface Treatment

Although TiAlN coating by PVD with coating thickness of 5 and substrates roughness of 0.4 have good stripping performance and high adhesion strength, it's still insufficient for LGP injection molding, which needs strong adhesion strength and extremely smooth surface of screw in order to prolong its service life (a typical intermittent, repeated, variable and long-term process). Therefore, multi-film was proposed to reinforce the adhesion strength between the PVD TiAlN coating and substrate (SKD61 screw). We chose Cr or Ti to be the cushion layer between TiAlN and substrate, whose hardness is lower than TiAlN coating but higher than SKD61, matching the crystal lattices.

FIGURE 5 shows the adhesion strength between substrate and coatings film, which indicated clearly that the adhesion strength of multi-films greatly exceeded the single film (TiN or TiAlN). Combined with FIGURE 6 and 7, we found that TiN coating owned the biggest size of crystal particles and heaves (R_{max}=189nm), and TiAlN coating was on the contrary (R_{max}=32.1nm), and TiAlN/Cr (73.0 nm) and TiAlN/Ti (69.9 nm). TiAlN coatings in FIGURE 7(b) presented the excellent surface quality as a result of its smoothness, continuum, and low surface wave, and its average roughness of surface R_a was only 4.38 nm. However, the R_{max} of TiAlN/Cr coating and TiAlN/Ti coating were slightly worse than TiAlN due to the deposition of Cr or Ti. But on the other hand, the R_{max} of TiAlN/Cr and TiAlN/Ti were much lower than TiN coating, and their R_a and R_q remained much close to TiAlN coating. Even so, owing to its better adhesion strength with substrate, lower stripping stress with PC and smoother surface of coatings, multi-films could be regarded as a desirable treatment approach to the special screw for LGP injection molding.

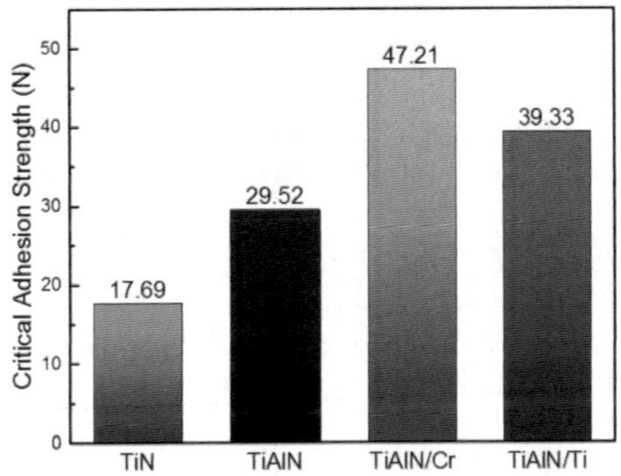

FIGURE 5. Adhesion strength between substrate and coatings

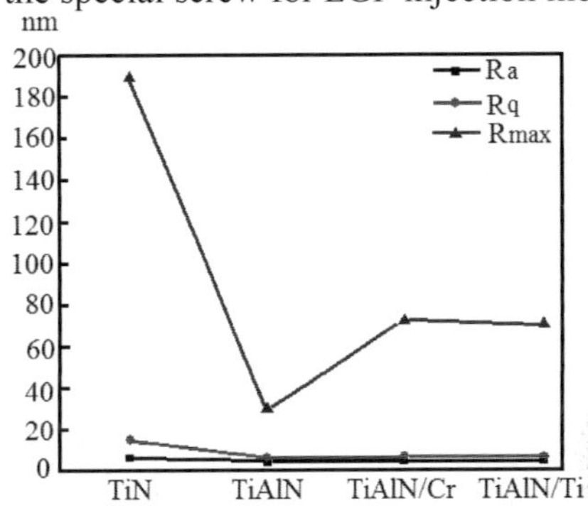

FIGURE 6. Comparisons of surface quality of different coatings

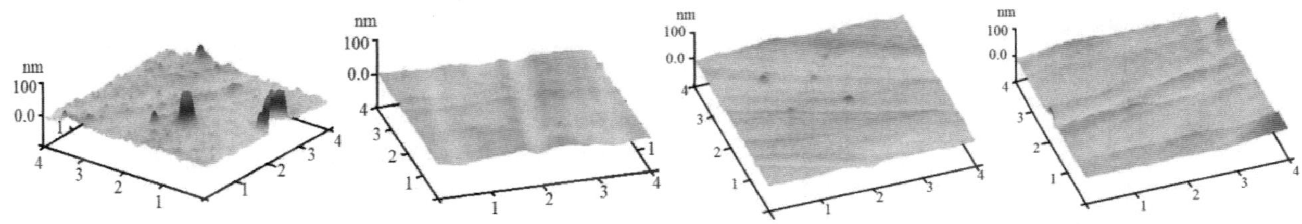

(a) TiN (b) TiAlN (c) TiAlN/Cr (d) TiAlN/Ti

FIGURE 7. Surface micro-morphology of various coatings film

Industrial Testing of Light Guide Plate Injection Molding

In order to verify the advantageous effects of multi-coating surface treatment on the screw, almost 10 thousand LGP parts were fabricated. Three aspects were determined to evaluate the overall performances of the various coatings film as follows: the sequence number of molded part that black discoloration appeared for the first time, the average defective percentage (the percentage of molded parts with black discolorations or yellow streaks, as shown in FIGURE 8), and the surface quality after testing including whether coatings were exfoliated or corroded or heavily worn.

(a) Black discoloration on light guide plate (b) Yellow streak on the screw

FIGURE 8. Defects in the process of LGP injection molding

Industry testing results, shown in TABLE (4), further verified the previous studies. TiAlN/Cr coating had the best performance with defects latest appear, lowest defective percentages and best surface quality of the post-test screw, much better than hard chrome coating. In addition, it was found that once the first defective part emerged, wear of coatings would become much fiercer and defect parts with black discoloration would continuously emerge in several adjacent molding cycles.

TABLE (4). Industrial test result of LGP injection molding

Coating Type	Sequence Number of first defect part	Average Defective Percentages (%)	Surface Quality of The Post-Test Screw
Hard chrome	410	19.21	Severe degradation, change in color
TiN	823	0.46	Tiny broken surface
TiAlN	1098	0.19	Intact and no degradation
TiAlN/Cr	2076	0.11	Intact and no degradation

CONCLUSIONS

An experimental study was carried out to investigate the molding quality of an optical component in precision-injection-molding, with a particular focus on the effects of surface treatment on the screw. Multi-coating was firstly proposed to apply in the surface treatment of screw. Meanwhile, the influence of the surface roughness of substrate, the thickness of coatings film on the surface quality and adhesion strength of coatings were also investigated. It can be found that the TiN and TiAlN coatings owned the improved surface quality and significantly reduced the bonding force between the surface and polycarbonate compared with hard chrome film. Besides, with increasing the roughness of substrate, the adhesion strength of coating swiftly decreased. However, the coating should be deposited within a certain thickness to maintain strong adhesion strength with substrates. What's more, multi-coatings including TiAlN/Cr and TiAlN/Ti greatly increases the adhesion strength between substrates and coatings due to the cushion coating Cr and Ti. And according to industry testing results, the average defective percentage of light guide plate molded with TiAlN/Cr coated screw decreased from 19.21% to 0.11% compared with conventional hard chrome coated screw.

ACKNOWLEDGMENTS

This study was financed by the National Natural Science Foundation of China (51203009 and 51576012). Haitian Plastics Machinery Ltd. is gratefully acknowledged for the industrial testing of injection molding light guide plate. Also acknowledgement and appreciation extended to anyone contributed to this study.

REFERENCES

1. C. FJ. Kuo, G. Dewantoro, C. C. Huang, *Journal of Intelligent Manufacturing* **26**, 677-690(2015).
2. B. Y. Joo, J. H. Ko, *Journal Of the Optical Society of Korea* **19**, 159-164 (2015).
3. S Hong, I. Min, K. Yoon, et al, *Journal of Micromechanics and Microengineering* **24**, 015009 (2014).
4. M. S. Huang, C. F. Chung, *Journal of Applied Polymer Science* **121**, 1151-1159(2011).
5. J. S. Kim, D. S. Kim, J. J. Kang, et al, *Polymer Engineering and Science* **50**, 1696-1704(2010).
6. S. T. Wang, *International Journal of Materials & Product Technology* **45**, 83-95(2012).
7. P. Xie, G. Gou, Z. Wen, et al, *Advanced Polymer Processing* **87-88**, 306-310(2010).
8. G. S. Fox-Rabinovich, B. D. Beake, K. Yamamoto, et al, *Surface and Coatings Technology* **204**, 21–22(2010).
9. G. S. Fox-Rabinovich, K. Yamamoto, M. H. Aguirre, et al, *Surface and Coatings Technology* **204**, 2465-2471(2010).
10. G. S. Fox-Rabinovich, K. Yamamoto, B. D. Beake, et al, *Surface and Coatings Technology* **204**, 3425-3435(2010).
11. G. S. Fox-Rabinovich, K. Yamamoto, A. I. Kovalev, et al, *Surface and Coatings Technology* **202**, 2015-2022(2008).

Production and Characterization of Rotomolded Hybrid Polymer Composites based on Natural Fibers

Fatima Ezzahra Hanana and Denis Rodrigue*

Department of Chemical Engineering and CERMA, Université Laval, Quebec, QC, G1V 0A6, Canada

Abstract. Today, rotational molding is still a growing sector of the general polymer production. But to further improve on this process, the design and production of new products with better properties must be performed. For this reason, the development of polymer composites is of interest, especially when natural fibers are used. In this case, the addition of rigid particles leads to higher mechanical properties (mostly rigidity and strength) compared to the base matrix. Unfortunately, very limited information on the production of rotomolded composites can be found in the literature. In this work, an overview on the development of rotomolded composites is made based on the addition of natural fibers (hemp and maple wood) in a thermoplastic resin (linear low density polyethylene). From the samples produced, tensile properties of the final parts was measured as a function of fiber concentration, fiber size, fiber ratio and surface treatment via maleic anhydride addition in solution. From the results obtained, the optimum processing conditions (rotational speed, oven temperature, and cycle time) and composition were obtained for the materials studied. The production of hybrid composites was also investigated by using a blend of both fibers with different ratios to further improve on the properties of the rotomolded parts. Finally, the composites produced are shown to be very efficient in terms of production characteristics by having reduced material costs and better looks.

Keywords: rotomolding, polyethylene, hybrid composites, hemp, maple.
PACS: 81

INTRODUCTION

Rotational molding (rotomolding or rotocasting) is a polymer processing technology able to produce complex hollow shapes over a wide range of sizes. The main advantage of rotomolding is the possibility to produce parts without weld lines which are weak points inside polymers, especially for polymer composites. Compared to other processes such as injection molding, rotational molding uses low-cost molds made from different materials and thicknesses since virtually no pressure is applied on the materials. Final part thickness can also be easily controlled by the amount of material loaded inside the mold, as well as by the biaxial rotational speeds and mold wall thickness. Nevertheless, rotational molding has some limitations like long cycle times and very few resins to choose from [1-3].

Today, the different grades of polyethylene represent about 90% of all the parts molded [4-6]. But more recently, interesting developments were associated with the possibility to mold parts with reinforcements/fillers to reduce costs and/or improve the mechanical properties. The main applications for these materials are in automotive, civil infrastructure, sporting goods, packaging, as well as different tanks/containers [7]. One main area of polymer composites development is the use of natural fibers (lignocellulosic) because they are biodegradable, CO_2 neutral, lightweight and low costs [8-9].

To rotomold a composite part based on a thermoplastic matrix and a reinforcing particle, two techniques can be used to introduce the particles inside the polymer matrix: melt blending or dry blending. Here, the latter was used as all the initial materials are in a powder form. So, high-shear mixers can be used to disperse the materials before being loaded in the mold. This method has several benefits as being fast, low cost and very limited material degradation occurs [10]. Nevertheless, this method has some limitations because poor adhesion (interfacial voids) at the particle-matrix interface is associated to the lack of compatibility and the absence of pressure to compact the materials. This usually leads to poor mechanical properties. Fortunately, this problem can be solved by performing a surface treatment on the particles as the interface between the fibers and the matrix is controlling the mechanical properties of composites. Poor adhesion is the results of low interfacial interaction limiting mechanical load transfer [11]. But the polar and hydrophilic natural fibers are incompatible with most polymers which are non-polar and hydrophobic. This incompatibility leads to poor interface quality and poor mechanical performances [9-12]. But good interfacial adhesion (compatibility) between natural fibers and polymer matrices can be achieved by using physical or chemical surface treatments [11-13]. This can also lead to other advantages like improved mechanical properties, moisture resistance and better fiber distribution allowing the possibility to use higher fiber contents in rotational molding [8-14]. To this purpose, maleated copolymers (coupling agents) were developed as these molecules provide efficient interaction with the functional surface of the fibers and the matrix. For example, in maleated polyethylene (MAPE), the maleic anhydride groups of MAPE can react with the hydroxyl groups (OH) in the amorphous region of cellulose, while physical entanglement can occur between the polyethylene molecules of the matrix with the PE part of MAPE. These interaction/reaction mechanisms were described in details elsewhere [15,16].

Another way to improve the properties of polymer composites is to use a blend of different reinforcements to produce hybrid composites [17]. So the main objective of this work is to study the effect of different natural fiber contents and sizes on the properties of rotomolded polyethylene.

EXPERIMENTAL

The polymer used in this study was linear low density polyethylene (LLDPE) powder (EXXON MOBIL LL 8555). The density of this LLDPE is 936 kg/m^3 with a melting temperature and melt index of 126°C and 6.8 g/10 min (190°C/2.16 kg), respectively. The solvent used for fiber surface treatment was xylene (Laboratory purity grade) from Fisher Chemicals (USA) and the coupling agent was maleic anhydride grafted polyethylene (MAPE): Epolene C26 from Westlake Chemicals (USA). The latter has an average molecular mass of 65 kg/mol, a MFI of 8 g/10 min (190°C/2.16 kg), an acid number of 8.0 mg KOH/g and a melting point of 121°C. As reinforcement, maple wood fibers were purchased from PWI Industries (Canada), while hemp fibers was obtained from the Canadian hemp trade alliance (Canada). The particles were sieved to keep only particles between 355 and 500 μm for maple and 125 and 250 μm for hemp. This represents about the same range of the LLDPE powder (90-500 μm) to improve dry blending and limit particle segregation while still having different particle sizes for the reinforcements.

The samples were produced using the following steps. First, the fibers and LLDPE were dry blended with a high shear mixer Skyfood LAR-15LMB (Skyfood, USA) at 3320 rpm for 4-5 min to get a homogeneous blend. Then, around 700 g was introduced in an aluminum mold having a cubic shape (20 cm each side) with an approximate wall thickness of 9.5 mm (3/8") for the lateral sides and 12.7 mm (1/2") for the top/bottom sides. Before loading, a demolding agent (Trasys 420) was applied in the mold. The mold was then closed and introduced into the oven at 280°C. Then, the mold was kept rotating for 20 min with a 4:1 speed ratio, followed by a cooling period of 38 min with forced air convection (blowing fan).

Tensile tests were performed to get the stress-strain curves. From the data obtained, Young's modulus, tensile strength and elongation at break were obtained. All the experiments were performed on an Instron model 5565 universal testing machine (Instron, USA) using a 500 N load cell. Type V dog bones were cut from the rotomolded parts (ASTM D638). The deformation speed was fixed at 10 mm/min and the tests were performed at room temperature. Five samples were tested to report on the average and standard deviation.

RESULTS AND DISCUSSION

Figure 1 presents typical rotomolded samples. It can be seen that homogeneous parts with good surface quality were obtained with optimization of the processing conditions, even if the simple dry blending technique was used to introduce the reinforcing fibers into the LLDPE matrix. It is also clear that both fibers produced samples with different color and texture: hemp is dark brown and maple is light brown. Nevertheless, good aesthetic properties can be obtained with different types of fibers and this can also be adjusted by fiber concentration.

FIGURE 1. Typical rotomolded samples for hemp (left) and maple (right) fibers at 20% wt.

Based on the samples produced, the tensile properties were obtained and Table 1 reports on the data obtained so far for different total fiber contents (composites) and fiber ratios (hybrid composites) with and without surface treatment. Samples coding was used to simplify notation as: total fiber content/hemp content/maple content. For example, sample 20/25/75 represents a sample having a total fiber content of 20% wt. for which 25% is composed of hemp and 75% is composed of maple. A star (*) denotes a samples made from surface treated fibers.

TABLE 1. Tensile properties of the samples produced.
(values in parentheses represent standard deviations).

Sample code	Tensile modulus (MPa)	Tensile strength (MPa)	Elongation at break (%)
0/0/0	**174 (5)**	**15.5 (1.6)**	**1315 (59)**
12.5/100/0	210 (9)	8.6 (0.2)	30 (1)
12.5/100/0*	217 (3)	16.4 (0.4)	43 (1)
12.5/25/75	218 (3)	11.8 (0.7)	32 (1)
12.5/25/75*	229 (2)	17.0 (0.5)	44 (2)
12.5/0/100	190 (3)	9.2 (0.1)	30 (2)
12.5/0/100*	200 (3)	16.5 (0.2)	40 (2)
20/100/0	222 (4)	9.2 (0.6)	25 (2)
20/100/0*	250 (6)	17.1 (0.5)	36 (2)
20/25/75	230 (7)	11.0 (0.6)	25 (1)
20/25/75*	**284 (4)**	**18.1 (0.4)**	**32 (1)**
20/50/50	206 (11)	8.5 (0.1)	22 (1)
20/75/25	202 (9)	6.2 (0.6)	21 (1)
20/0/100	219 (3)	9.2 (0.6)	26 (1)
20/0/100*	246 (8)	16.6 (0.2)	38 (1)
27.5/25/75	193 (7)	8.7 (0.7)	18 (1)
27.5/25/75*	222 (7)	14.4 (0.5)	26 (2)
27.5/0/100	240 (2)	10.8 (0.5)	16 (1)
27.5/0/100*	221 (9)	16.0 (0.3)	27 (1)

The results of Table 1 show that the general effect of adding natural fibers is to increase tensile modulus and decrease elongation at break. For tensile strength, the trend depends on the type of fiber used: surface treated or not.

For the tensile modulus, the addition of natural fibers was very effective. For example, adding 20% of maple increased the LLDPE modulus by 26% (from 174 to 219 MPa). When using treated maple fibers, the improvement increased to 41% (246 MPa). Similar results were obtained for 20% hemp. In this case, hybridization was found to be very efficient as the highest tensile modulus obtained was 284 MPa for a 25/75 blend of hemp/maple treated fibers at a total content of 20%. On the other hand, the results at higher fiber content (27.5%) are not that good because of a lack of polymer matrix (too much fiber surface area to cover) and limited adhesion since there is no pressure applied on the compounds to achieve good contact between all the phases present. Nevertheless, all the values are still higher than the neat matrix.

For tensile strength, different trends were obtained. For untreated fibers, the value decreases with increasing fiber content. This behavior is associated to poor interfacial adhesion between the fiber and the matrix. For example, the tensile strength decreases from 15.5 MPa for LLDPE to 9.2 MPa at 20% untreated maple. On the other hand, tensile strength was found to increase when using treated fibers. For treated hemp fibers, the values increased from 15.5 to 16.4 and 17.1 MPa for 0, 12.5 and 20%, respectively. This behavior is expected as better adhesion is produced with the addition of a coupling agent on the fiber's surface. In this case, the tensile strength increased to 16.6 MPa at 20% of treated maple. Once again, the hybrid effect was important here as the highest value (18.1 MPa) was obtained for the 25/75 hemp/maple sample at 20% content. This increase in tensile strength can be related to combined reinforcing mechanisms in hybrid materials, with smaller reinforcements (hemp 125-250 μm) being able to accommodate between larger ones (maple 355-500 μm) to produce higher and more efficient specific surface area between the matrix and all the hemp/maple particles (good dispersion and orientation without particle-particle contact) [18].

In the case of elongation at break, all the values were below the neat polymer (1315%). This is associated to the low elasticity of the reinforcing fibers and the presence of voids/defects in the composites. As expected, the value decreases with increasing fiber content. For untreated maple, a decrease from 30 to 26 and 16% was observed for 12.5, 20 and 27.5%, respectively. But using surface treated fibers was found to give higher elongation at break. For example, the value increased from 26 to 38% at 20% maple. Finally, hybridization was less effective for this property. Although it did not produced the highest value, it is still in the highest ones for a fixed total fiber content.

CONCLUSION

In this work, rotational molding was used to produce polymer composites with natural fibers. In particular, linear low density polyethylene (LLDPE) was used as the matrix and maple/hemp fibers were used are reinforcement alone (composites) or together (hybrid composites). The effect of total fiber content (0, 12.5, 20 and 27.5% wt.) and fiber ratio (0/100, 25/75, 50/50, 75/25 and 100/0) was investigated using fibers with or without surface treatment (maleic anhydride grafted polyethylene in solution). From the samples produced, the mechanical properties in tension (modulus, strength and elongation at break) were measured and analyzed. Based on the results obtained, several conclusions can be made:

The simple dry blending technique was found to be sufficient to get good samples: homogeneous parts with good surface quality. This methodology is fast, cost-effective and limits polymer/fiber degradation.

The fiber surface treatment was very effective in increasing all the tensile properties for the range of conditions tested. The addition of maleic anhydride grafted polyethylene (MAPE) on the fibers' surface acted as a coupling agent improving fiber dispersion and mechanical load transfer.

The effect of hybridization was found to be very efficient to improve tensile modulus and tensile strength. The effect on elongation at break was not significant.

For the conditions tested, the best tensile properties were obtained for the sample having a total fiber content of 20% wt. composed of a 25/75 blend of hemp (125 - 250 μm)/maple (355 - 500 μm): 63% increase in tensile modulus and 17% increase in tensile strength when compared to neat LLDPE (174 MPa and 15.5 MPa, respectively).

Although interesting results were obtained, more work needs to be done to investigate the effect of the type of matrix and reinforcement, as well as to complete the characterization for the samples produced. A wider range of experimental conditions is currently under investigation.

ACKNOWLEDGMENTS

The authors would like to thank the financial support of the Natural Sciences and Engineering Research Council of Canada (NSERC), the Research center for high performance polymer and composite systems (CREPEC), as well as Centre de Recherches sur les Matériaux Avancés (CERMA) and Centre de Recherche sur les Matériaux Renouvelables (CRMR) of Université Laval for financial and technical support. The technical help of Mr. Yann Giroux was also much appreciated.

REFERENCES

1. S. Jung and K. Fu, *Polym. Test.* **27**, 209-220 (2008).
2. G. Gogos, L. Olson, X. Liu and V. Pashman, *Polym. Eng. Sci.* **38**, 1387-1398 (1998).
3. R. Crawford, *Mater. Proc. Technol.* **56**, 263-271 (1996).
4. R. Crawford. "Rotational molding of plastic." Second Edition, British Library Cataloguing In Publication Data, Great Britain, 1997, pp. 2-5.
5. R. Crawford and J. Throne. "Rotational molding technology." Plastic Design Library William Andrew Publishing, Norwich, New York, 2002, pp. 1-11.
6. F. Ezzahra Hanana and D. Rodrigue, *Plast. Eng.* **71**, 28-31 (2015).
7. S. Mazumdar. "Strategic growth opportunity in composites industry" in *Composites Executives Forum*. Washington DC, 2014, pp. 1-3.
8. A. Bendahou, H. Kaddami, H. Sautereau, M. Raihane, F. Erchiqui and A. Dufresne, *Macromol. Mater. Eng.* **293**, 140-148 (2008).
9. J. Juntaro, M. Pommet, A. Mantalaris, M. Shaffer and A. Bismarck, *Comp. Interf.* **14**, 753-762 (2007).
10. A. Mahfoudh, A. Cloutier and D. Rodrigue, *Polym. Comp.* **34**, 510-516 (2013).
11. M. Kabir, H. Wang, K. Lau and F. Cardona, *Composites part B.* **43**, 2883-2892 (2012).
12. K-Y. Lee, K.C.C. Ho, K. Schlufter, J. Hodgkinson and A. Bismarck, "Bacterial cellulose as the binder for natural fibre performs to produce green nanocomposites" in *European conference on composite materials*, Venice, Italy. (2012), pp. 24-28.
13. B. Wang, S. Panigrahi, L. Tabil and W. Cerar, *J. Reinf. Plast. Comp.* **26**, 447-463 (2007).
14. E.O. Cisneros-López, M.E. González-López, A.A. Pérez-Fonseca, R. González-Núñez, D. Rodrigue and J.R. Robledo-Ortíz, *Comp. Interf.* DOI: 10.1080/09276440.2016.1184556 (2016).
15. D. Chimeni Yomeni, J. L. Toupe, C. Dubois and D. Rodrigue, *Comp. Interf.* **23**, 405-421 (2016).
16. S. Mohanty, S. Verma and S. Nayak, *Comp. Sci. Tech.* **66**, 538-547 (2006).
17. A.A. Pérez-Fonseca, J.R. Robledo-Ortíz, F.J. Moscoso-Sánchez, F.J. Fuentes-Talavera, D. Rodrigue and R. González-Núñez, *J. Polym. Environ.* **23**, 126-136 (2015).
18. A. Ramezani Kakroodi, S. Leduc and D. Rodrigue. *J. Appl, Polym. Sci.* **124**, 2494-2500 (2012).

Dynamic Mechanical Analysis And Dynamic Fatigue of Bead Foams

Jan-Hendrik Keller [a,*], Volker Altstädt[a,*]

[a] *University of Bayreuth, Universitätsstr. 30, D-95447 Bayreuth, Germany*
E-Mail: altstaedt@uni-bayreuth.de

Abstract. Technical components, such as crash absorbers in automotive industry or running shoes in sports products, made by bead foams are today exposed very often to complex large scale cyclic loading conditions in combination with different environmental loadings. In order to characterize the time and temperature dependent mechanical behavior during fatigue loading, the principles of in-situ mid-curve and subsequent hysteresis evaluation are best suited. These recently further-developed tools can provide instantaneous feedbacks about the state of fatigue in the specimen or the component. In addition, using these methods, complementary information about the prevailing deformation mechanisms of the sample can be derived (e.g. linear, non-linear, failure) at any time during the test. Within this paper, the results of commercially available EPS under several fatigue loading conditions are discussed, where non-linear mechanical behavior could be seen even at low stress amplitudes. Also the influence of different loading conditions on the long term mechanical behavior is discussed

Keywords: Hyteresis, bead foams

INTRODUCTION

Energy and consequently CO2-savings play a big role in todays engineering jobs. Therefore several strategies have been developed to build a light part with enough structural support. Naturally, the use of foams is preferred, as they can be tailor made according to the application. In addition with attached face sheets, these so called sandwich structures for example, are used to increase the bending stiffness of wind blades [1]. The use of foams and lightweight structures is not new since nature shows us some nice examples.

Polymer foams can be subdivided into several fields, such as extrusion foams, batch foams, injection molded foams, or bead foams. In this study, bead foams are examined. Parts made of bead foams consist of several hundred individual beads, which are fused together using heat [2]. The main advantage of bead foams are the freedom to create complex geometries with thick walls. Through the choice of the bead density itself, one can also tailor made the final density of the part and therefore determine the mechanical properties to ones needs.

Bead foams are mainly found in insulating applications such as insulation of houses or thermal sensitive foods. Today, new applications can be found, where not only the thermal behavior but also the mechanical behavior is important. Bead foams can be found nowadays as sun visors or other more complex parts in the automotive industry [3]. While the car is driving, varying loading conditions have to be endured by these parts. To mimic these loading conditions, modified testing methods are needed to unveil the mechanical behavior of bead foams under such large amplitude loadings.

The dynamic mechanical analysis is preferred for testing polymers with small deformation amplitudes. This test can be used to get an idea, how the material reacts under frequency changes or temperature changes. However, when it comes to larger amplitude loading, polymers and especially polymer foams behave quite non-linear. The basic principles of DMA is not valid anymore. Therefore the midcurve and hysteresis tools, developed by B. Lazan [4], can be used instead. Therefore the servo-hydraulic testing machine is used to describe the mechanical behavior under larger amplitude loading situation.

In this work, EPS is subjected to small amplitude loading in the DMA to unveil the general mechanical behavior in common. Additionally, EPS is subjected to large amplitude loading conditions to detect the border of critical and non-critical cyclic creeping and other fatigue related parameters.

MATERIAL & METHODS

Bead Foams

Commercially available expandable polystyrene (EPS) for mainly insulation purposes was bought from a local supplier in form of 20 mm thick sheets. The specified density is 30 g/l. The specimens were cut into cubes of 20 mm edge length using a circular saw. The faces were sanded by hand after the skin was cut off by a sharp razor blade.

Compression Testing

Prior to the dynamic testing, simple compression tests were done on an universal testing machine (Z7050, Zwick Roell, Ulm, Germany). The velocity of the crossbar was set to a constant speed of 100 mm/min. The displacement was determined using the crossbar, while the force was measured using a load cell (500 N, Zwick Roell). The rather high speed of the testing machine mimics the test circumstances in the dynamic experiments. As known from literature, the stress-strain behavior of foams is dependent from the testing speed (e. g. [5]). Within this test, the critical stress is detected as follows: The end of the linear region, where the stress-strain behavior is in transition from the linear to the plateau phase, is taken as the critical level.

Dynamic Mechanical Analysis

A dynamic mechanical analysis test machine (Eplexor 500 N, Gabo/Netzsch, Selb, Germany) was used in compression mode for detecting the mechanical properties under varying temperature and frequency conditions. Qubic specimens were placed inside the temperature controlled chamber and loaded according to the testing plan.

The machine was running both a temperature sweep and a strain sweep. The amplitude of loading was set to a very low value, as well as the pre-force. Details for the test procedures can be seen from table 1.

TABLE 1). Dynamic mechanic analysis test parameters for testing both, temperature and frequency sweep.

Experiment	Set point 1	Set point 2
Temp. sweep	30 - 100 °C	1 Hz = const.
Freq. sweep	30 °C = const.	0.1 - 100 Hz

Servo-hydraulic Testing

A servo-hydraulic testing machine (PSA 010, Instron Schenck, Darmstadt, Germany) was used in pure compression mode for detecting the long-term mechanical properties. The displacement was taken from the position sensor of the piston, while the force was taken by a load cell (1kN, Type S2, HBN, Germany). Additionally, the hysteresis loops were obtained with this testing device using a sophisticated software. The evaluation of these loops shows the linear or non-linear stage of the sample while testing. The machine was operated in both pure stress-stress controlled and stress-strain controlled mode, depending on the type of the experiment.

The test frequency was set to 5 Hz constant. This ensures, that the core temperature of the specimens does not increase more than 5 °C compared to the environment. The load levels were set to identify the critical load level. Five specimens were tested using the same loading conditions. Mean values were calculated from these experiments. Table 2 shows the loading levels in details.

In the stress-stress controlled experiments, the upper and the lower load levels are set to a constant stress value. Due to cyclic creeping, the upper and lower position limits of the piston changes continuously.

In contrast, the stress-strain controlled experiments are more complex. The loading level was set to a constant deformation while the un-loading level was set to a minimum stress. This ensures, that the piston has always contact to the specimens. Also, this pre-stress is needed to qualify the cyclic creep rate upon unloading. Within this test, the piston level while unloading increases to match the pre-force. This is needed, because the sample will be deformed plastically throughout the testing.

TABLE 2). Servo-hydraulic test parameters for testing both, stress-stress and stress-strain controlled load level.

Experiment	Set point 1	Set point 2
Stress-Strain	Stress: 20 kPa	Strain: 20, 30, 40 %
Stress-Stress	Mid-stress: 50, 80, 100 kPa	Amplitude: 30 kPa = const.

RESULTS

Dynamic Mechanical Analysis

The results for the DMA frequency sweep are displayed in figure 1. The frequency was changed between 0.1 and 100 Hz in a logarithmic way. At least 5 datasets per decade were recorded. On the left hand side, the store modulus (E') and loss modulus (E") are shown. The tan(d) is shown on the right hand side. Up to 5 Hz, there is no change in the E' and only a slight increase in the E". These frequencies (0,1 - 5 Hz) can be used for further testing in the servo-hydraulic testing machine as less cyclic heating is expected here. However, increasing the testing frequency to 10 Hz and further will lead to an increase in the E" and consequently in the tan(d). Around 20-30 Hz, the maximum of the tan(d) can be seen. Within this frequency, the EPS foam acts more and more as a damping device. Higher frequencies of more than 80 Hz lead to an increase in the E' value. Within this range, a hardening effect of the modulus can be detected. This might be attributed to the air inside the foam, which acts as an additional spring device.

Interestingly, the temperature sweep results might be odd on the first look. Figure 2 shows the change of pre-force and pre-strain over the temperature. Within this experiment, the pre-force was set to a value of 200 mN ± 20 mN and is applied after every measurement point. Starting at 30 °C and heating with 0.1 k / min, the experiment run until the final temperature was reached. The pre-force rises during the test until 80 °C, but within this range never exceeding the limits. The pre-strain, which follows from the pre-force, decreases until 80 °C slightly. This might be attributed to the positive temperature expansion ratio of the EPS foam. While the trapped gas inside is constantly building up pressure, less pre-strain is needed to put the pre-force on the sample. Shortly after 80 °C, the pre-force suddenly decreases rapidly and the pre-strain is doing the opposite. At this temperature, the EPS foam is deformed plastically as the T_G level is reached. While the pre-force is rising throughout the experiment, so does the E' (not shown). Reasons for this can be found in inhomogeneous deformation of the specimen due to the small loadings. Other than that, the trapped air inside the foam expands and creates a counter pressure, which is an explanation for the increase in the E'.

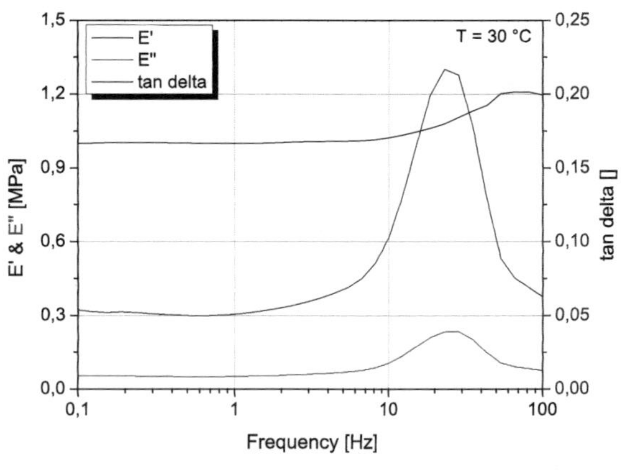

FIGURE 1. Frequency behavior of EPS30 at 30 °C. The store modulus (black) increases with increasing frequency. The loss modulus has a maximum at about 20 Hz. The damping of EPS in this region is maximized.

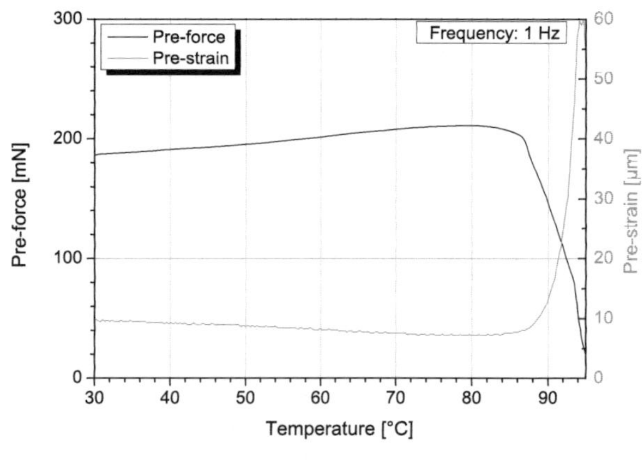

FIGURE 2. Temperature dependent behavior of EPS30. With increasing temperature, the pre-force is rising constantly. This might be due to the positive thermal expansion coefficient. At 85 °C, the specimen liquefies.

Servo-hydraulic Testing

The stress-strain controlled experiments show, that cyclic creeping is only existent, if a certain force level is exceeded. Oscillations generally take place with a given amplitude around a mid-point. The mid-stress is the mid-point for the force oscillation in this case. Unloading took place to a fixed stress level (i. e. 20 kPa). The piston of the testing machine was moving in a sinus of 5 Hz between those aforementioned two set points.

For EPS, the mid-stress is given in figure 3 (solid) for the different compression levels (i. e. 20 %, 30 %, 40 % in relation to the height). As the sample is compressed to a fixed length based value, it will soften and deform plastically over time. Softening can be seen in a decrease in the mid-stress over the cycles.

Figure 3 shows also the mid-strain (dashed) for the test setup. As the sample is compressed it will deform plastically and therefore the height of the sample will decrease subsequently. The mid-strain is a measure, similar to the mid-stress, which describes the strain value, which is "seen" by the sample in average. More interestingly in this case, the mid-strain changes only within the first 20k cycles. Afterwards it stays more or less constant.

The results for the servo-hydraulic testing with large amplitudes unveil, that the EPS is subjected to cyclic creeping only, if a certain stress level is passed. This level can be identified to about 120 kPa force maximum. So if the force exceeds this critical value while loading the specimen, it will deform plastically within this cycle. If the force is lower than this value, no additional cyclic creeping can be observed. The change in the deformation after 20 k cycles can be compared to the values found for quasi-static creep [6]. Exact models will be published elsewhere. As for stress-strain controlled experiments the loading amplitude decreases over time in this case, less parts of the specimen is compressed. Damaged parts build shear band like structures, but they are not further collapsing, if the loading force is lower than this critical level.

To prove this working hypothesis, stress-stress controlled experiments were conducted while changing the mid-stress and keeping the amplitude constant. The results can be seen in figure 4. Up to 80 ± 30 kPa, nearly no creeping is visible. However, if the critical level of 120 kPa is exceeded (s. figure 4, blue curve), a sudden change in the mechanical behavior is visible. The mid-strain curve suddenly drops and the sample is compressed up to 15 % after the experiment.

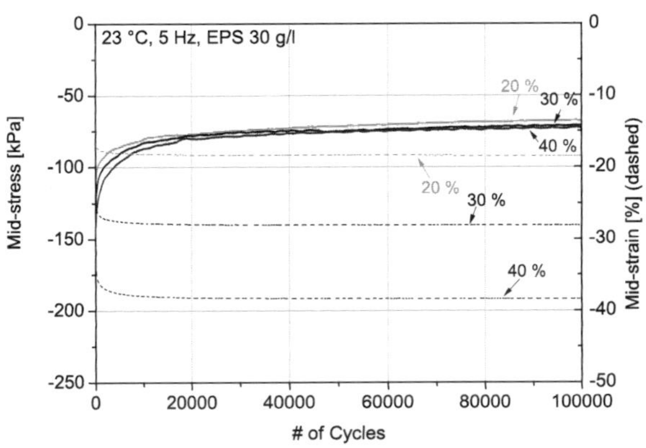

FIGURE 3. Mid-stress (solid) and mid-strain (dashed) for EPS30 under varying loading conditions. Load levels were set to 20 %, 30 %, and 40 % compression and 20 kPa unloading.

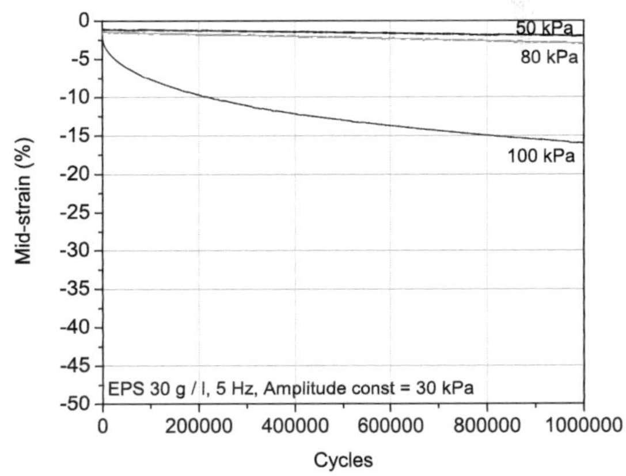

FIGURE 4. Influence of the mid-stress on the mid-strain for EPS30. The amplitude was set constant to 30 kPa and the mid-stress was varied. Below 80 ± 30 kPa nearly no cyclic creeping can be seen. If the critical level of 120 is exceeded, sudden creep occurs.

DISCUSSION

Dynamic mechanical analyses and servo-hydraulic testing show, that EPS30 has a complex deformation behavior. Loading with small amplitudes lead to an inhomogeneous deformation and temperature dependent effects can be detected, such as thermal expansion or melting. However, EPS can be used up to 80 °C when no loading is present.

Cyclic loading with larger amplitudes is even more complex. If the critical load level, which was identified here to 120 kPa is exceeded, severe cyclic creeping and plastic deformation takes place. In this case, this seems to be the limit of the material itself. When loading to values smaller than 120 kPa, whereby any combinations of stress amplitude and mid-stress can be used, only little cyclic creeping is visible.

Further investigations are needed to fully understand what is happening at the border of the critical load level. Therefore, in-situ SEM compression studies will be published elsewhere.

ACKNOWLEDGMENTS

The financial support of the University of Bayreuth Graduate School is highly acknowledged. The scientific input of the members of the department of Polymer Engineering and New Materials Bayreuth GmbH is highly appreciated.

REFERENCES

1. A. Fathi, J.-H. Keller, V. Altstädt, *Composites: Part B* ***70,*** 2015, pp. 156-166
2. D. Raps, N. Hossieny, C. B. Park, V. Altstädt, *Polymer* ***56***, 2015, pp. 5-19
3. Ch. Trassl, V. Altstädt, "Leichtbau- und Designwerkstoffe der Zukunft" in *Kunststoffe* ***2*** 2014, KU111605, pp. 73-76
4. B. J. Lazan, "Damping of materials and members in structural mechanics" in Pergamon Press Inc., New York, 1968
5. W. Chen, H. Hao, D. Hughes, Y. Shi, J. Cui, Z.-X. Li, *Materials and Design* ***69***, 2015, pp. 170-180
6. J. S. Horvath, Manhatten College Research Report No. CE/GE-98-3, 1998, p. 12

The Evaluation of Stretchability and Its Applications for Biaxially Oriented Polypropylene Film

Toshitaka Kanai [a,*], Tomoaki Takebe[b]

[a] *KT Polyme: 5-7-14 Kuranamidai, Sodegaura, Chiba, Japan;* toshitaka.kanai@ktpolymer.com
[b] *Performance Materials Laboratories, Idemitsu Kosan Co., Ltd. Anesaki, Ichihara ,Chiba, Japan*

Abstract: In recent years, the plastic film is essential in our daily lives. Among them, polypropylene has good physical properties, for example, tensile strength, impact strength and transparency. Recently, it was found that the stretchability of polypropylene (PP) is improved by blending a small amount of low modulus polypropylene (LMPP). The LMPP is a low tacticity polypropylene. The purpose of this research is to investigate the effect of blending LMPP, which has different characteristic of resin such as tacticity and a molecular weight, on PP and to study the mechanism of the structure formation in a biaxial stretching process. The blending LMPP to PP reduces the spherulite size and increases the thickness of amorphous phase in the lamella without changing the thickness of crystal phase, namely melting temperature. LMPP blends also reduced the yield values during the stretching process and increased the stress build up ratio. From these reasons, they improved film thickness uniformity of biaxially stretched film and reduced film break during the high production line. Furthermore, as the stretchable temperature window was widened to low temperature side, the thickness uniformity was improved in the wide stretchable temperature. As a result, improving the stretchability and the film thickness uniformity of PP were achieved by blending LMPP to PP.

Keywords: low modulus PP, biaxial stretching, tacticity, crystallinity, stretchable temperature width
PACS: 68.55.-a

1. INTORODUCTION

Polypropylene (PP) is widely used for the biaxially oriented film using sequentially biaxal oriented process and the supply capacity in 2014 reached 10 million tons/year. This film is mainly used in the field of food package and industrial usage because of high transparency, heat resistance and chemical resistance. According to the increase of the demand, high production speed and ultra thin film such as capacitor film have been required. In case of the high speed production, as the high stress is applied during the biaxially oriented process, the film break and the problem of uneven film thickness occur. Low tacticity polypropylene (LMPP) [1)] has low crystallinity, low melt temperature and low density.

Our previous research reported that the addition of LMPP to PP which reduced the size of sherulites and decreased the yield value of Stress-Strain curve expected the better stretchability from the results of uniaxial stretching [2), 3)].The purpose of this research is to improve the stretchability of PP and to investigate the mechanism of stretchability in terms of both S-S curve and super structure when a small amount of LMPP is blended to PP.

2. EXPERIMENT

2.1 Samples

The samples which were used are listed in Table 1. Isotactic polypropylene was BOPP grade A produced by Prime Polymer. L1, L2 and L3 are LMPP grades and L1 is standard grade of LMPP and L2 is lower tacticity than L1 and L3 is higher molecular weight than L1. Iso-polypropylenes A blended with 5 wt% of LMPP, namely L1, L2 and L3 were formed 500μm sheets of A/L1, A/L2 and A/L3 respectively under the resin temperature 250℃ and chill roll temperature 30℃. The samples were cut in 58mm x

58mm square size. The previous research showed that as 5 wt % LMPP blended to PP had the best stretchablility, 5 wt % blended various samples were prepared.

Table1 Characteristics of resins

Sample	MFR (g/10min)	Mw $\times 10^5$	Tm (℃)
A	3	3.8	161
L1	60	1.3	70
L2	60	1.3	60
L3	7	2.3	70

2.2 Experimental Equipment

The biaxial stretcher SDR-527K which was designed by our group and made by Etoh Corporation was used. The schematic view of this equipment is shown in Fig.1. Various data could be obtained at the same time such as stress - strain curve (S-S), refractive indexes in three directions, phase difference distribution and light scattering data. The stretching conditions are shown in Table 2.

Table2 Stretching conditions

Stretching mode	Sequential biaxial stretching
Stretching speed	10mm/s
Stretching ratio	MD5times×TD7times
Stretching temp.	159℃
Preheating time	7min

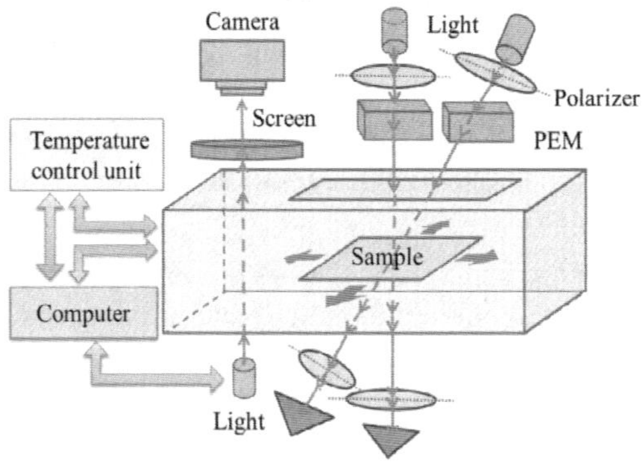

Fig.1 Schematic diagram of biaxial stretching machine

2.3 Stress- Strain Curve during the Stretching

Fig.3 shows an image pattern of stress- strain curve during the sequentially biaxial stretching. As the tenter process is usually used in the TD stretching and the final thickness uniformity is considered to be determined in TD stretching, TD stretching was focused. From the previous research, good and not good stretchable patterns are obtained by the stress-strain curve. The stress pattern which has a high yield point at the initial stretching stage and then reducing stress shows undesired film thickness uniformity.

On the contrary, the stress pattern which has a low yield point at the initial stretching stage and then increasing stress shows good film thickness uniformity. The reduction of yield value is important to obtain good thickness uniformity [2)]

3. RESULTS AND DISCUSSION

3.1 Stress- strain Curve

Fig.2 shows stress- total stretching ratio (MD stretching ratio x TD stretching ratio) curves for various samples. Sample A has high yield value and after this point passes, the stress decreases. This means the necking deformation during the stretching process occurs and is undesirable for thickness uniformity.

The A/L1 and A/L2 which have about 40% lower yield values than PP A show stress hardening. Although A/L3 shows almost same yield value as PP A, the stress of A/L3 increases rapidly in the field of the high stretching ratio. L3 is high molecular weight LMPP and has molecular entanglement in the amorphous phase during the stretching process.

From these results, it is expected that LMPP blended samples create better film thickness uniformity than PP A sample and furthermore film break during the stretching process can be reduced.

3.2 Film Thickness Uniformity of Biaxially oriented film

Table3 shows the yield value σ_{yield} , maximum stress σ_{max} , stress build up ratio $\sigma_{max}/\sigma_{yield}$ and standard deviation of film thickness σ_t when the samples are stretched in TD at stretching temperature 159℃. The thickness uniformity was measured at 60 points, 1 cm interval by Toyo Seiki Thickness Meter B-1. Fig.3 shows that the relationship between stress build up ratio $\sigma_{max}/\sigma_{yield}$ and standard deviation of thickness uniformity σ_t. This result shows good correlation and the increasing the stress build up ratio $\sigma_{max}/\sigma_{yield}$ has a potential of improving thickness uniformity. The reduction of yield value increases stress build up ratio $\sigma_{max}/\sigma_{yield}$ and reduces the necking deformation. It makes uniform stretching and reduces standard deviation of film thickness.

3.3 Relationship between stretchable temperature window width and film thickness uniformity

The stretchable temperature window width was defined as the stretchable temperature range. Fig.4 shows the relationship between the stretchable temperature window width and the standard deviation of film thickness uniformity.

The stretchable temperature range of PP A is 158~163℃ (6℃) and the ones of A/L1, A/L2and A/L3 which have lower stretcable temperature windows are 156~163℃ (8℃).

As the blending LMPP to PP increases lower melting components and decreases crystallization speed during the stretching process, the stretching at a lower temperature is possible.

As a result, A/L1 and A/L2 improved the film thickness uniformity and the standard deviation is about half at 159 ℃, compared with PP A. A/L3 also improved thicknes uniformity, but it is not so good as A/L1 and A/L2. A/L3 which has high molecular component shows molecuar chain entanglement and strain hardening, but does not show the reduction of yield value at the initial stretching stage and stress build up ratio does not become so large.

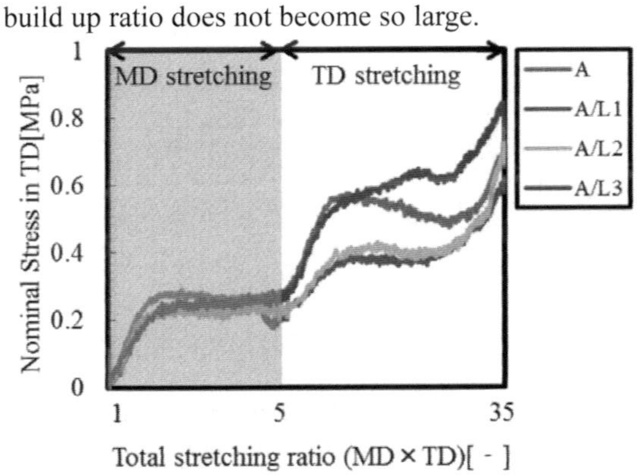

Fig.2 Nominal Stress-Total stretching ratio curves

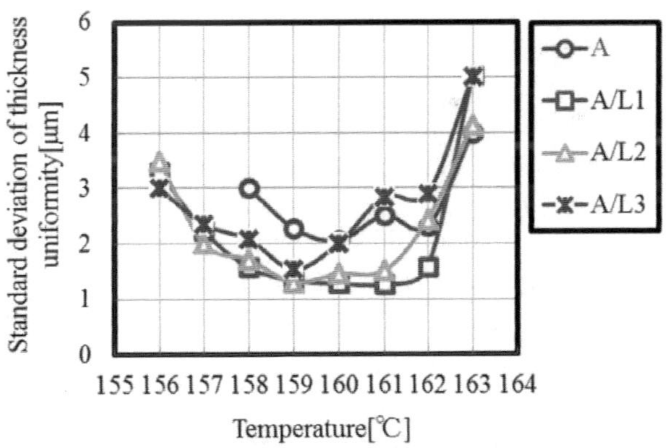

Fig.4 Standard deviation of thickness uniformity and temperature

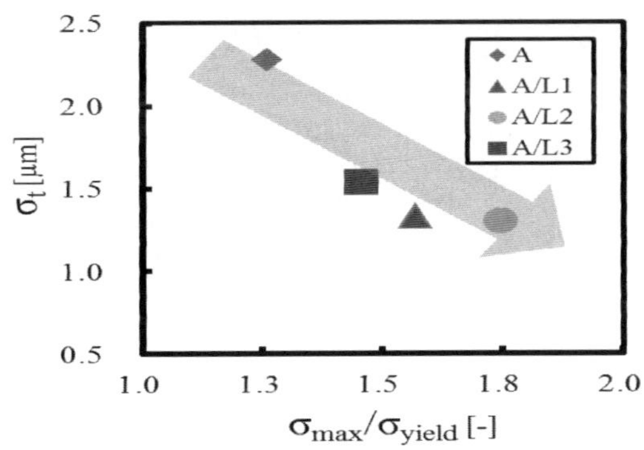

Fig.3 Stress build up ratio $\sigma_{max}/\sigma_{yield}$ and standard deviation of thickness uniformity σt

Table3 Stretching properties of PP samples in TD(159℃)

Sample	σ_{yield} [MPa]	σ_{max} [MPa]	$\sigma_{max}/\sigma_{yield}$ [-]	σt [μm]
A	0.57	0.72	1.26	2.28
A/L1	0.35	0.55	1.57	1.33
A/L2	0.40	0.70	1.75	1.30
A/L3	0.56	0.83	1.46	1.53

3.4 Three Dimensional Refractive Indexes

The orientation during the stretching can be obtained by measuring the three dimensional refractive indexes. Fig. 5 shows the refractive indexes of PP and PP/ LMPP1 blend samples under the stretching temperature 166℃. MD refractive index Nx increases during MD stretching ratio 5 and then TD refractive index Ny increases and becomes larger refractive index Ny than Nx over TD stretching ratio 5. ND refractive index decreases during the MD and TD stretching. The PP/ L1 (PP/LMPP1) blend sample reduces neck like stretching and decreases fluctuation and fluctuating range, compared with PP sample.

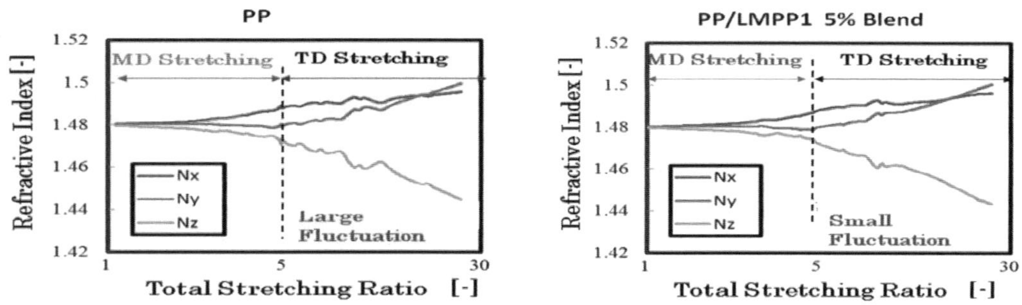

Fig. 5 Refractive indexes and total stretching ratio in PP and PP/LMPP1 5% blend

3.5 Light Scattering（LS）

The spherulite size of sheet sample was measured by the light scattering. The existence of spherulites in a sheet makes the strechablity worsen [3]. He-Ne laser (wave length λ=632.8nm) was used as a light source and the polarizer and the analyzer were mutually set perpendicularly and Hv pattern was observed. Hv patterns measured by the light scattering are shown in Fig.6. The radius of shpherulite size U_{max} was calculated by Eq.(1).

$$U_{max}= 4.09/\ [(4\pi/\lambda)\sin(\theta_{SALS,max}/2)] \quad (1)$$

$\theta_{SALS,max}$:Maximum scattering angle showing maximum value 45 degree to polarizing direction

Table 4 shows the spherulite size of various samples. As Hv scattering pattern of sample A has small, the spherulite size is large and reaches about 10.5μm. On the contrary, Hv scattering pattern of samples A/L1 and A/L2 are large and the spherulite size is reduced and about half, compared with PP A. As the crystallization speed of LMPP is much slower than PP, it is considered that the blending of LMPP to PP reduces the crystallization speed and the growth speed of spherulite. As the spherulite size of A/L3 is almost same as PP A, the blending effect of L3 is weak.

Table4 Spherulite sizes of PP samples

Sample	Sperulitesizes[μm]
A	10.5
A/L1	5.0
A/L2	4.7
A/L3	10.4

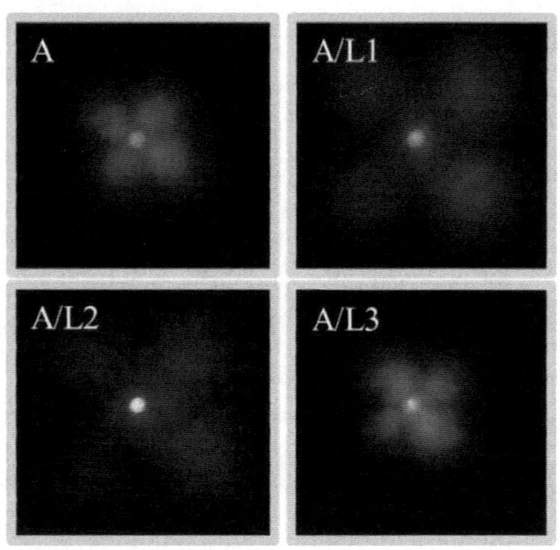

Fig.6 Light scattering images of PP samples

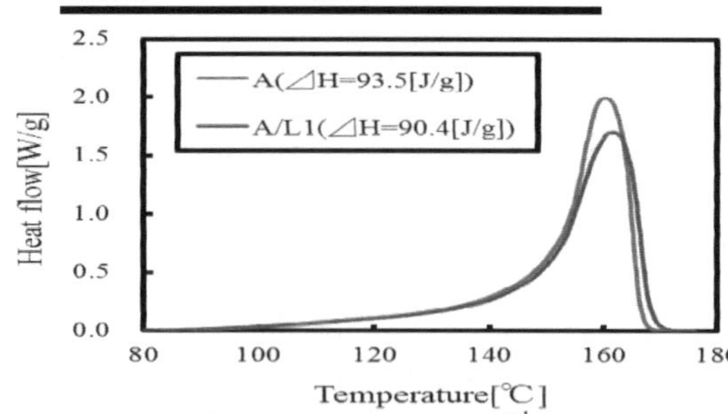

Fig.7 DSC curves on 2nd heat

3.6 Differential Scanning Calorimetry（DSC）

The enthalpy of fusion ΔH and the melting temperature were measured by DSC-7 produced by Perkin Elmer. Fig.7 shows DSC curves. The blending LMPP to PP A reduces ΔH by about 4%. The melting temperature of LMPP blends which keeps constant is the same as PP A.

3.7 Small Angle X-ray Scattering（SAXS）

The long period L was measured for 30 minutes using Rigaku X-ray SAXS at voltage 50kV, Ampere 280mA, CuKα (λ=1.5418Å) and calculated by the maximum value of scattering pattern $2\theta_{max}$ and Bragg equation Eq.(2).

$$2L\sin\theta_{max}=\lambda \quad (2)$$

Fig.8 shows the relationship between the relative intensity and the scattering angle $2\theta_{max}$. From these results, the scattering peak of A/L1 and A/L2 are shifted to the small angle side. It means the long period increases, but A/L3 shows a little shift.

From the long period and ΔH obtained by DSC, the thickness of crystal phase was obtained. Table5 shows the long period, the thickness of crystal phase and amorphous phase. The increase of LMPP blend samples' long period is caused by penetrating LMPP into amorphous phase, not crystal phase. LMPP reduces the crystallinity of blend samples.Fig.9 shows that the superstructure is changed by blending LMPP into PP.

Table5 Higher-order structures of PP samples

Sample	Long period [μm]	Crystalline Phase [nm]	Amorphous Phase [nm]
A	12.6	5.6	7.0
A/L1	13.9	6.0	7.9
A/L2	13.4	5.8	7.6
A/L3	13.0	5.6	7.4

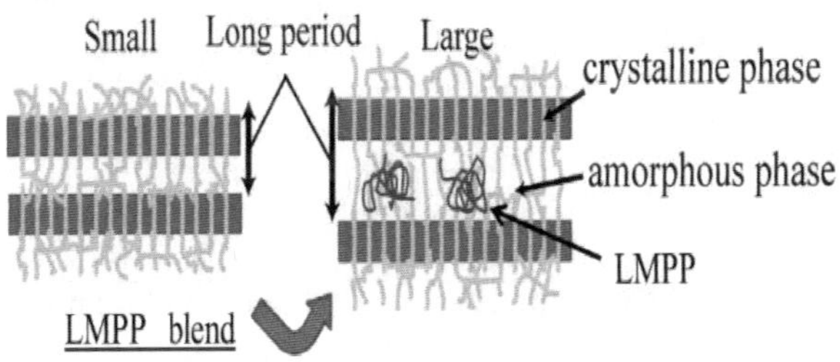

Fig.9 High order structure model by LMPP blend

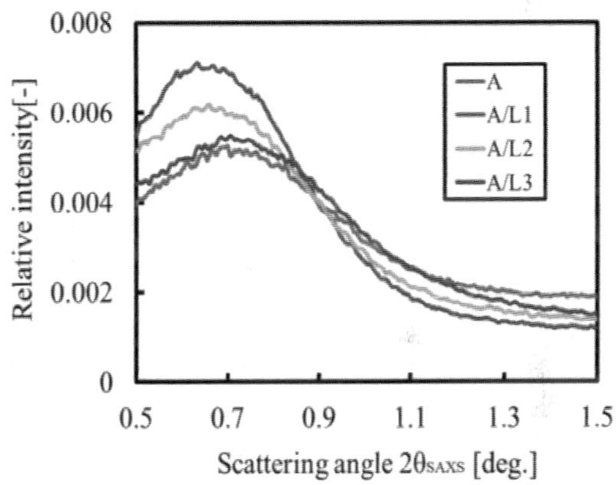

Fig.8 Relative intensity vs scattering angle 2θ

4. CONCLUSIONS

Blending LMPP into PP was investigated in order to improve the biaxial stretchability of PP. LMPP blends reduced the yield values during the stretching process and increased the stress build up ratio. As a result, they improved film thickness uniformity of biaxially stretched film and reduced film break during the production line. Furthermore, as the stretchable temperature window was widened to low temperature side, the thickness uniformity was improved in the wide stretchable temperature. The blending LMPP to PP reduces the spherulite size and increases the thickness of amorphous phase in the lamella without changing the thickness of crystal phase, namely melting temperature. As a result, reducing crystallinity and improving the stretchability and the film thickness uniformity of PP were achieved.

REFERENCES

1. T.Takebe,Y.Minami, T.Kanai; *Seikeikakou,* Vol. 21, (4),202-207(2006)
2. S.Tamura, ,I. Kuramoto,T. Kanai; *Polym. Eng. Sci.*, Vol. 52,(6),1383-1393 (2012)
3. A.Funaki,S.Kuratani,T.Yamada,T.Kanai; *Seikeikakou,* Vol. 23,(5),229-235(2011)

New Trends in Preparation and Characterization of Nanofibre Filters

Wannes Sambaer[a], Martin Zatloukal[a,*], Dušan Kimmer[b], Tomas Sedlacek[c] and Petr Sáha[a,c]

[a] *Polymer Centre, Faculty of Technology, Tomas Bata University in Zlin, Vavreckova 275,760 01, Zlin, Czech Republic*
[b] *SPUR a.s., T. Bati 299, 764 22 Zlin, Czech Republic*
[c] *?Centre of Polymer Systems, Tomas Bata University in Zlin, 760 01 Zlin, Czech Republic*

Abstract. Full 3D polydisperse particle filtration modeling at low pressures has been performed for a polyurethane nanofiber based filter prepared via electrospinning process in order to more deeply understand the filter clogging and the cake formation. In this work, realistic SEM image based 3D filter model, transition/free molecular flow regime, Brownian diffusion, aerodynamic slip, particle-fiber and particle-particle interactions together with a Euclidian distance map based methodology to calculate the pressure drop have been utilized. Model predictions have been compared with relevant experimental data in order to validate the used assumptions, methodologies and numerical scheme. The effect of particle-particle as well as particle-fiber interactions on the nanofiber based filter efficiency, pressure drop and the quality factor during the filter clogging has been investigated in more detail.

Keywords: Nanostructure, Electrospinning, 3D filter clogging modeling, Polymers, Porous media, Filtration.
PACS: 81.07.-b, 81.05.Qk, 81.05.Rm, 47.56.+r

INTRODUCTION

Fibrous based air filters are widely used in many applications in area of pharmaceutical, medical, biology, food, semi-conductor industries and therefore it is not surprising that both experimental and theoretical studies dealing with their basic characteristics (such as the pressure drop and particle collection efficiency). Nowadays, nanofiber nonwoven based filters becomes of high interest due to their ability to reach high filtration efficiency for ultrafine particles with a low pressure drop due to aerodynamic slip around the nanofibers [1-5]. Unfortunately, till now, the detailed theoretical 3D analysis of the nanofiber based filters clogging has not been performed yet and thus, the full understanding of the filtration cake formation on the nanofiber based filters and its role on the final filter efficiency is not fully understood. In order to provide a better understanding of the nanofiber based filters loading process with respect to pressure drop and filtration efficiency evolution, a realistic SEM image based 3D structure model of the filter, transition/free molecular flow regime, Brownian diffusion, particle-fiber interactions, aerodynamic slip and sieve has been utilized in this work.

EXPERIMENTAL

Materials

A polyurethane solution in dimethylformamide (DMF) based on 4,4'methylenebis(phenylisocyanate) (MDI), poly(3-methyl-1,5-pentanediol)-alt-(adipic, isophtalic acid) (PAIM) and 1,4 butanediol (BD) was synthesized in molar ratio 9:1:8 at 90°C for 5 hours (per partes way of synthesis starting with preparation of prepolymer from MDI and PAIM and followed by addition of BD and remaining quantity of MDI). The prepared solution was suitable for electrospinning process and had a PU concentration of 24.1 wt%, viscosity of 1.45 Pa.s and conductivity of 157 μS/cm (adjusted by 3 parts of citric acid and 1 part of borax).

Nanofiber Layers Preparation by Electrospinning Process

In this work, a commercially available NanoSpider™ machine (Elmarco s.r.o. Liberec, Czech Republic, http://www.elmarco.com/) equipped with patented (PCT/CZ2010/000042) rotating electrode with 3 cotton cords spinning elements was used to prepare nanofiber nonwoven based filter sample from polyurethane solution. The experimental conditions were: relative humidity 25 %, temperature 23.5 °C, high voltage applied into PU solution 75 kV, distance between electrodes 210 mm, rotational electrode speed 7 rpm and speed of polyester nonwoven supporting textile collecting nanofibers was 0.4 m/min. Samples for further analyses were taken from the middle part of the 40 cm wide fabric coated with layer of nanofibers.

Nanofiber Filter Characterization

A nanofiber based filter, prepared via electrospinning process, has been characterized by Scanning Electron Microscope technique (Vega 3, Tescan, Czech Republic), as visible in Figure 1a.

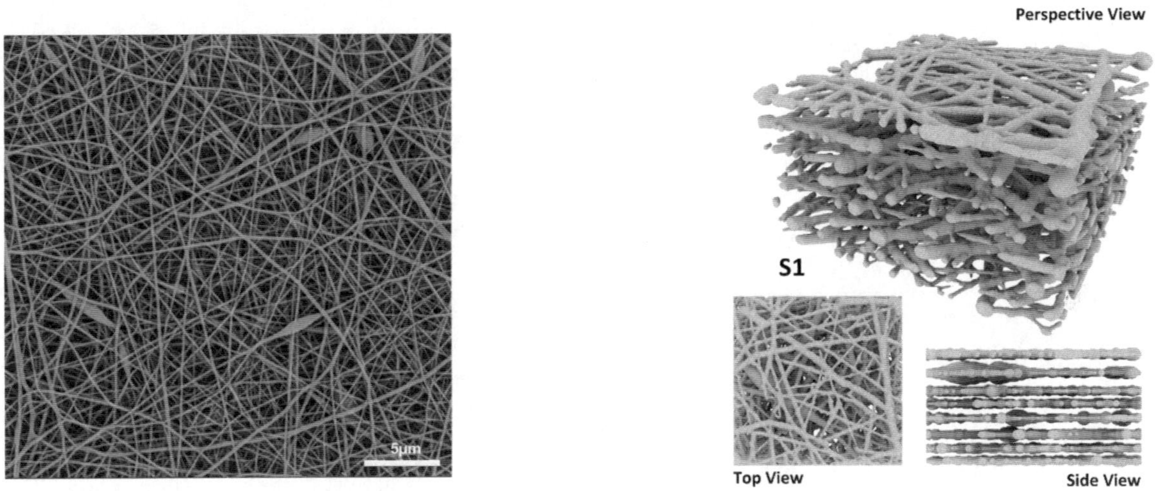

FIGURE 1. Manufactured nanofiber based nonwoven filter. **1a)** SEM image. **1b)** Full 3D structure model of the manufactured filter.

The obtained SEM picture has consequently been used for the fiber diameter/pore size distribution determination by using recently proposed digital image analysis technique [6]. The filter has an average fiber diameter and average pore size equal to 133 nm and 156 nm, respectively, thickness (3.59 µm) and mass area (0.54 g/m²). The electric potential generated on the surface of the filter was measured by electrostatic fieldmeter Simco, model FMX-003, the Netherlands. It has been found that the surface charge potential of the filter was practically negligible, equal to 70 V, i.e. there are practically no electrostatic forces, which could affect its filtration characteristics. The filtration efficiency for the prepared nanofiber based filter was determined experimentally in the ultrafine particle range Note that the filtration efficiency and pressure drop measurements were repeated four times and the following filtration efficiency characteristics of the filter were experimentally determined: the Maximum Penetration Pore Size (MPPS) 70 nm, the minimum filtration efficiency 98.32 %, the average pressure drop 148.5 Pa and the quality factor 27.5 kPa^{-1}, which is defined according to Eq. 1.

$$QF = -\ln(E)/\Delta p \tag{1}$$

where Δp is the pressure drop and E is the filtration efficiency at MPPS.

THEORETICAL

All calculations performed in this work have been obtained with in-house developed software (UTBsoft Filtration) which has been ran on a PC utilizing Windows 7 Professional with the following configuration: Intel® Core™ I7-980X processor equipped with 6 cores able to run 12 threads, having each a clock speed of 3.33 GHz, and 24 GB DDR3 memory and a NVIDEA Quadro 2000 graphical card. The utilized filtration model is provided in [3-4]. In this work, the fully sticking (single touching point) boundary condition was applied (similarly to other theoretical studies, which can be justified for the 10nm-5µm particle size range as showed experimentally by Wang and Kasper

[7]. Under this condition, dendrites (formed by the accumulated particles during the filtration process) cannot bend or break down.

Evaluation of Model Prediction Capabilities for Clean Filter

The filtration efficiency of the filter is theoretically determined by the number of particles, which can be captured by the filter [3]:

$$E = \frac{N_{in} - N_{out}}{N_{in}}, \tag{2}$$

where N_{in} and N_{out} are the numbers of entering and exiting particles, respectively. Twenty different groups of 20 000 identical particles having diameter linearly varying from 20 to 500 nm on a logarithmic scale were injected in the large modeled clean filter inlet having dimensions of 7.2 μm on 7.2 μm. Polyurethane nanofiber 3D structure model (see Figure 1b), generated from the SEM image, was utilized in the filtration modeling. The comparison between the predicted and experimentally determined filtration efficiency, pressure drop and quality factor data for the tested filter samples is provided in Figure 2. It is visible, that firstly, the model predictions are in a fairly good agreement with the corresponding experimental data, which justifies the used model assumptions.

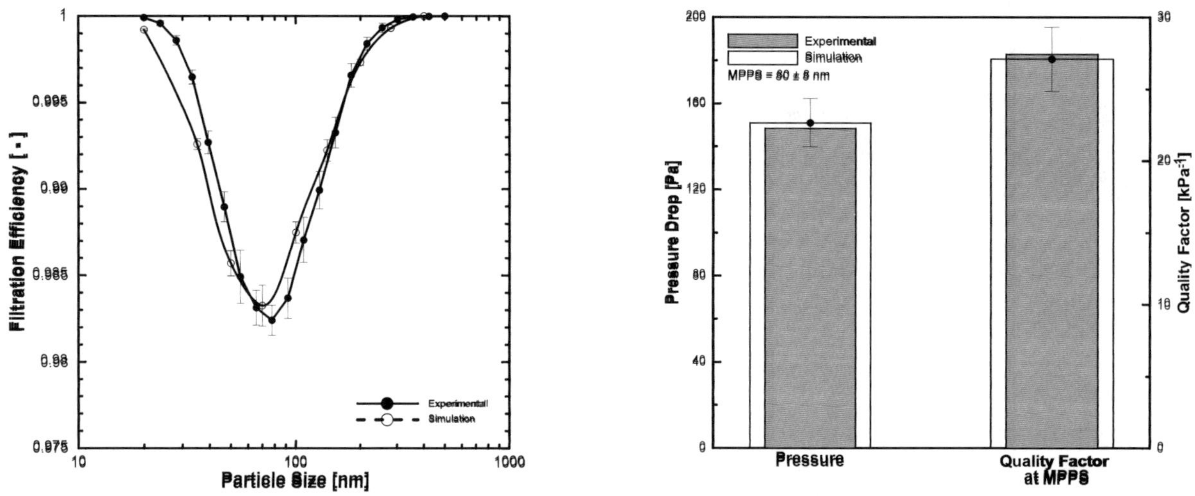

FIGURE 2. Filtration model predictions vs. experimental data. **2a)** Filtration efficiency. **2b)** Pressure drop and quality factor.

Deposition Dynamics vs. Filtration Characteristics

In order to understand deposition dynamics for nanofiber based filter, the 3D structure model of the filter depicted in Figure 1b was randomly penetrated with 50000 particles, having the particle size distribution the same as the nebulized solution used in the experimental filtration test, considering both particle-fiber and particle-particle interactions (see Figure 3). Calculated pressure drop as the function of the depth direction is provided in Figure 4. The closer analysis revealed existence of four regions with different pressure gradients entitled here as A, B, C and D. The region A describes top cake active layer open to the ambient air at which particle attachment takes predominantly the place. Due to the openness of this region the pressure gradient is the smallest here. The region B represents consolidated layer of the cake characterized by a constant mean particles density where the probability that the new particle arrives from the surrounding gas is minimal. Thus, the pressure gradient in region B is higher than in region A. In region C (near-fiber wall region), the particle density is the highest due to presence of nanofibers having high collecting surface area. As the result, the pressure gradient in region C is the highest. It is interesting that even if the near-fiber wall region is very thin, it consumes considerable amount of pressure. The region D consists of the clogged filter at which particle density decreases with the filter depth. The pressure gradient in region D is comparable with region B due to structural similarity in both regions leading to equivalent resistance against the air flow. In this particular case, the overall structural similarity of D region (containing low number of non-consolidated particles) with region B (containing high number of consolidated particles) is reached by the presence of nanofiber based filter having very small average pore size diameter (156 nm in this case). In numbers, the overall packing factor for region D (determined as the volume fraction occupied by particles and fibers) is 0.1305 whereas the packing factor in region B is 0.1207. Note that the packing factor of the clean filter (i.e. the volume fraction occupied by fibers only) is 0.1162.

FIGURE 3. Side view of predicted 50 000 particles deposit.

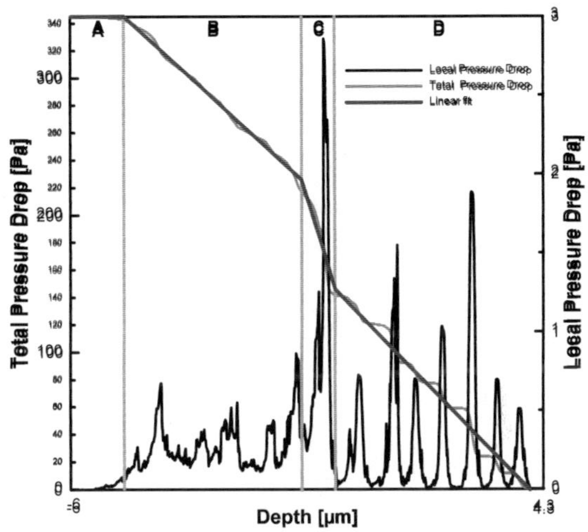

FIGURE 4. Calculated total and local pressure drop as the function of the depth direction for top cake active layer (region A), consolidated layer of the cake (region B), near-fiber wall region (region C) and clogged filter region (region D).

The role of Particle-Fiber Friction Coefficient

In order to more deeply understand the particle-fiber friction coefficient effect on the cake build up, filtration efficiency, pressure drop and quality factor of the nanofiber based filters, its value was varied in the utilized filtration model. The 3D filter model was randomly penetrated with 50000 particles, with an experimentally defined particle size distribution. The filtration efficiency, the pressure drop and quality factor (calculated by Eq. 1) was recorded during the filter clogging as a function of the captured particle mass for different particle-fiber frictions and the results are depicted in Figure 5a. Clearly, the increase in the fiber-particle friction coefficient leads to higher quality factor. On the other hand, the filter samples with higher fiber-particle friction coefficient shows much higher quality factor reduction with the increased collected particle mass in comparison with the filter samples having lower fiber-particle friction coefficient. These theoretical conclusions are supported by the recent experimental data showing that roughness structure of nanofibers, created by nano protrusions, shows more enhances filtration properties (the quality factor 30.8 kPa^{-1}) than circular cross section nanofibers (the quality factor 21.8 kPa^{-1}) for given polymer type and air filtration conditions [8]. In order to understand these findings in more detail, pressure and packing profiles for two different particle-fiber friction coefficients (0 and 0.033) were generated and plotted in Figure 5b and Figure 5c, respectively. As it can be seen, particle-fiber friction coefficient influences pressure and packing profiles only at the near-fiber wall region C and clogged filter region D. In more detail, increase in friction coefficient increases the pressure drop and packing factor in region C considerably. On the other hand, in region D, the trend is switched and the effect of friction coefficient on the pressure drop is very small due to the fact that the packing factor of the particles is one order of magnitude lower than packing factor of the clean filter, i.e. the variation of the particle packing factor due to change in friction coefficient has only a small effect on the pressure drop here. This suggests that fiber-particle friction coefficient controls the overall pressure drop predominantly via the packing factor at the near-fiber wall region C at which the cake build up is initiated. In this particular case, increase of particle-fiber friction coefficient from 0 to 0.033 increases the overall pressure drop by 15.6%.

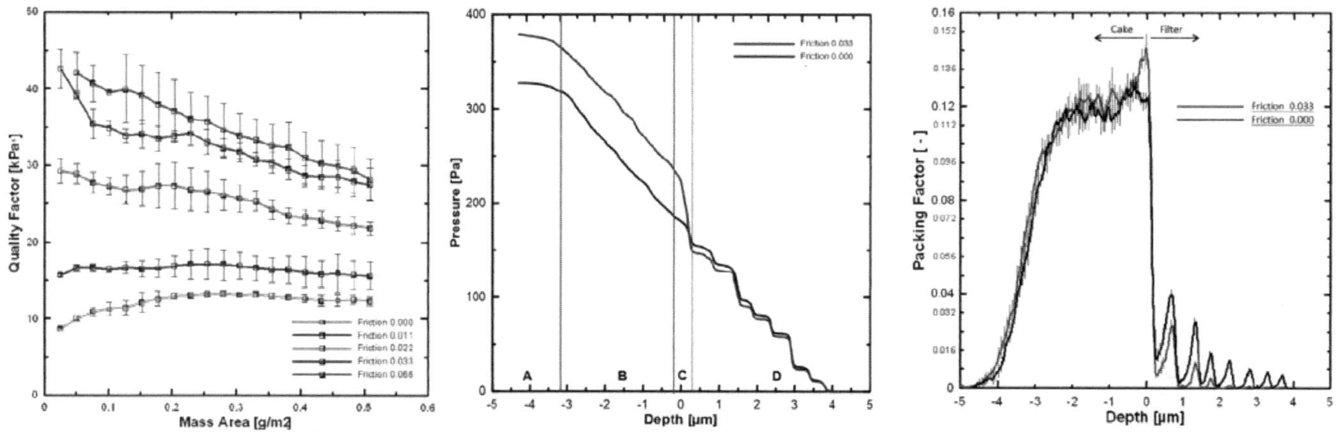

FIGURE 5. Theoretically predicted effect of fiber-particle friction coefficient during filter clogging. **5a)** Quality factor. **5b)** Pressure drop. **5c)** Packing factor.

CONCLUSIONS

Polyurethane nanofiber based filter has been prepared by electrospinning process having average fiber diameter and average pore size equal to 133nm and 156 nm, respectively, thickness 3.59 µm, mass area 0.54 g/m^2, maximum penetration pore size 70 nm, minimum filtration efficiency 98.32 %, average pressure drop 148.5 Pa and quality factor 27.5 kPa^{-1}. The realistic SEM image based 3D filter model, transition/free molecular flow regime, Brownian diffusion, aerodynamic slip, particle-fiber and particle-particle interactions have been utilized for the prepared polyurethane nanofiber based filter in order to more deeply understand the filtration cake formation and its role on the final filter efficiency and the pressure drop. Model predictions have been found to be in a fairly good agreement with the performed filtration efficiency, pressure drop and quality factor measurements for the clean filter, which justifies the used assumptions and numerical scheme for the given filter type and filtration conditions. Based on the performed theoretical analysis, the following conclusion remarks can be formulated. Increase in the fiber-particle friction coefficient leads to filter quality factor increase due to the creation of highly packed and thin cake layer occurring just at the top filter nanofibers, which causes much higher increase in the filtration efficiency in comparison with the pressure drop. On the other hand, filters with higher fiber-particle friction coefficients show much higher quality factor reduction with the increased collected particle mass than filters having lower fiber-particle friction coefficients.

ACKNOWLEDGMENTS

The authors wish to acknowledge the Grant Agency of the Czech Republic (grant No. 16-05886S) for the financial support.

REFERENCES

1. Sun, B., Long, Y.Z., Zhang, H.D., Li, M.M., Duvail, J.L., Jiang, X.Y., Yin, H.L., *Progress in Polymer Science*, 39(5), 862-890 (2014).
2. Hosseini, S.A., Tafreshi, H.V., *Chem. Eng. Sci.*, 65, 2249-2254 (2010).
3. Maze, B., Tafreshi, H.V., Wang, Q., Pourdeyhimi, B., *J. Aerosol Sci.* 38(5), 550-571 (2007)
4. Sambaer, W., Zatloukal, M., Kimmer, D., *Chem. Eng. Sci.*, 66, 613-623 (2011).
5. Sambaer, W., Zatloukal, M., Kimmer, D., *Chem. Eng. Sci.*, 82, 299-311 (2012).
6. Sambaer, W., Zatloukal, M., Kimmer, D., *Polym. Test.* 29, 82–94 (2010).
7. Wang, H.C., Kasper, G., *J. Aerosol Sci.*, **22(1)**, 31–41 (1991).
8. Wang, N., Si, Y., Wang, N., Sun, G., El-Newehy, M., Al-Deyab, S.S., Ding, B., *Separation and Purification Technology*, 126, 44-51 (2014).

Effect of Long-Chain Branches on Fabrication of High Expansion Microcellular Injection Molding of Polypropylene Foams

Long Wang, Shota Ishihara, Atsushi Minato, Masahiro Ohshima*

Department of Chemical Engineering, Kyoto University, Katsura, Kyoto 615-8510, Japan

Abstract. Nowadays, intensive studies have been conducted on foam injection molding due to its reduction of material consumption and less deterioration in mechanical properties. However, the main drawback of this technology is its low volume expansion ratio and maximum volume expansion ratio is usually smaller than 5-fold. Here well-defined cell structures of long chain branched polypropylene (LCBPP) foams with 10-fold volume expansion ratio were successfully fabricated by a core-back foam injection molding with nitrogen. Comparing with isotactic polypropylene (iPP), the addition of long-chain branches imparts LCBPP with remarkably higher elongational viscosity and obviously strain hardening characteristics. The long-chain branched structures can stabilize cell growth and reduce cell coalescence or rupture, which can be authenticated from the lower open cell content of LCBPP foams (5/10-fold) relative to that of iPP foams as well as the fibrillated cell structures of iPP foams. Furthermore, the operating window of manufacturing high expansion ratio foams was notably widened for LCBPP foams due to the presence of long-chain branches. This study demonstrates that the introduce of long-chain branches is needed to fabricate high volume expansion ratio microcellular PP foams and in great potential capacity for even more than 10 times of expansion microcellular foams.

Keywords: Long-Chain Branches, High Expansion Foams, Foam Injection Molding
PACS: 82.35.-x, 61.41.+e, 61.20.Lc, 61.25.Hq, 81.05.Rm

INTRODUCTION

Polypropylene (PP) is a fast growing commercial polymer since it has many desirable and beneficial properties such as, good thermal stability, resistance to corrosive chemical, good mechanical properties, and easy recycling. These outstanding properties have made PP more promising in foaming products than other thermoplastics such as polystyrene (PS) and polyethylene (PE). Therefore, a lot of research has been conducted on PP foams [1]. However, conventional PP has poor melt properties due to its linear molecular structure, such as low melt strength, and this greatly restricts its application. Thus, it is urgently needed to prepare high melt strength PP and introducing long chain branches (LCB) into PP's backbone is the most efficient method to increase the melt strength. To achieve this, a commercial LCBPP was adopted in this study.

Microcellular foam injection molding (FIM) is an extension of injection molding with foaming and is a powerful technique for producing lightweight products without sacrifice or obviously decrease its mechanical properties. There are many advantages of foam injection molding include absence of the sink mark on the part surface, less residual mark, better geometric accuracy, weight reduction, less shrinkage of product, high stiffness-to-weight ratio, etc. However, comparing with batch foaming or extrusion foaming products, one main drawback of FIM is its low weight reduction (expansion ratio), which is restricted to the fixed mold in the conventional foam injection molding machine. To overcome this, a new foam injection molding machine with core-back operation is fabricated, which can greatly increase the expansion ratio of the FIM products [2]. As far as the authors' concern, the reported maximum expansion ratio was 5 fold, i.e., 80 wt.% weight reduction [3]. In the present study, LCBPP was utilized to prepare high expansion foams by a core-back foam injection molding technique and the effect of LCB in producing high expansion microcellular foams was discussed.

EXPERIMENTAL

Materials

A commercial available high-tacticity (97 %) isotactic polypropylene, iPP (F133A, Prime polymer) with a weight-averaged molecular weight (Mw) of 379 kg/mol and a melt flow index (MFI) of 3.0 g/10min (230 °C/2.16 kg) was used as a linear PP source. A metalloscene-based LCBPP (MFX6) with a MFI of 3.0 g/10min (230 °C/2.16 kg) was kindly provided by the Japan Polypropylene Corporation. Nitrogen (N_2) with a purity of 99.98 % supplied by Izumi Sangyo, Japan was used as the physical blowing agent for foaming experiments.

*E-main address: wangl.kevin@cheme.kyoto-u.ac.jp (Long Wang) ohshima@cheme.kyoto-u.ac.jp (Masahiro Ohshima)

Foam Injection Molding with Core-back Operation

A 35-ton clamping force foam injection molding machine (J35EL III-F, Japan Steel Work) installed with a gas injection device (Trexel SCF system SII TRJ-10-A-MPD) were used for the foaming experiments (Figure 1). The materials were used as the received and the detailed processing conditions were shown in Table 1. The nitrogen was pressurized to 24 MPa by the Trexel SCF system and then injected into the barrel of the injection molding machine at 0.2 wt.% of polymer. The injected gas was mixed with the molten polymer at a high temperature and pressure with a screw which enhances the gas dispersion and dissolution and creates a single-phase N_2/polymer solution. The mold was a box-type cavity with a gate of 1.5 mm in diameter. The cavity dimension is 70 mm in length, 50 mm in width and 1 mm in the original thickness.

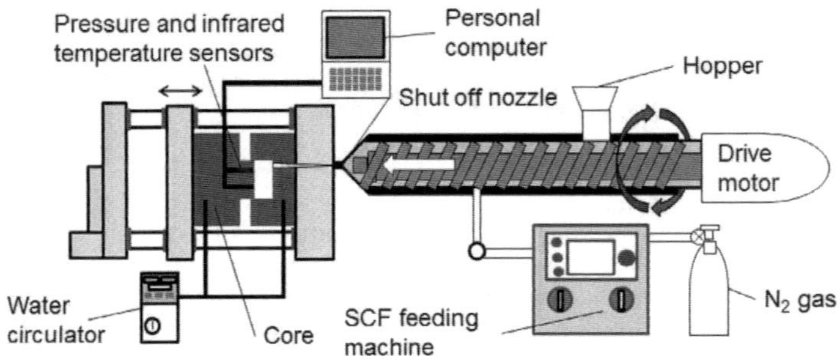

FIGURE 1. Schematic of the foam injection molding machine with core-back operation and a gas injection device.

Unlike the conventional structural foam injection molding process, the core-back foam injection molding process can greatly increase the cavity volume through the following process: after filling up the cavity with gas/polymer solution, a portion of the mold can be moved at a designated core-back timing (the end of the dwelling time), which can significantly increase the mold cavity volume, rapidly decrease the cavity pressure and meanwhile induce the foaming process. By manipulating the core-back rate and distance, the expansion ratio of the cavity volume and the depression rate of the cavity pressure can be controlled. In this study, the core-back rate were fixed at 20 mm/s and the core-back distance were set to four different distances, 1, 4, 6, and 9 mm. More details about the core-back operation process can be found in [3].

Table (1). Processing conditions for foam injection experiments

Parameters	Values
Melt temperature (°C)	195
Mold temperature (°C)	40
Cooling water (°C)	20
Injection speed (mm/s)	100
Injection pressure (MPa)	160
Dwelling time (s)	2.6~5.0
Shot size (mm)	35
Packing pressure (MPa)	60
Core-back rate (mm/s)	20
Core-back distance (mm)	1, 4, 6, 9
N_2 content (%)	0.2

Characterization of Materials and Foams

The dynamic rheological properties and the elongational viscosity were measured using a rheometer (AREX-G2, TA Instruments). The oscillation frequency sweep tests were conducted at 180 °C at a strain of 1 % in the linear viscoelastic regime and the temperature sweep experiments were measured at the cooling rate of 2 °C/min after the polymer melting at a temperature of 210 °C. The thermal stability was ensured by time sweep for all specimens. Uniaxial elongational experiments were carried out at a temperature of 180 °C with an extensional viscosity fixture under several elongational rates between 0.05 and 0.5 s^{-1}. Before experiment, a prestretch of 10 s was conducted to ensure that no slipping occurred between the specimen and the fixture.

The foamed specimens were cryogenically fractured in liquid nitrogen. After the surfaces of the specimens were covered with a thin layer of gold, the foam morphology was observed from both views in perpendicular and parallel to core-back direction by a scanning electron microscopy (Tiny-SEM Mighty-8, Technex), at an accelerating voltage of 17 kV. The image processing was analyzing using an ImageJ software, developed by National Institutes Health, US. The cell-population density (N_0) was calculated as the number of cells per unit volume with respect to the un-foamed polymer and was determined by:

$$N_0 = (\frac{n}{A})^{3/2} \frac{\rho_p}{\rho_f} \quad (1)$$

where n is the number of cells in the micrograph, A is the area of the micrograph, ρ_p and ρ_f are the densities of un-foamed and foamed specimens, respectively.

RESULTS AND DISCUSSION

Oscillatory Shear Flow Properties

Linear viscoelastic properties are very sensitive to the structural change in the materials. Figure 2 shown the results of temperature sweep of the complex viscosities as a function of temperature for linear PP and long-chain branched PP. The drastic increase in viscosity was caused by the crystallization. It can be observed that LCBPP had a larger crystallization temperature than the iPP sample, which might be due to the heterogeneous nucleation effects of the long-chain branched points. Neat polymers follow the typical terminal flow characterized by storage modulus $G' \sim \omega^2$ and loss modulus $G'' \sim \omega^1$ at the low frequency region. The storage modulus (G') of both samples plotted as a function of frequency was shown in Figure 3. iPP exhibited the typical terminal behavior, indicating that it was linear polymer. However, the LCBPP sample was deviated from the terminal behavior. With the addition of long-chain branching structures, G′ increased at low frequency and the terminal slope of G′ decreased from 1.52 of iPP to 0.70 of LCBPP. The non-terminal behavior of LCBPP indicated that there was a longer relaxation mechanism, which was beneficial for the foaming process.

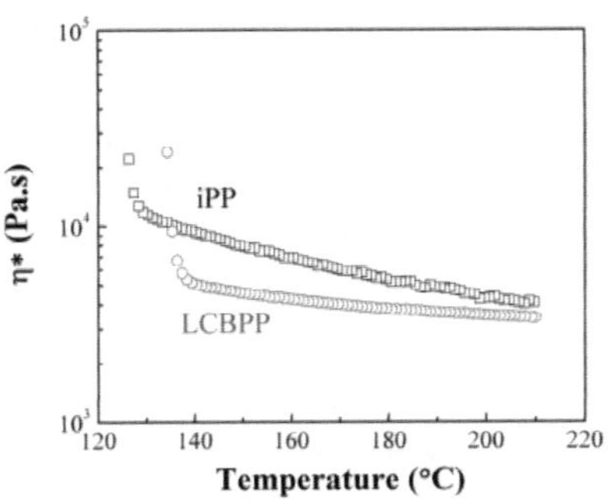

FIGURE 2. Complex viscosity of iPP and LCBPP samples.

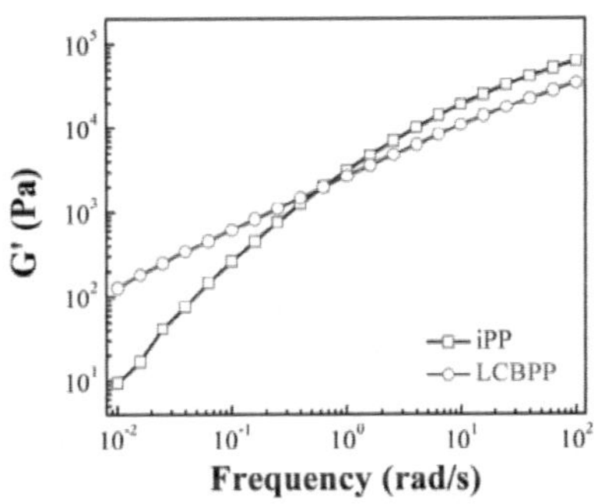

FIGURE 3. Storage modulus of iPP and LCBPP samples.

Uniaxial Elongational Flow Behaviors

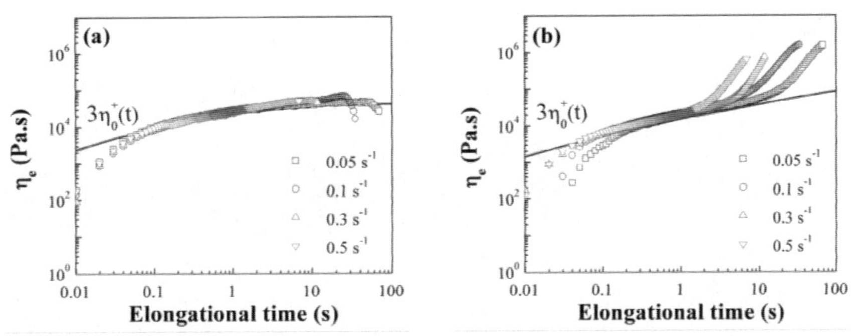

FIGURE 4. Changes in elongational viscosity as a function of elongational time at 180 °C for (a) iPP and (b) LCBPP.

The uniaxial elongational flow experiments at various elongational rates were conducted at 180 °C for the iPP and LCBPP samples. When the elongational viscosity is recorded as a function of specimen strain, a sudden increase in viscosity at a certain strain is called strain-hardening behavior. Such a strain hardening is important for polymer processing where high melt strength is required, such as high-speed fiber spinning, foaming and film blowing. Figure 4 shown the time-dependent elongational viscosity (η_e) at elongational rates ranging from 0.05 to 0.5 s^{-1}. As shown in Figure 4a, sample iPP did not exhibit any strain hardening, and all curves measured at different elongational rates superimposed on the linear viscoelastic curve which corresponded very well with the threefold value of the linear start-up viscosity in shear $\eta_0^+(t)$ as it should be in the range of linear deformation. In contrast, the LCBPP sample displayed obvious strain hardening behavior at all measured elongational rates.

Cell Morphology and Structures

Figure 5 shown the SEM images of the cross sectional areas of LCBPP sample foamed at various expansion ratio parallel to the core-back direction. It can be observed that the cells were highly elongated in the core-back direction with an increase of expansion ratio. In the following discussion, the cell diameter was calculated from the SEM images taken in a direction perpendicular to the core-back direction with the assumption that the cell shape was spherical.

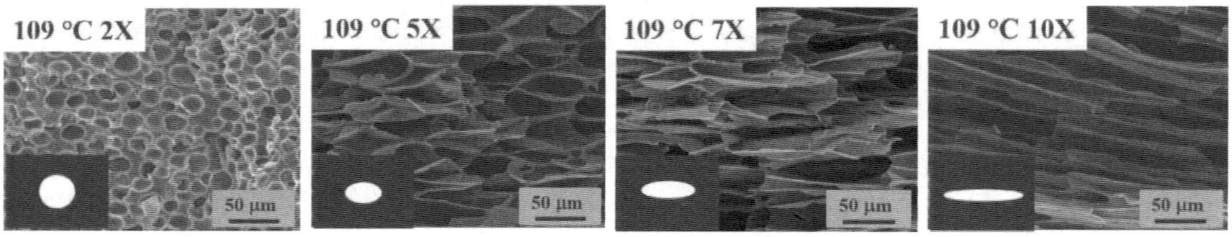

FIGURE 5. SEM images of LCBPP samples at different expansion ratio.

Figure 6 shown the SEM images of iPP and LCBPP five-fold expansion ratio foams under different foaming temperature. The foaming temperature was controlled by tuning the core-back timing and then the cell size can be minimized as much as possible. For iPP foams, it can be observed that the cell sizes were reduced by decreasing the foaming temperature, which might be due to the increasing melt strength as well as the proceeding crystallization process. Furthermore, with the inclusion of long-chain branching structures, the cell sizes of LCBPP were notably reduced when compared with the iPP samples. The cell densities as a function of foaming temperature for iPP and LCBPP foams were shown in Figure 7a. It shown that the cell density of iPP was about 10^7 cells/cm^3, and it was slightly improved by decreasing the foaming temperature. However, by introducing the long-chain branching structures, the cell density of BPP foams were significantly increased to 10^8 cells/cm^3 due to the increasing melt strength. In other words, the cell density of the LCBPP foam was nearly ten times of that of the iPP foam, which indicated that the presence of long-chain branches in PP enhanced the melt strength and obviously suppressed cell coalescence during the foaming process, resulting in enhancing cell density. This improvement in foaming process by introducing the long-chain branches can also be further authenticated by the variation of cell sizes as shown in Figure 7b. Over the entire foaming temperature range investigated, the average cell diameters of LCBPP were much smaller than that of iPP foam. For iPP foam, the cell sizes were very large (around 40~70 μm) due to the low melt strength. With the presence of long-chain branches, the average cell diameters of LCBPP foam were significantly reduced to 20~30 μm due to the increasing melt strength and thus the reducing cell coalescence and cell rapture.

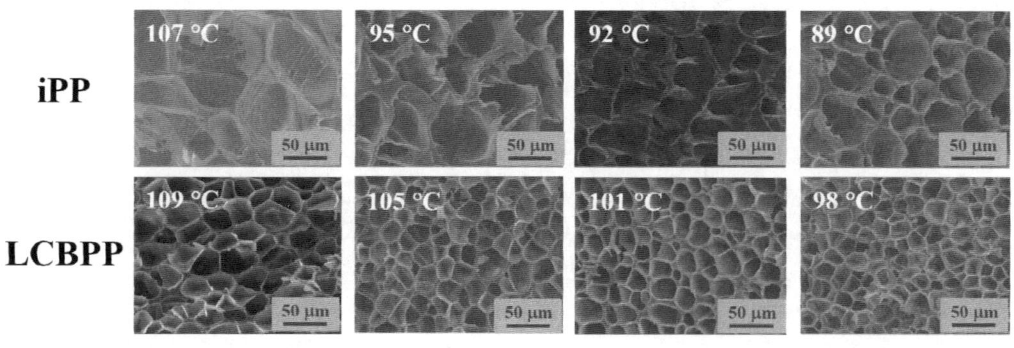

FIGURE 6. SEM micrographs of iPP and LCBPP five-fold expansion foams at various foaming temperature.

Figure 8 shown the SEM images of iPP and LCBPP at the expansion ratio of 10 times under different foaming temperature. For iPP foam, only fibrillary cell structures can be observed even at the cross sectional area perpendicular to the core-back direction, indicated that the cells were highly deformed and seriously cell coalescence occurred during the foaming process under such a high expansion ratio. In contrast, the cell morphology of LCBPP 10-fold expansion foams was still spherical, indicated that the long-chain branches can stabilize the cell grow and reduce cell coalescence as well as the cell rupture. Furthermore, LCBPP had a much larger foaming processing windows than that of iPP.

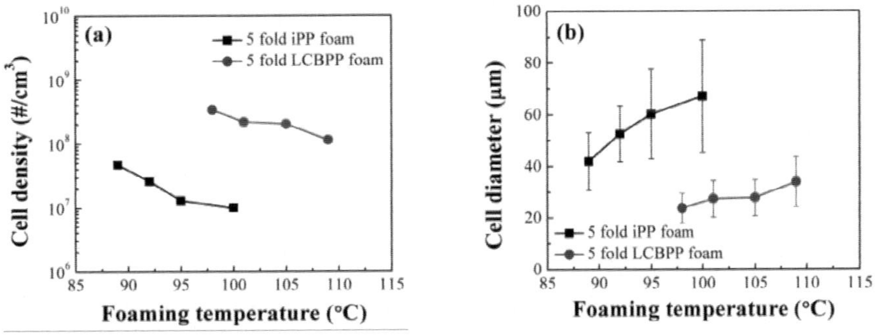

FIGURE 7. (a) Cell density and (b) average cell diameter as a function of foaming temperature of iPP and LCBPP foam.

Figure 9 shown the open cell content of iPP and LCBPP foams at various expansion ratio measured by the pycnometer. The open cell content was obviously increased by increasing the expansion ratio. Furthermore, when comparing with linear iPP, the long-chain branching structures reduced the open cell content. It means that long-chain branches can stabilize the cell morphology and keep the cell structure even at high expansion ratio.

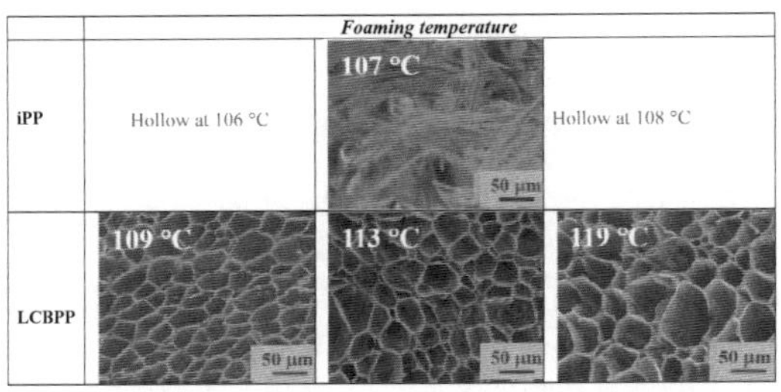

FIGURE 8. SEM of iPP and LCBPP with 10-fold expansion.

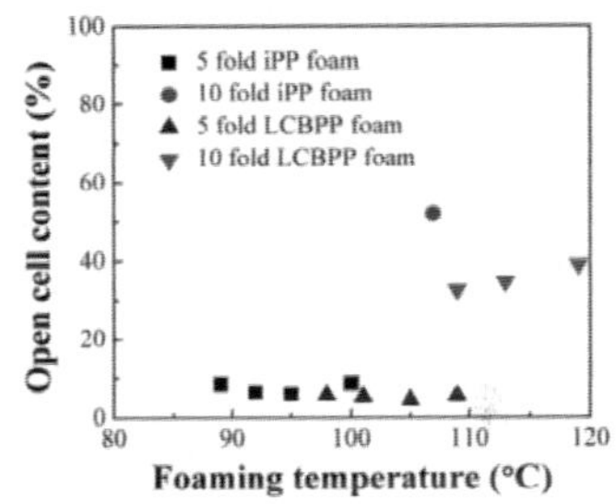

FIGURE 9. Open cell content of iPP and LCBPP foams.

CONCLUSIONS

Here, high expansion microcellular polypropylene foams were successfully fabricated using a foam injection molding machine with core-back operation. Long-chain branches imparted PP with remarkably higher elongational viscosity and obviously strain hardening characteristics. The long-chain branched structures can stabilize cell growth and reduce cell coalescence as well as cell rupture, which can be evidiced from the lower open cell content of LCBPP foams (5/10-fold) relative to that of iPP foams as well as the fibrillated cell structures of iPP 10-fold foams. In addition, the operating window of manufacturing high volume expansion ratio foams was notably widened for LCBPP due to the presence of long-chain branches. This study demonstrates that LCBPP is needed to fabricate high expansion microcellular foams and is in great potential capacity for even more than 10 times of expansion ratio microcellular foams.

ACKNOWLEDGMENTS

The research was conducted in associated with Grants-in-Aid for Scientific Research (B), Number:26289289 of Japan Society for Promotion of Science (JSPS).

REFERENCES

1. K. W. Suh, C. P. Park, M. J. Maurer, M. H. Tusim, R. D. Genova, R. Broos and D. P. Sophiea, *Adv. Mater* **12**, 1779-1789 (2000).
2. T. Ishikawa and M. Ohshima, *Polym. Eng. Sci.*, **51**, 1617-1625 (2011).
3. R. Miyamoto, S. Yasuhara, H. Shikuma and M. Ohshima, *Polym. Eng. Sci.*, **54**, 2075–2085 (2014).

Needleless Melt Electrospinning for Nanofiber Preparation and Its Scaling-up

Li Haoyi, Tan Jing, Ma Xiaolu, Chen Hongbo, Yang Weimin

College of Mechanical and Electrical Engineering, Beijing University of Chemical Technology, Beijing 100029, China

Abstract: Many kinds of electrospinning methods have emerged and been well investigated in the past two decades in the pursuer of nanofiber preparation and its scaling-up. Among the electrospinning methods, melt electrospinning, as a non-solvent route, has draw great attention in recent years since its high yield and safety. However, there was few work has been done on the mass production of nanofiber and yarn production by this polymer melt based method. On this basis, a melt differential electrospinning method without needles was proposed, by which the single nozzle can yield two orders of magnitude more than the capillaries did and nanofiber production in kilo scale was realized. A simple solution for yarn production was also introduced with narrow diameter distribution. Larger scale production can be easily performed by modular designing and assembling. For example, Polypropylene nanofiber with diameter around 500nm and Poly Lactic Acid (PLA) nanofiber with diameter around 800nm can be achieved by this method. Based on this principle, complete commercial techniques and equipments have been developed, and applications of as-spun fibers in air filtration, oil spill recovery and water treatment, etc, are promising.

Key words: melt electrospinning, needleless, nanofiber

Since Formhals[1] invented the electrospinningtechnique in 1934, electrospun fibers have been attracting a lot of interest due to their high specific surface area, good mechanical properties, and functional applications. To develop polymer melt electrospinning technique, there are three serious problems that should be considered: high voltage insulation and related safety issues of electrospinning system, fiber refinement, and productivity.

Aiming to solve the above mentioned shortage of traditional equipment using needle-like spinneret, weproposed a melt differential electrospinning (MD-ESP) method [2] for ultrafine fiber preparation, through which fiber smaller than one micrometer can be produced and a yield of 10-20g/h can be achieved by a needleless nozzle. We termed this process as MD-ESP because of a melt flow was divided into tens of minor Taylor-cones as a result of a self-organizing process. The design of MD-ESP spinning nozzle includes the inner-cone type and the outer- cone type [3-4] as shown in figure 1.

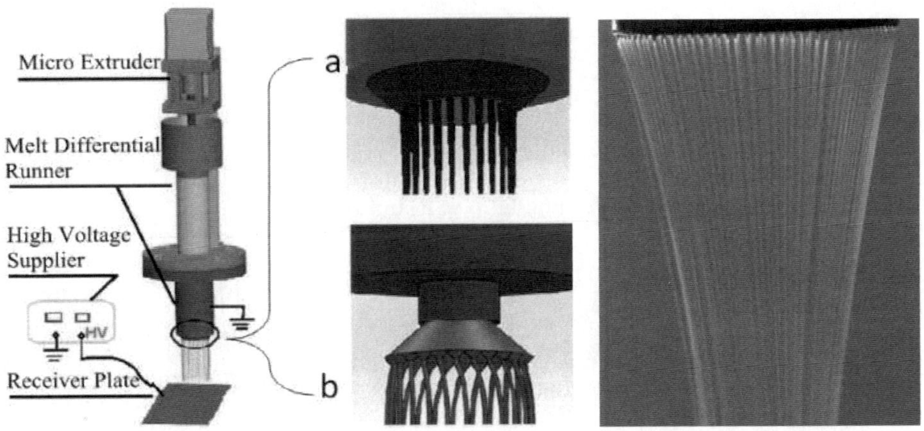

Figure 1. A schematic diagram of the melt differential electrospinning (inner-cone and out-cone nozzle)

Based on the principle of polymer MD-ESP, a prototype equipment of pilot line was also designed and operated. This machine can realize mass production of polymer nanofiber by melt electrospinning route. As shown in figure 2, the equipment includes nine components: a small twin screw extruder, a melt filter, a melt metering pump, a melt differential electrospinning box, air auxiliary equipment, suction device, a lapping machine, and rolling-coiling machine.

Figure 2. The photo of the pilot prototype of polymer melt differential electrospinning device.

The performance parameter of pilot line are as follows:

1. Integration of the 32 nozzles;

2. The jets number ejected from each nozzle (diameter of 26mm) can range from 50- 100;

3. The defined width is 0.8m, output is 300- 600g/h, and the averagediameter is 500-800nm;

4. The thickness of non-woven fabrics is adjustable from 10μm to 1000μm;

5. Continuous supply of polymer melt or polymer blend melt can be realize on line;

6. Modularly extendable to the 6kg/h production line;

The melt differential electrospinning (MD-ESP) process and resultant diagram were compared with that of solution electrospinning(S-ESP) in figure 3.

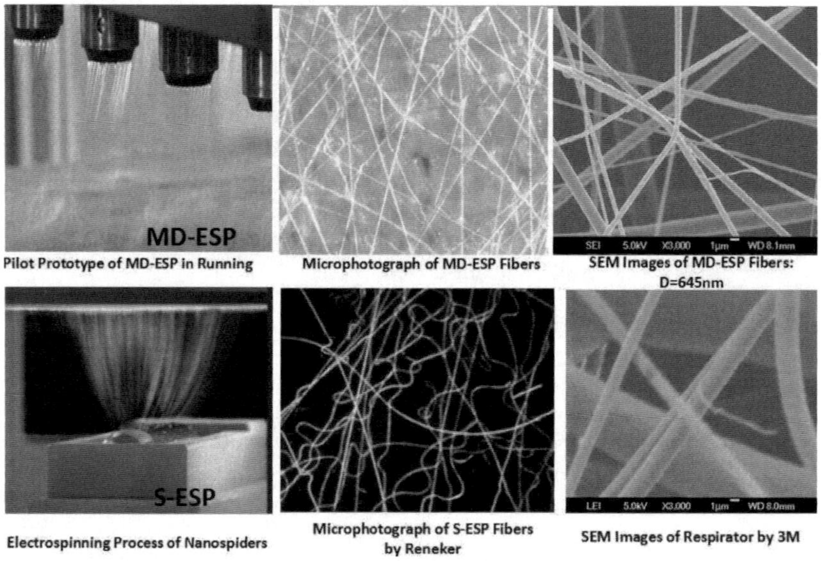

Figure 3. The spinning process of pilot line and the diagram comparison.

Table 1 shows the material and process that we have used, and average diameter that we have achieved.

Table1 Material, process and average diameter of melt differential electrospinning

Material	Process	Average diameter
PA6	Outer-cone nozzle	8.85μm
PET	Outer-cone nozzle	6.37μm

PLA	Outer-cone nozzle	1.61μm
PCL	Outer-cone nozzle	1.35μm
PP	Inner-cone nozzle	645nm
PLA(2002D)/ PEG(200)	Inner-cone nozzle	860nm

References

[1] A. Formhals, Process and apparatus for preparing artificial threads [P]. US, 1975504, 1934,10,2.

[2] Weimin Y, Haoyi L. Principle and equipment of polymer melt differential electrospinning preparing ultrafine fiber[C]//IOP Conference Series: Materials Science and Engineering. IOP Publishing, 2014, 64(1): 012-013.

[3] Li H Y, Bubakir M M, Xia T, et al. Mass production of ultra-fine fibre by melt electrospinning method using umbellate spinneret[J]. Materials Research Innovations, 2014, 18(S4): S4-921-S4-925.

[4] Li H, Chen H, Zhong X, et al. Interjet distance in needleless melt differential electrospinning with umbellate nozzles [J]. Journal of Applied Polymer Science, 2014, 131(15).

Improving the Transparency of the Highly-Oriented Melt-Crystallized Polyethylenes: Toward High-Impact Window Application

ABSTRACT: Highly transparent, ultra-drawn high-density polyethylene (HDPE) films were successfully prepared using compression molding and solid-state drawing techniques. The low optical transmittance (< 50%) of the pure drawn HDPE films can be drastically improved (> 90 %) by incorporating a small amount (> 1% wt/wt) of specific additives to HDPE materials prior to drawing. It is shown that additives with relatively high refractive index result in an increased optical transmittance in the visible light wavelength which illustrates that the improvement in optical characteristics probably originates from refractive index matching between the crystalline and non-crystalline regions in the drawn films. Moreover, the optically transparent drawn HDPE films containing additives maintain their physical and mechanical properties, especially their high modulus and high strength, which make these films potentially useful in a variety of applications such as high-impact windows.

Fiber structure development and crystallization behaviors of PHBH in high-speed melt spinning process

Qing Qin, Wataru Takarada, Takeshi Kikutani

Department of Organic and Polymeric Materials, Graduate School of Science and Engineering, Tokyo Institute of technology, 2-12-1, O-okayama, Meguro-ku, Tokyo 152-8552, Japan

Abstract: High-speed melt spinning of poly(3-hydroxybutyrate-co-3-hydroxyhexanoate) (PHBH) with the 3-hydroxyhexanoate compositions of 5.4 and 10.4 mol% (X131A and X151A) were carried out. It has been reported that PHBH fibers can be prepared through the melt spinning process without sticking each other on the winding bobbin only when extrusion temperature is lower than the melting temperature of pure PHB (176 °C). The high-speed melt spinning experiment in this study revealed that the crystallization of X131A proceeded on the spin-line at high take-up velocities even when the extrusion temperature was higher than the melting temperature of pure PHB. On the other hand, the existence of a small amount of β-form crystals was detected for the fibers prepared at high take-up velocities and the tensile strength of the fibers showed positive correlation with the β-form fraction. However, for the melt spinning of X151A fibers, crystallization did not occur because of suppressed crystallizability and crystallization rate.

Keywords: high-speed melt spinning / crystalline orientation / 3-hydroxyhexanoate /β-form crystals

INTRODUCTION

Polyhydroxyalkanoates (PHAs), which are generated by a wide range of microorganisms [1], are one of the families of typical eco-friendly polymers. It attracts a great deal of attention for its biodegradability and biocompatibility. Bacterial poly (3-hydroxybutyrate) (PHB) is one of the well characterized members of PHA family, which is synthesized from renewable carbon sources such as sugars and vegetable oil by a wide variety of bacteria [2-3]. It is known that PHB is brittle at room temperature because of its high degree of crystallinity. In addition, its melting temperature is close to the temperature where thermal degradation starts. Therefore, melt-processing of PHB is rather difficult and its applications are very limited. To overcome such drawback, various PHB based co-polymers have been developed. Poly (3-hydroxybutyrate-co-3-hydroxyhexanoate) (PHBH) is one of those polymers and has been known as a melt-processable aliphatic co-polyester. Nevertheless, processing of this polymer is still not easy because the glass transition temperature of PHBH is below the room temperature whereas the crystallization rate of this polymer is extremely low. Especially, the introduction of 3-hydroxyhexanoate (3HH) results in reduction of crystal growth rate and degree of crystallinity [4]. Accordingly, it is difficult to form fibers, films and other products from PHBH through the melt processing because the polymer is still in a rubbery state after it is cooled down to the room temperature.

In terms of the crystalline structure of PHB, it has been reported that the PHB crystallizes into orthorhombic crystals with the unit cell parameters of a =0.576 nm, b=1.320 nm, and c (fiber axis) = 0.596 nm (α-form) with their chains in the left-handed 2_1 helix [5]. In addition, β-form crystals with planar zig-zag conformation can be observed in the highly drawn PHB fibers. The β-form crystal is represented by a crystalline reflection appearing on the equator at 2θ=19.7° in the WAXD measurement. Orts et al. suggested that the β-form was introduced by the orientation of free chains in the amorphous regions between α-form lamellar crystals [6]. Furuhashi et al. proposed another formation mechanism that the β-form is formed as the crystals of pseudohexagonal structure [7]. In any case, most reports indicated that the β-form exists only in the highly oriented fibers and films of PHB and its copolymers.

In this research, attempts were made to overcome the difficulty of low crystallizability of PHBH for the melt processing by applying the high-speed melt spinning process. When take-up velocity is increased, high tensile stress can be applied to the spinning line. Development of molecular orientation and orientation-induced crystallization are expected to occur under such high tensile stress.

EXPERIMENTAL

PHBH with the 3-hydroxyhexanoate content of 5.4 mol% (X131A) and 10.4 mol% (X151A) supplied by Kaneka Co. Ltd. were used in this study. Molecular weight of the polymer was $5.3x10^5$ and $6.5x10^5$, respectively. The melt spinning experiment was carried out using an extrusion system consisting of a single-screw extruder and a gear pump. A spinneret with four holes of 0.5 mm diameter was attached to the spinning head. The polymer was extruded at temperatures of 160,170,180 and 190 °C. The total throughput rate was controlled to 5.0, 10.0, and 20.0 g/min. To evaluate the spinnability of the polymer, take-up velocity was increased from 0.5 km/min to the attainable highest speed. Thermal analysis of the polymers was conducted with the differential scanning calorimetry (DSC Q-100; TA Instruments) at the heating and cooling rates of 10 K/min in the temperature range of -30°C to 200°C. Crystalline structure of the as-spun fibers were investigated through the measurements of the two dimensional wide-angle X-ray diffraction (WAXD) patterns using an X-ray generator equipped with a CCD detector (RINT-III, Rigaku Co., Ltd). Mechanical properties of the fibers were measured using a tensile testing machine (SHIMADZU, AG-I). The gauge length was 20 mm, and the tensile speed was 10 mm/min.

RESULTS AND DISCUSSION

Crystallization behavior of PHBH polymer chips

Figure 1 shows the DSC thermograms of PHBH polymer chips of X131A and X151A measured at a cooling rate of 10 K/min. It is noted that a crystallization peak with large enthalpy was observed for X131A whereas X151A did not show a crystallization peak. This result coincides with the report which describes that the introduction of 3-hydroxyhexanoate (3HH) results in reduction of crystal growth rate and degree of crystallinity [4]. It is expected to be more difficult for X151A to form fibers because of its lower crystallizability.

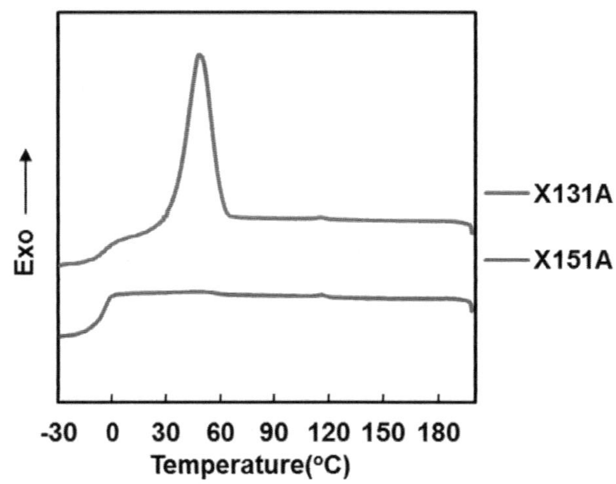

FIGURE 1. DSC cooling curves for PHBH polymer chips of X131A and X151A measured at cooling rate of 10 K/min.

Appearance of prepared fibers

Appearance of as-spun fibers cut-off from the bobbin is summarized in Figure 2. At the take-up velocity of 0.5 km/min, the fibers stuck together at 180 °C. This result coincides with the result reported by Yamane et. al. [8]. It can be considered that when the extrusion temperature was high, crystallization rate of PHBH was low and molten polymer in the spinning line reached the take-up bobbin in a rubbery state without crystallization. It should be noted that the glass transition temperature of PHBH of around 0 °C is lower than the ambient temperature. On the other hand, for the X131A, well-separated fibers could be obtained at high take-up velocities. It is well known that the crystallization can be accelerated in the high-speed melt spinning process because of the high tensile stress and the development of molecular orientation in the spinning line. It appears that such mechanism is also applicable for the high-speed melt spinning of PHBH. However, for the X151A, the fibers stuck together even at the high take-up velocity of 5 km/min,

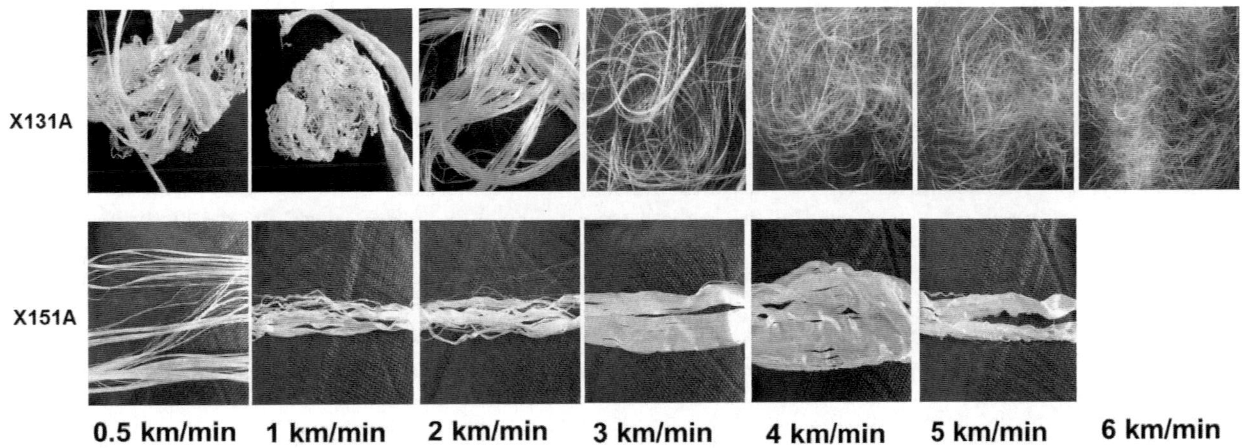

FIGURE 2. Photographs of as-spun PHBH fibers of X131A and X151A prepared at extrusion temperature 180 °C and total through put 5 g/min.

Structure of as-spun fibers

WAXD patterns of as-spun PHBH fibers of X131A and X151A prepared at extrusion temperature 180°C and through-put 5 g/min are shown in Figure 3. It can be noted that the as-spun fibers prepared at 0.5 km/min showed α-form crystals of no orientation. For the X131A fibers, degree of crystalline orientation increased gradually with an increase in the take-up velocity, and the crystalline reflections from (020) and (110) (2θ=13.4° and 16.9°, both on the equator) were clearly detected from 2 km/min. However, for the X151A fibers, there was only a slight increase of crystalline orientation with the increase of take-up velocity. Even at 5 km/min, the WAXD pattern indicated quite low orientation ofα-form crystals. Appearance of the fibers after the spinning process shown in Figure 2 suggested that crystallization of X151A fibers proceeded after wound up on the bobbin.

In addition, for the WAXD patterns of the as-spun X131A fibers prepared at high take-up velocities, a small diffraction peak was detected on the equator at around 2θ=19.7°. This reflection corresponds to the highly oriented β-form crystals. The β-form crystals were detected for the X131A fibers from the take-up velocity of 3 km/min and the reflections from the β-form crystals became more distinct with the increase of take-up velocity. Formation of the β-form crystals suggested that the crystallization proceeded under the effect of high tensile stress. On the other hand, existence of the β-form crystals could not be confirmed for the X151A fibers even at 5 km/min because of its lower crystallizability.

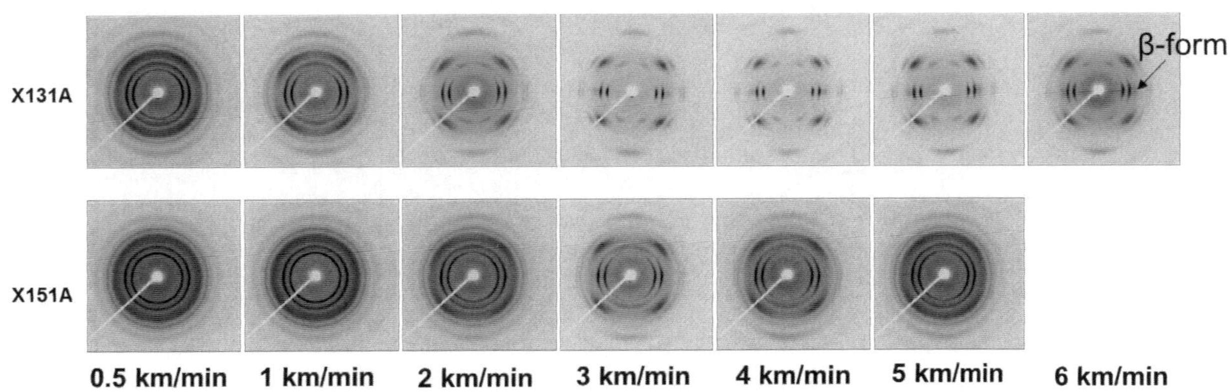

FIGURE 3. WAXD patterns of as-spun fibers of X131A and X151A prepared at extrusion temperature 180°C and total through put 5 g/min.

Figure 4 shows the variation of the calculated crystalline orientation of as-spun PHBH fibers of X131A and X151A prepared at the extrusion temperature of 180°C. Crystalline orientation was analyzed utilizing azimuthal intensity distribution of (020) reflection. For the X131A fibers, the crystalline orientation was improved with the increase of take-up velocity. In addition, the crystalline orientation was found to increase with the increase of through-put. This result was speculated to be due to the higher spine-line tension originated from higher viscosity. It was analyzed that a significant degree of thermal degradation proceeds in the extrusion process of PHBH. Residence time of polymer in the extrusion system becomes shorter with the increase of through-put. Eventually, thermal degradation was prevented that resulted in the higher viscosity. For the X151A fibers, crystalline orientation was quite low, and the effect of the take-up velocity for the improvement of orientation was found to be minimal.

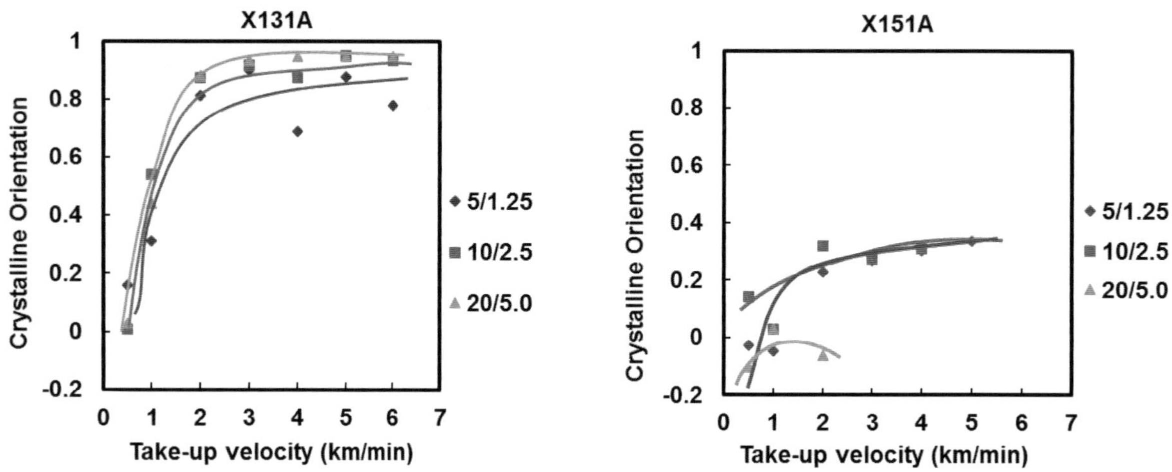

FIGURE 4. Comparison of the crystalline orientation of the PHBH fibers of X131A and X151A prepared at extrusion temperature of 180°C and three different through-puts. Crystalline orientation was analyzed from the azimuthal intensity distribution of (020) reflections appearing at 2θ=13.4°

Relative proportions of the two crystalline forms were analyzed from the integrated intensities of α-form and β-form reflections. The results for the X131A fibers are shown in Figure 5. The fibers prepared with the through-put of 20 g/min showed the highest β-fraction up to 4 km/min. This is another indication of the higher viscosity at high through-put caused by the shorter resident time. On the other hand, the fibers prepared with the through-put of 10 g/min started to show higher β-fraction from 5 km/min. It can be speculated that the effect of the increment of air-friction stress with the reduction of through-put because of larger specific surface area exceeded the effect of the reduction of viscosity.

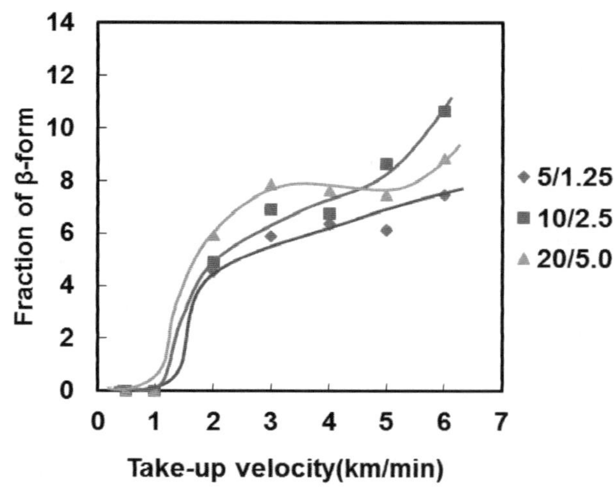

FIGURE 5. Changes of fraction of β-form with take-up velocity for as-spun PHBH fibers prepared at extrusion temperatures of 180°C and three different through-puts.

Mechanical Properties of as-spun fibers

Variations of tensile strength and β-form fraction with take-up velocity for the as-spun X131A fibers prepared with three different through-puts were compared in Figure 6. Tensile strength increased with an increase in the take-up velocity. It can be noted that the tensile strength and β-form fraction showed a similar tendency indicating that the tensile strength of the fibers generally shows positive correlation with the β-form fraction.

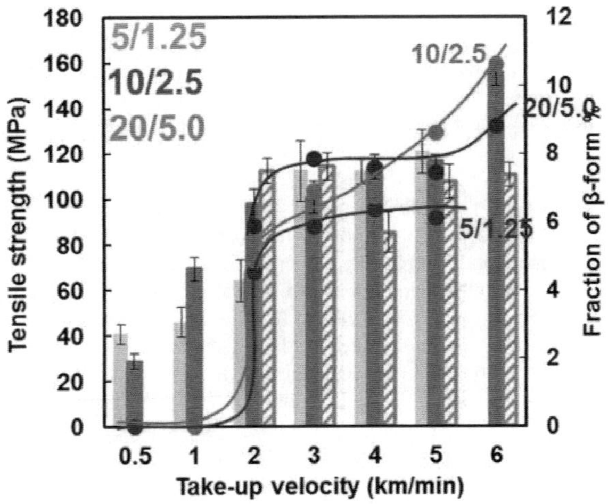

FIGURE 6. Changes of tensile strength and fraction of β-form with take-up velocity for as-spun PHBH fibers prepared at extrusion temperatures of 180°C and three different through-puts.

CONCLUSIONS

Enhancement of crystallization at high take-up velocities was confirmed in the high speed melt spinning of X131A fibers. The fibers spun at higher take-up velocities showed higher crystalline orientation, higher content of the fraction of β-form crystals. The fraction of β-form crystals appeared to increase under the spinning conditions with higher spin-line stress. In addition, β-form fraction showed positive correlation with the tensile strength of the as-spun fibers. On the other hand, for the X151A fibers, low oriented α-form crystals were formed at high take-up velocities, however sticking among fibers could not be avoided. This was caused by the higher 3HH composition, which resulted in the reductions of crystallization rate and degree of crystallinity of the as-spun fibers.

REFERENCES

1. A. Steinbuchel and H. E.Valentin, *FEMS, Microbiol. Lett.* **128**, 219-228 (1995).
2. T. Tanaka, T. Yabe, S. Teramachi, and T. Iwata, *Poly. Deg. and Stab,* **92,** 1016-1024 (2007).
3. A. J. Anderson and E. A. Dawes, *Microbiol Rev.*, **54**, 450 (1990).
4. H. Abe, Y. Doi, *Biomacromolecules*, **3**, 133–138(2001).
5. M. Yokouchi, Y. Chatani, H. Tadokoro, K. Teranishi, and H. Tani, Polymer, **14**, 267-272 (1973).
6. W.J. Orts, R.H. Marchessault, T.L. Bluhm, and G.K. Hamer, *Macromolecules,* **23**, 5368-5370 (1990).
7. Y. Furuhashi, H. Ito, T. Kikutani, T. Yamamoto, M. Kimizu, and M. Cakmak, *J Polym Sci: Part B: Polym Phys.*, **36**, 2471-2482 (1998).
8. J. C. Lee, Y. Furuhashi, Y. Kawahara, and H. Yamane, Sen'i Gakkaishi, **62**, 115-122 (2006).

Preparation and Magnetic Property of Electrosprayed PES/SPES Magnetic Microspheres by In-situ Reaction

Ting Yu[a], Minle Peng[a], Gang Zhang[b], Shengru Long[b], Xiaojun Wang[b] and Jie Yang[b,c,*]

[a] *College of Polymer Science and Engineering, Sichuan University, Chengdu, China*
[b] *Institute of Materials Science and Technology, Sichuan University, Chengdu, China*
[c] *State Key Laboratory of Polymer Materials Engineering, Sichuan University, Chengdu, China*

Abstract. In this paper, Polyethersulfone/Sulfonated Polyethersulfone (PES/SPES) magnetic microspheres were prepared for the hydrophilic modification of PES magnetic microspheres. And the effects of SPES on the preparation and magnetic property of magnetic microspheres were investigated. The images of scanning electron microscope showed that the ferrite nanoparticles were embedded into the surface of the PES/SPES electrosprayed microspheres. The vibrating sample magnetometer test revealed that the PES/SPES in-situ magnetic microspheres showed superparamagnetism and had extremely small hysteresis loop as well as low coercivity. The magnetization of PES/SPES microspheres could reach 20.95 emu/g, increased by 17.8% compared with PES microspheres. And the PES/SPES microspheres can be directional separated faster than PES microspheres under magnetic field. PES/SPES magnetic microspheres had better magnetic property compared with PES magnetic microspheres. The results proved that the improvement of the microspheres' hydrophilicity was advantageous for the generation of magnetic nanoparticles and the adhesion onto the surface.

Keywords: Electrspray; In-situ reaction; Sulfonated; Magnetic microspheres

Magnetic polymer microspheres not only have remarkable characteristics of polymer microspheres but also have magnetic responsiveness, so that they are able to directional separate quickly under the applied magnetic field. With this special characteristic, magnetic microspheres have attracted much attention in recent years and will be currently and potentially used in many fields, such as wastewater treatment, enzyme immobilization, controlled drug delivery, drug targeting, magnetic resonance imaging, cell labeling and separation. As an advanced method to prepare magnetic polymer microspheres, electrospraying combined with in-situ reaction can be easily conducted, and the product has small diameter, large specific surface area and excellent properties.

In our previous studies, electrosprayed PES in-situ magnetic microspheres have been already prepared successfully. However, magnetic particles are hydrophilic, and PES is a hydrophobic material. Therefore, Sulfonated Polyethersulfone (SPES) was added into electrospraying Polyethersulfone (PES) solution, and the effect of the hydrophilic modification of electrosprayed microspheres on the characters of in-situ magnetic microspheres was investigated.

PREPARATION OF PES/SPES MAGNETIC MICROSPHERES

First, PES/SPES (8.5:1.5wt) was dissolved in Dimethyl Formamide (DMF) at 70°C for 4h until the solution became homogeneous, and the concentration of it is 0.09g/ml. For preparing microspheres, the solution was placed in a 50 mL syringe, to which a capillary tip of 0.5mm inner diameter was attached. A direct current high-voltage generator was used to provide a voltage of 10kV. The positive electrode of high-voltage power supply was connected to the capillary tip. The grounded electrode was connected to a metallic collector immersed in a water bath. The distance between the tip and the collector was maintained at 6cm and the flow rate was constant. The spraying stream is broken up after moving a certain distance, and then electrosprayed microspheres were collected in the water bath. Then PES/SPES electrosprayed microspheres were immersed in deionized water for 48h at room temperature to remove the solvent. After suction filtration and drying, semi-finished products were obtained.

Then 0.01g PES/SPES electrosprayed microspheres and 1g $FeCl_2.4H_2O$ were put into a centrifuge tubes, and 20mL deionized water was added. Next the tube was sealed with plastic wrap and vibrated with constant temperature water bath at 40°C for 48h. Subsequently, the materials in the centrifugal tube were centrifugated at 5000rpm for 10min, and the supernatant liquid was poured out. And then, 20mL NaOH aqueous solution with a concentration of 0.02mol/L was added into the tube, and the tube was put into constant temperature water bath at 60°C for 4h. After that, 10mL H_2O_2 (1%) was added, and put into constant temperature water bath at 30°C for 4h. Finally, microspheres were isolated, and dried for 24h. According to the above steps, PES in-suit magnetic microspheres were obtained.

STRUCTURE OF PES/SPES MAGNETIC MICROSPHERES

4mg magnetic microspheres were dispersed in 10mL deionized water. The color of the microspheres was similar to PES magnetic microsphere as shown in FIGURE 1. It can be proved preliminarily that the dark brown nanoparticles generated on the surface of microspheres were ferrite.

FIGURE 1. Photograph of the (a) PES/SPES in situ magnetic microsphere dispersed in water and (b) PES in situ magnetic microsphere dispersed in water.

Scanning electron microscope (SEM) was employed to investigate the surface of the microspheres, as shown in FIGURE 2. The inserts display the morphology of PES/SPES electrosprayed microspheres and PES/SPES in-situ magnetic microspheres, indicating their spherical shape. As shown in FIGURE 2b, microspheres exhibited relatively denser surfaces than that of PES/SPES electrosprayed microspheres. It further confirmed that iron oxide particles could be readily embedded into the surface of the electrosprayed PES/SPES microspheres. And there were a few free ferrite nanoparticles scattered between microspheres. The results proved that the improvement of the microspheres' hydrophilicity was advantageous for the generation of magnetic nanoparticles and the adhesion onto the surface.

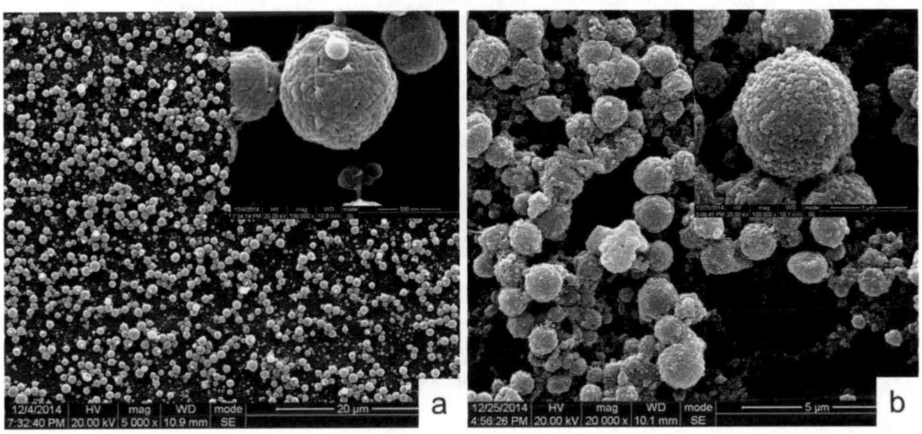

FIGURE 2. SEM images of the surface of the PES/SPES electrosprayed microspheres (a) and PES/SPES in-situ magnetic microspheres (b). The insert illustrates the high-magnification morphology of them.

MAGNETIC PROPERTIES

To investigate the magnetic property of microspheres, vibrating sample magnetometer (VSM) test was conducted as shown in FIGURE 3. Their magnetization increased with the increase of the intensity of magnetic field. They exhibited an extremely small hysteresis loop and low coercivity. This was typically characteristic of superparamagnetic particles. The saturation magnetization obtained from the hysteresis loop was 20.95emu/g increased by 17.8% compared with PES microspheres whose was 17.78emu/g. It further proved that the improvement of the microspheres' hydrophilicity was advantageous for the generation of magnetic nanoparticles and the adhesion onto the surface.

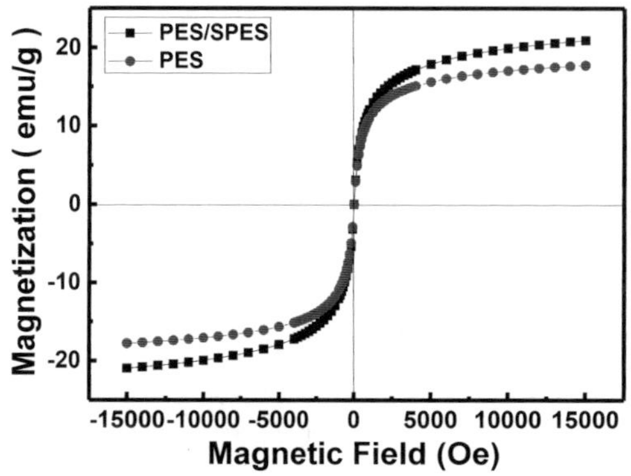

FIGURE 3. The magnetic hysteresis curves for the PES/SPES in situ magentic microspheres.

Besides, separation performance was also studied. From FIGURE 4, the photo showed that magnetic microspheres could align quickly in the magnetic field within a short time. Almost all microspheres separated completely from water environment within five minutes with imposed magnetic field. It revealed that magnetic microspheres possessed a sensitive magnetic response. In order to compare the separation performance of PES and PES/SPES magnetic microspheres, absorbence of supernatant after separation for 5min was tested by an ultraviolet-visible spectrophotometer. Absorbance of PES supernatant was 0.484, while PES/SPES was 0.310. The reduction of absorbance revealed that less residual microspheres or less free ferrite nanoparticles existed in the supernatant. It demonstrated that separation performance of PES/SPES magnetic microspheres was better.

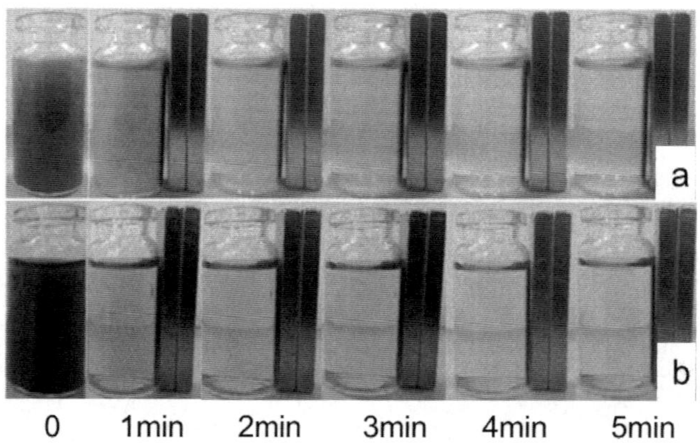

FIGURE 4. The photograph of the magnetic separation of the (a) PES/SPES and (b) PES in situ magentic microspheres

In conclusion, all the results revealed PES/SPES magnetic microspheres had better magnetic property compared with PES magnetic microspheres. It proved that the improvement of the microspheres' hydrophilicity was advantageous for the generation of magnetic nanoparticles and the adhesion onto the surface.

REFERENCES

1. B. L. Droumaguet, G. Mantovani, D. M. Haddleton and K. Velonia, J. Mater. Chem., 2007, 17, 1916.
2. A. Dyal, K. Loos, M. Noto, S. W. Chang, C. Pagnoli, K. V. P. M. Sha, A. Ulman, M. Cowman and R. A. Gross, J. Am. Chem. Soc., 2003, 125, 1684.
3. J. C. Kang and D. L. Wei, Acta Pharmaceutica Sinica., 1997, 32, 536.
4. R. S. Molday, S. P. S. Yen and A. Rembaum, Nature, 1977, 268, 437.

5. L. O. Kolarik, N. J. Anderson, B. A. Bolto, C. T. Chin and A. J. Priestley, J. Water Suppl., 1994, 12, 253.
6. Min-le Peng, et al. Structure Controlling and Adsorption Application of Polyethersulfone Porous Microspheres Prepared via Electrospraying. Chinese Journal of Polymer Science, 2014， 32(10): 1390-1399.
7. Doina Hritcu,Gianina Dodi,Mihaela Silion.et al. Composite magnetic chitosan microspheres:In situ preparation and characterization[J].Polym. Bull. 2011(67):177-186.

Preparation and Characteristic of Inflammatory Response Type Guided Tissue Regeneration Membrane

Jingjing Ye [a], Rui Shi [b] and Liqun Zhang [a,*]

[a] Beijing Laboratory of Biomedical Materials, Beijing University of Chemical Technology, Beijng 100029,China

[b] Laboratory of Bone Tissue Engineering of Beijing Research Institute of Traumatology and Orthopaedics, Beijing 100035,China

Abstract:

Biomimetic surface modification of electrospun PCL nanofiber membrane was demonstrated. Drug was covalently grafted on the surface of the nanofiber membrane to fabricate enzyme sensitive drug release. We used dopamine self-polymerization to form thin poly(dopamine) film on the surface of PCL nanofibers in a dopamine solution. Hydroxyl groups were introduced on the surface of the nanofibers. Silane(KH550) grafting as the carried out on the surface of the poly(dopamine)-coated PCL nanofibers to introduce amine groups. A ester has been synthesized from acryloyl chloride and metronidazole. Then drug was grafted on the surface of the nanofiber from C=C on the eater and the amine groups on the nanofibers by means of Michael addition reaction.The modification process of the nanofibers was characterized by SEM,IR,EDS and XPS. Drug released from the surface of the nanofibers by the action of enzyme. The membranes showed enzyme sensitive drug release kinetics with no cytotoxicity.

Key words: Guided tissue regeneration membrane, Inflammation responsive, Dopamine, Silane coupling agent KH550, Cholesterol esterase

***Corresponding Author**: Liqun Zhang, zhanglq@mail.buct.edu.cn

Modeling and Experimental Validation of Anomalous Moisture Absorption of Micro and Nanocomposite Laminates

Youssef K. Hamidi [a,b], Gorkem E. Guloglu [a], Jacob P. Anderson [a] and M. Cengiz Altan [a,*]

[a] *School of Aerospace and Mechanical Engineering, University of Oklahoma, Norman, OK 73019 USA*

[b] *Ecole Nationale Supérieure des Mines de Rabat, Morocco*

* *Corresponding author. E-mail: altan@ou.edu*

Susceptibility of polymeric composites to moisture has been well known for several decades. Most high performance epoxy or bismaleimide (BMI) resins and their fiber-reinforced composites may absorb up to 5 wt% moisture which could lead to 10-30% reduction in various mechanical properties, including flexural strength, stiffness, impact resistance, and interlaminar shear strength (ILSS). In particular, fiber-matrix interface and process-induced defects such as microvoids often act as moisture storage sites, thus increasing the maximum intake level. It has been common practice to use a Fickian model to characterize the diffusion of moisture into polymeric composites. However, in several high-performance and mission critical applications, more sophisticated models accounting for the edge effects, anisotropy of absorption, molecular interactions, and interfacial storage are required to fully describe the long- and short-term moisture absorption dynamics. In this article, a model that combines the classical Fickian behavior and diffusion hindrance due to molecular bonding is used to explain anomalous absorption. The hindered diffusion model (HDM) is shown to predict both short-term Fickian and long-term anomalous, non-Fickian absorption behavior often observed in structural composites. The total amount of absorption is shown to be the sum of bound and unbound liquids, which are coupled through a differential diffusion and a temporal storage model. The accuracy of the model predictions is discussed by comparing the model predictions with the experimentally measured mass gain of graphite/epoxy laminates and clay/epoxy nanocomposites. It is shown that the anomalous moisture absorption dynamics observed in these laminates can be accurately predicted by the hindered diffusion model.

Keywords: Composites, Moisture Absorption, Anomalous Diffusion, Hindered Diffusion Model, Anisotropy.
PACS: 81.05.Qk.

INTRODUCTION

Due to their high performance and design flexibility, polymer composites have been widely used in commercial, military, and space applications [1-2]. Nevertheless, susceptibility of polymeric composites to moisture, to which they are typically exposed during service, has been well known for several decades [3-4]. In addition, the degradation observed in flexural strength, stiffness, impact resistance, and ILSS often depends on the moisture intake and absorption kinetics [5-6]. Consequently, preventing in-service failure of composite structures would require full characterization of the moisture absorption behavior prior to the design phase.

The most commonly used model to describe moisture absorption in polymers and their composites is the isotropic, one-dimensional Fickian model, where the moisture concentration in the composite is expected to follow Fick's second law. However, considerable evidence for non-Fickian, or anomalous, moisture diffusion in polymer-based materials has been reported [7-9]. Consequently, a number of researchers have proposed a variety of non-Fickian models for moisture absorption. These models include two-phase polymer dual-diffusivity models [10-11]; dual-diffusivity models [12]; coupled diffusion-relaxation models [13]; as well as dual-mode sorption models, referred to as the Langmuir-type or hindered diffusion model (HDM) [14].

The Langmuir model has been successfully applied to describe the non-Fickian absorption behavior of numerous polymers [15-17] and their composites [18-20]. An additional advantage of the Langmuir model lies in its ability to simplify to Fickian behavior, which removes the constraint of a priori knowledge of the absorption behavior. Moreover, a comparative study of the effectiveness of these models reported the Langmuir model to be particularly effective in representing the non-Fickian behavior of moisture absorption in an epoxy system [22]. Further evidence supporting this approach includes measurements of bound and unbound water molecules in epoxy resin and epoxy composites [15-16,22-23] using nuclear magnetic resonance and Fourier transform infrared spectroscopy.

Recently, Grace and Altan [24] proposed a three-dimensional Hindered Diffusion Model (HDM). Building on the established Langmuir framework, the HDM offers the possibility to describe both anisotropic diffusion in three dimensions and the interactions between water molecules and polymer chains, while accounting for finite sample size. The HDM stipulates that moisture absorption in polymer composites is related to both the polymer/water affinity and the availability of free volume, or empty space, within the polymer. Polymer/water affinity depends on the presence of hydrophilic functional groups within polymer chains. The availability of free volume within the polymer, on the other hand, depends on its microstructural morphology and crosslink density. Water molecules that can diffuse into the free volume of the polymer are often referred to as unbound. Some absorbed water molecules cease to diffuse once they couple with hydrogen bonding sites or polar groups of the polymer. These molecules, referred to as bound, often interrupt the interchain hydrogen bonding of the polymer [25].

In this article, a comprehensive absorption model that combines the anomalous diffusion hindrance due to molecular bonding and three-dimensional anisotropy is presented. The ability of the HDM to predict both short-term Fickian and long-term anomalous, non-Fickian absorption behavior often observed in structural composites is investigated for clay/epoxy nanocomposites and graphite/epoxy laminates. In addition, the accuracy of the model is discussed by comparing the model predictions with the experimentally measured total moisture intake.

THEORETICAL BACKGROUND

The Langmuir model, proposed by Carter and Kibler [14], expresses the total moisture concentration as the sum of two components: bound moisture concentration N and unbound moisture concentration n, coupled through a differential diffusion and a temporal storage model. Building on the Langmuir approach, Grace and Altan [24] recently proposed the anisotropic, three-dimensional Hindered Diffusion Model (HDM) that simplifies to the Langmuir model for one-dimensional cases. The HDM governing equations along the principal axes of diffusion are written as:

$$D_x \frac{\partial^2 n}{\partial x^2} + D_y \frac{\partial^2 n}{\partial y^2} + D_z \frac{\partial^2 n}{\partial z^2} = \frac{\partial n}{\partial t} + \frac{\partial N}{\partial t}, \tag{1a}$$

$$\frac{\partial N}{\partial t} = \gamma n - \beta N. \tag{1b}$$

where γ is the rate at which unbound molecules become bound, and β is the rate of bound molecules becoming unbound, and D_x, D_y and D_z are the diffusion coefficients in the x, y and z directions, respectively. For both the Fickian and hindered diffusion models, a set of absorption parameters is needed to effectively characterize the penetrant absorption into the polymer. The set includes: (i) diffusion coefficients D_x, D_y and D_z (ii) the maximum amount of penetrant that material can absorb, M_∞, often given in terms of weight percentage, and (iii) the hindrance parameters γ and β. For one-dimensional cases, this set reduces to four parameters as planar diffusion along the x and y directions are not considered. In this case, Equation (1) simplifies to:

$$D_x \frac{\partial^2 n}{\partial x^2} = \frac{\partial n}{\partial t} + \frac{\partial N}{\partial t}, \tag{2a}$$

$$\frac{\partial N}{\partial t} = \gamma n - \beta N. \tag{2b}$$

Moisture absorption continues until no more water molecules diffuse into the composite. Thus, equilibrium is reached when the ratio of the bound moisture concentration, N_∞, to unbound moisture concentration, n_∞, is a constant regardless of time or location within the composite sample. The equation describing this equilibrium condition can be written as:

$$N_\infty / n_\infty = \gamma / \beta \tag{3}$$

The total mass gain, $M(t)$, at time t, can is given by the exact analytical solution as [14]:

$$M(t) = M_\infty \left\{ \begin{array}{l} 1 - \frac{8}{\pi^2} \sum_{i=1}^{\infty(odd)} \frac{r_i^+ exp(-r_i^- t) - r_i^- exp(-r_i^+ t)}{i^2 (r_i^+ - r_i^-)} \\ + \frac{8}{\pi^2} \left(\frac{K\beta}{\beta + \gamma} \right) \sum_{i=1}^{\infty(odd)} \frac{exp(-r_i^- t) - exp(-r_i^+ t)}{(r_i^+ - r_i^-)} \end{array} \right\} \tag{4}$$

where:

$$r_i^{\pm} = \frac{1}{2} [(Ki^2 + \gamma + \beta) \pm \sqrt{(Ki^2 + \gamma + \beta)^2 - 4K\beta i^2}],$$

$$K = \frac{\pi^2 D}{h^2},$$

and h is the thickness of the composite sample.

USE OF ANALYTICAL SOLUTION AND EXPERIMENTAL VALIDATION

One of the most accurate methods available for simultaneously recovering the absorption parameters was proposed by Aktas et al. [26]. In this technique, based on a modified version of the steepest descent method and described in detail elsewhere [26-27], the recovery of the absorption parameters set is obtained by iteratively matching the experimental data to the analytical exact solution of the HDM given in Eq.4. To start the error minimization algorithm, initial values are assigned to unknown absorption parameters. Careful selection of the initial guess is essential as these values highly affect the number of iterations, and thus the computational effort needed, before convergence.

In order to validate the accuracy of the HDM in recovering the absorption parameters, we use experimental data obtained from immersion of the two materials systems in water at 25°C: (i) nanoclay/epoxy composites, fabricated with 1wt% nanoclay, and (ii) graphite/epoxy laminates (66% fiber volume fraction). A later experimental measurement of moisture intake for each material system is used to validate the HDM predictions.

Characterizing Absorption Behavior of a Nanoclay/Epoxy Laminate

In order to manufacture the thin nanoclay/epoxy laminate, Nanomer® I.30E nanoclay and EPON 862 epoxy resin/EPIKURE W curing agent system are used at 1 wt% nanoclay content. Eight 31.8 x 31.8 mm x 0.97 mm - specimens are cut from a larger laminate, dried in a vacuum oven, then immersed in a distilled water bath at 25°C. The experimental data obtained from the moisture conditioning experiment, shown in Figure 1, are used to recover the absorption parameters of this laminate. The recovered absorption parameters are presented in Table 1. These values agree well with the absorption parameters reported for similar materials [28].

TABLE 1. Recovered Absorption Parameters of 1 wt% Nanoclay/Epoxy Composite.

Absorption Parameters	Values	Units
Diffusion Coefficient, D	7.97	10^{-10} m²/hr
Bound → Unbound, β	6.61	10^{-4} hr^{-1}
Unbound → Bound, γ	1.05	10^{-4} hr^{-1}
Maximum Moisture, M_∞	2.22	wt%
RMS error/data point	2.63	10^{-3}
Number of iterations	728	-

Figure 1 depicts the moisture absorption behavior predicted by the recovered absorption parameters. Figure 1 also shows the separate contributions of bound and unbound moisture to the total moisture uptake as analyzing the bound and unbound moisture contents can provide more insight into understanding the anomalous behavior of moisture absorption in nanoclay/epoxy composites.

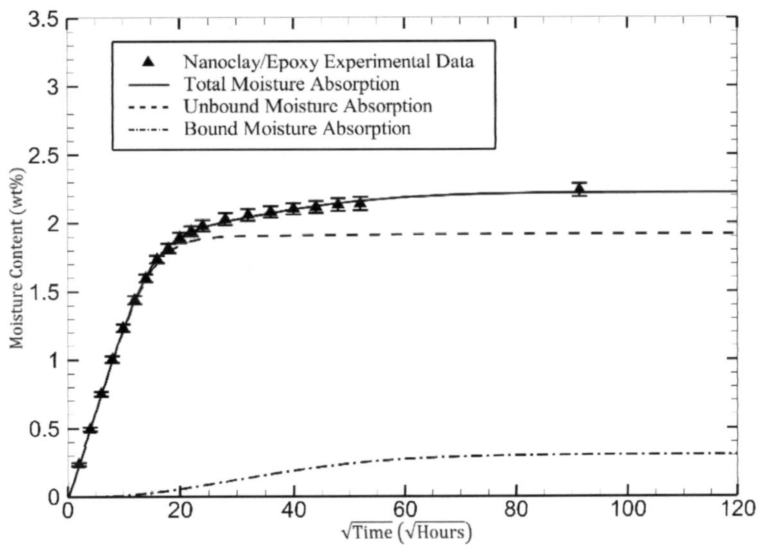

FIGURE 1. Moisture Absorption Kinetics of the 1 wt% Nanoclay/Epoxy Composite.

The predictions of the HDM show excellent agreement with experimental data. The short-term moisture absorption is observed to originate mainly from the unbound moisture, and seems to follow a purely Fickian behavior. As water molecules start to interact and bond with the polymer chains, bound moisture contribution causes the total moisture to deviate from a Fickian behavior. It is worth noting that the last experimental data point is shown only for validating the longer-term predictive ability of the hindered diffusion model, and is not part of the data used in the recovery process. Data collected for almost 2,750 hours allowed the accurate prediction of the moisture content after almost 8100 hours of immersion with an RMS error per data point in the order of 10^{-3}. The maximum moisture intake predicted by the model of 2.22% also agrees well with the later measurement. The number of iterations needed before convergence is moderately high at 728 iterations. This number can be reduced using other techniques discussed elsewhere [27].

Characterizing Absorption Behavior of a Graphite/Epoxy Laminate

The laminate investigated in this study is fabricated using eight-ply of the TenCate® IM7/EX-1522 plain weave graphite/epoxy prepregs cured at a temperature of 177°C and a pressure of 483 kPa. The average thickness of the laminate is 1.58 mm, and the fiber volume fraction is 66%. Six 31.7 x 32.0 mm-specimens are dried in a vacuum oven, then immersed in a distilled water bath at 25°C. The experimental data obtained from the moisture conditioning experiment, shown in Figure 2, are used to recover the absorption parameters of the graphite/epoxy composite. The recovered absorption parameters for the graphite/epoxy laminate are presented in Table 2. The recovered diffusivity and maximum moisture are comparable to those reported in the literature for similar materials [29].

TABLE 2. Recovered Absorption Parameters of 66.1% Fiber Content Graphite/Epoxy Laminate.

Absorption Parameters	Values	Units
Diffusion Coefficient, D	12.10	10^{-10} m²/hr
Bound → Unbound, β	0.31	10^{-4} hr^{-1}
Unbound → Bound, γ	1.52	10^{-4} hr^{-1}
Maximum Moisture, M_∞	1.88	wt%
RMS error/data point	1.56	10^{-3}
Number of iterations	185	-

Figures 2 depict the moisture absorption behavior predicted by the recovered absorption parameters. Again, the contributions of bound and unbound moisture to the total moisture absorption are shown separately in Figure 2 to help analyze the moisture absorption kinetics behind the anomalous behavior of graphite/epoxy laminates.

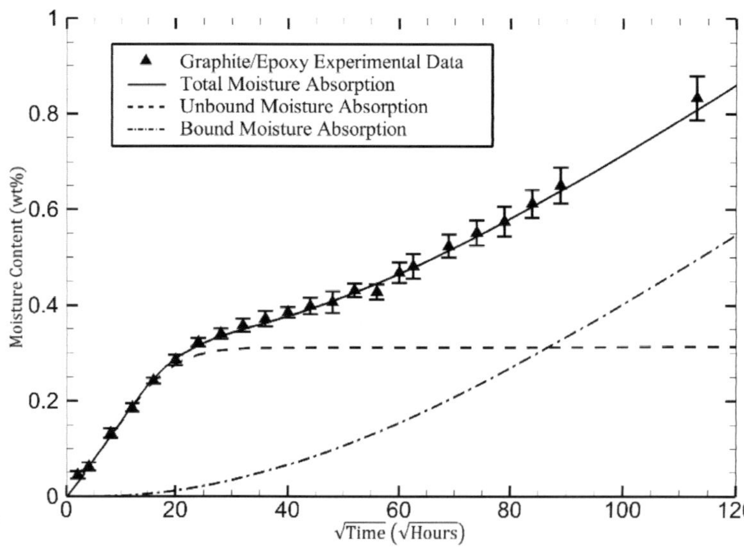

FIGURE 2. Moisture Absorption Kinetics of the 66% Graphite/Epoxy Laminate.

Once again, the model shows excellent agreement with experimental data. The short-term Fickian moisture absorption is due primarily to the unbound moisture. Around 400 hours into the moisture conditioning, water molecules start to interact the polymer chains. The bound moisture accelerates rapidly and causes the total moisture to significantly deviate from a Fickian behavior. The bound moisture contribution even surpasses its unbound counterpart after around 2,700 hours of immersion. The model predicts that at equilibrium, the amount of bound moisture would be almost five-fold of the bound. This phenomenon of higher ratio of bound moisture is not commonly reported in epoxy composites, and may originate from excessive storage sites in the laminates, such as microvoids and hydrophilic fiber/matrix interface. The last experimental data point is once more shown only for validating the longer-term predictive ability of the HDM, and is not part of the data used in the recovery process. It is worth mentioning here that the moisture conditioning in this case is still in its early stages, as the equilibrium moisture intake of 1.88% is predicted to be reached after approximately 8 years. Yet, data collected for almost 7,500 hours allowed the accurate prediction of the moisture content after almost 12,600 hours (last experimental data point) with an RMS error per data point in the order of 10^{-3}. As presented in Table 2, the number of iterations needed before convergence is reasonable at 185 iterations, and is believed to result from a careful selection of the initial guess values.

CONCULDING REMARKS

In this article, a comprehensive absorption model that combines the anomalous diffusion hindrance due to molecular bonding and three-dimensional anisotropy was presented. Unlike the widely used one-dimensional Fickian model, the hindered diffusion model (HDM) offers the possibility to account for the anisotropy of absorption and molecular interactions or storage, characteristic of anomalous absorption behaviors often observed in structural composites.

The ability of the HDM to accurately characterize both short-term Fickian as well as the long-term anomalous diffusion behaviors was demonstrated using experimental data from immersed samples of nanoclay/epoxy composites and graphite/epoxy laminates. For both material systems, the absorption parameters were recovered simultaneously using a modified version of the steepest descent search algorithm by minimizing the error between the experimental data and the analytical HDM solution. The absorption behaviors predicted by the recovered parameters for both nanoclay/epoxy composites and graphite/epoxy laminates were then validated by the long-term experimental data points not used in the recovery process. In addition, both predictive models showed high precision, as the RMS errors per data point were found to be in the 10^{-3} range. Furthermore, unfolding the total moisture absorption into bound and unbound contributions helps further explain the anomalous absorption kinetics.

REFERENCES

1. F. Rodriguez, C. Cohen, C. K. Ober, and L. A. Archer, "Introduction," **in** *Principles of Polymer Systems*. CRC Press, New York, 2015, p. 1-19.
2. Y. K. Hamidi, M. C. Altan, and B. P. Grady, "Polymer Composites," **in** *Encyclopedia of Chemical Processing*., Taylor and Francis, New York, 2005, p. 2313-2322.
3. L. Aktas, Y. K. Hamidi and M.C. Altan, *J. Mat. Proc. & Man. Sci.* **10**, 239-254 (2002).
4. L. Aktas, Y. K. Hamidi and M. C. Altan, *J. Mat. Proc. & Man. Sci.* **10**, 255-267 (2002).
5. E. Perez-Pacheco, J. I. Cauich-Cupul, A. Valadez-Gonzalez and P. J. Herrera-Franco, *J. Mat. Sci.* **48**, 1873-1882 (2013).
6. E. G. Wolff, *SAMPE J.* **29**(3), 11- 19 (1993).
7. M. T. Aronhime, X. Peng and J. K. Gillham, *J. Appl. Polym. Sci.* **32**, 3589-3626 (1986).
8. M.C. Lee and N.A. Peppa, *Prog. Polym. Sci.* **18**, 947-961, (1993).
9. S. Roy, W. X. Xu, S. J. Park and K. M. Liechti, *J. Appl. Mech.* **67**, 391-396 (2000).
10. P. M. Jacobs and F. R. Jones, *J. Mat. Sci.* **24**, 2331-336. (1989).
11. C. Maggana and P. Pissis, *J. Polym. Sci. B* **37**, 1165-182 (1999).
12. L.-W. Cai and Y. Weitsman, *J. Comp. Mat.* **28**,130-154 (1994).
13. A. R. Berens and H. B. Hopfenberg, *Polymer* **19**, 489-496 (1978).
14. H.G. Carter and K.G. Kibler, *J. Comp. Mat.* **12**, 118-31 (1978).
15. M. R. Vanlandingham, R. F. Eduljee and J. W. Gillespie, Jr, *J. of Appl. Polym. Sci.* **71**, 787-798 (1999).
16. S. Popineau, C. Rondeau-Mouro, C. Sulpice-Gaillet and M. E. R. Shanahan, *Polymer* **46**, 10733-10740 (2005).
17. G. LaPlante, A. V. Ouriadov, P. Lee-Sullivan and B. J. Balcom, *J. Appl. Polym. Sci.* **109**, 1350-359 (2008).
18. P. Scott and J. M. Lees, **130**, 1898-1908 (2013).
19. V. La Saponara, *Comp. Struct.* **93**, 2180-2195 (2011).
20. Y. Joliff, L. Belec and J. F. Chailan, *Comp. Struct.* **97**, 296–303 (2013).
21. F. B. Boukhoulda, L. Guillaumat, J. L. Lataillade, E. Adda-Bedia and A. Lousdad, Mat. & Des. **32**, 4080-4087 (2011).
22. T. I. Glaskova, R. M. Guedes, J. J. Morais and A. N. Aniskevich, *Mech. Comp. Mat.* **43**, 377-388 (2007).
23. J. Zhou and J. P. Lucas, *Polymer*, **40**, 5505–5512 (1999).
24. L. R.Grace and M. C. Altan, *Polym. Comp.* **34**, 1144-1157 (2013).
25. M.J. Adamson. *J. Mat. Sci.* **15**, 1736-1745 (1980).
26. L. Aktas, Y. K. Hamidi and M. C. Altan, *J. Eng. Mat. & Tech.* **126**, 427-435 (2004).
27. G. E. Guloglu, Y. K. Hamidi and M. C. Altan, *Submitted to Polym. Eng. & Sci.* (2016).
28. R. Gopalan, R. M. V. G. K. Rao, M. V. V. Murthy and B. Dattaguru, *J. Reinf. Plas. & Compos.* **5**, 51-61 (1986).

Blown film extrusion and compression moulding of high density polyethylene/multi-walled carbon nanotube composites: structure, mechanical and electrical properties

Dong Xiang [a,*], Eileen Harkin-Jones[b]

[a] *School of Materials Science and Engineering, Southwest Petroleum University, 610500, China*

[b] *School of Engineering, University of Ulster, BT37 0QB, UK*

Abstract. The structure, mechanical and electrical properties of melt mixed HDPE/MWCNT composites processed by compression moulding and blown film extrusion were investigated in this study. The addition of MWCNTs leads to a more uniform thickness distribution of the blown films due to the deformation of the material being more stable in film blowing. Compared with the compression moulded composites, the blown composite films exhibit enhanced mechanical properties due to the orientation and disentanglement of MWCNTs. At a blow up ratio (BUR) of 3, the breaking strength and elongation in the machine direction of the blown film with 4 wt% MWCNTs are respectively 239% and 1054% higher than those of compression moulded samples. The compression moulded composite containing only 2 wt% MWCNTs exhibits a decrease of 13 orders of magnitude in resistivity relative to the pure HDPE sheet, indicating the formation of lots of conductive pathways. However, the resistivity of the composite films increases significantly with increasing BUR due to the destruction of conductive pathways. There is no significant increase in the resistivity of the composite films containing 8 wt% MWCNTs due to a sufficient density of nanotubes forming a robust conductive network.

Keywords: polyethylene; carbon nanotubes; compression moulding; blown film extrusion; processing.
PACS: 81.07.De; 81.20.Hy; 81.05.Qk

INTRODUCTION

Carbon nanotubes (CNTs) filled polymer composites are increasingly finding applications in aerospace and automotive industries for antistatic, electrostatic dissipation (ESD) and electromagnetic interference (EMI) shielding when light weight, flexibility and low cost are required. In different polymer processing operations, the polymer/CNT composites will experience different deformational and thermal regimes, resulting in different structure and properties in the final products[1-3]. In Choong's study[4] of polycarbonate (PC)/multi-walled carbon nanotube (MWCNT) composites processed by compression moulding and injection moulding, it was observed that the mechanical and electrical properties of the composites were affected by the processing history. Kasaliwal et al.[5] investigated the influence of compression moulding temperature and time on the electrical resistivity of PC/MWCNT composites near the percolation threshold. It was observed that the resistivity decreased with increasing moulding temperature or moulding time due to the presence of more nanotube secondary agglomerates which enhanced conductive pathways. Although the blown film extrusion process of pure polymers has been investigated in some previous papers,[6-8] there are very few studies of the blown film extrusion of polymer/CNT composites. Russo et al.[9] prepared thermoplastic polyurethane (TPU)/MWCNT thin composite films using blown film extrusion and cast film extrusion. In their study, it was shown that cast film extrusion can achieve a better CNT dispersion compared to blown film extrusion at the same loading and extrusion temperatures. Tensile testing indicated that the blown films exhibited lower modulus and higher elongation at break compared to the cast films. However, the electrical properties of the blown and cast films were not taken into account in their study.

In this study we are interested in examining the differences in structure, mechanical and electrical properties of melt mixed HDPE/MWCNT composites that arise due to two processes of interest (1) compression moulding because it is the main method to produce samples for laboratory studies of nanocomposites and (2) blown film extrusion because it is a major commercial method to produce thin plastic films that are used in areas such as packaging. This work will be significant to better realise the influence of different processing routes on the structure and properties of polymer/CNT nanocomposites.

EXPERIMENTAL

Materials and Preparation

HDPE (HTA108) was purchased from ExxonMobil. It has a density of 0.961g/cm^3 and a high intrinsic viscosity (MFI = 0.7g/10min). MWCNTs (NC7000) with a purity of 90% were kindly supplied by Nanocyl. The MWCNTs with a density of 1.85 g/cm^3 have a nominal diameter of 9.5 nm and an average length of 1.5 μm.

The HDPE pellets were ground into a powder using a Wedco 12″ plant grinder. The HDPE powder was premixed with the MWCNTs using a PRISM Pilot 3 high speed mixer. The blends, containing 2, 4 and 8 wt% MWCNT loadings, were melt mixed in a Collin ZK 25 co-rotating twin-screw extruder with a temperature profile of 175, 220, 220, 215, 210, 200 °C from zones 1 to 6 and a screw speed of 150 rpm. The extruded strand was cooled in a water bath and then pelletized.

Processing and Characterization

The extruded pellets were compression moulded (CM) in a steel mould of 1 mm thickness at 200 °C and 100 bar for 5 min followed by a slow cooling at approximate 20 °C/min using a Collin P200P platen press machine in order to produce sheet samples. Blown film extrusion of the pure HDPE and HDPE/MWCNT composite pellets with 2, 4 and 8 wt% MWCNTs was conducted using a 38mm Killion single-screw extruder equipped with a film blowing die and cooling tower. The barrel temperature from zone 1 to 8 was 190, 230, 230, 230, 250, 250, 250 and 250 °C respectively. The screw speed was 35.6 rpm, and the take-up rate was 6.9 m/min. The materials were blown into films with blow up ratios (BURs) of 2, 2.5 and 3. The average thickness of blown films was obtained by measuring the thickness at 10 points distributed evenly along the film circumference using a Mitutoyo 293-821 micrometer with a resolution of 1 μm.

The morphology of MWCNTs in the compression moulded sheets and blown films was investigated using a JEOL 6500F SEM with an operating voltage of 5.0 kV. All samples for SEM analysis were plasma etched for 60 seconds at an etching power of 100 W using a reactive ion etching system (STS Cluster C005). Then the etched samples were gold sputtered prior to imaging. Tensile tests were conducted for the compression moulded sheets and blown films using an Instron 5564 Universal Tester at room temperature (BS EN ISO 527:1996). Five samples were tested for each sheet and average values were calculated. Volume resistivity testing was carried out for the compression moulded sheets and blown films. For the samples with a high resistivity ($>10^8$ Ω), volume resistivity testing was performed using a Keithley 6517A high resistivity electrometer equipped with a Keithley 8009 resistivity test fixture according to ASTM-D257. In this method, the circular samples cut from the compression moulded and blown films have a diameter of 70 mm. For more conductive samples ($\leq 10^8$ Ω), two-point probe measurements were conducted using a Keithley DMM 2000 multimeter. The strips with dimensions of 50 mm × 10 mm were cut from the compression moulded and blown films. Silver paste was used to minimise contact resistance effects in the two-point probe measurements. Two specimens for each type were tested and the average values were calculated.

RESULTS AND DISCUSSION

Blown Film Extrusion

Blown film extrusion of the pure HDPE and HDPE/MWCNT composite pellets with 2, 4 and 8 wt% MWCNTs was carried out at BURs of 2, 2.5 and 3. The average circumferential thickness (ACT) of the blown films with increasing BURs was measured and is shown in Figure 1. The coefficients of variation (CV), the ratio of the standard deviation to the corresponding average circumferential thickness, for the ACT of thin films were calculated and are also shown in Figure 1. The data points are horizontally offset for clarity. It can be seen in Figure 1 that the ACT of the thin films of both pure HDPE and HDPE/MWCNT composites nonlinearly decreases from around 30 to 20 μm with increasing BUR. The inclusion of MWCNTs does not influence the ACT of the thin films under the same processing conditions. The CVs increase for both pure HDPE and the composite films with increasing BUR. However, the CVs for the composite thin films appear to be smaller compared with the pure HDPE. This is most likely to be due to the biaxial deformation of the material during film blowing being more stable with the addition of MWCNTs.[1,2]

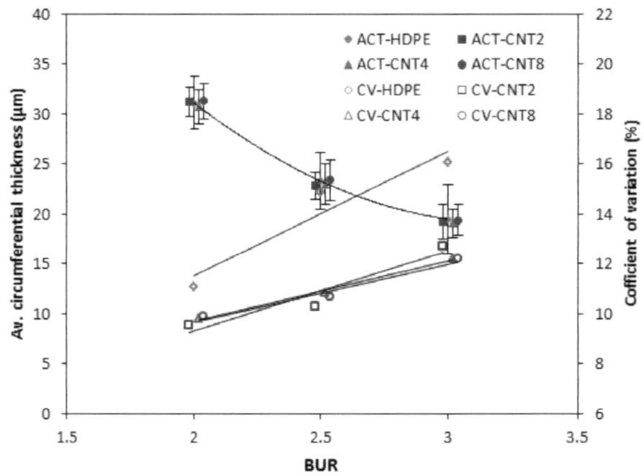

FIGURE 1. Average circumferential thickness of the blown films and the coefficients of variation for the circumferential thickness with increasing BURs.

Morphology

The morphology of MWCNTs in the compression moulded sheets and blown films containing 4 wt% MWCNTs was examined using SEM, as shown in Figure 2. Figure 2a shows that the MWCNTs are randomly oriented and a network-like structure of nanotubes can be observed in the compression moulded composites. The SEM images for the blown films of HDPE/MWCNT composite at BURs of 2 and 3 are shown in Figure 2b and Figure 2c respectively. In Figure 2b, MWCNTs oriented along the flow direction and randomly oriented can be observed at a BUR of 2. Figure 2c shows that there is less clear MWCNT orientation along the machine direction (MD) and fewer MWCNT agglomerates due to the increased deformation in the transverse direction (TD) at a BUR of 3.

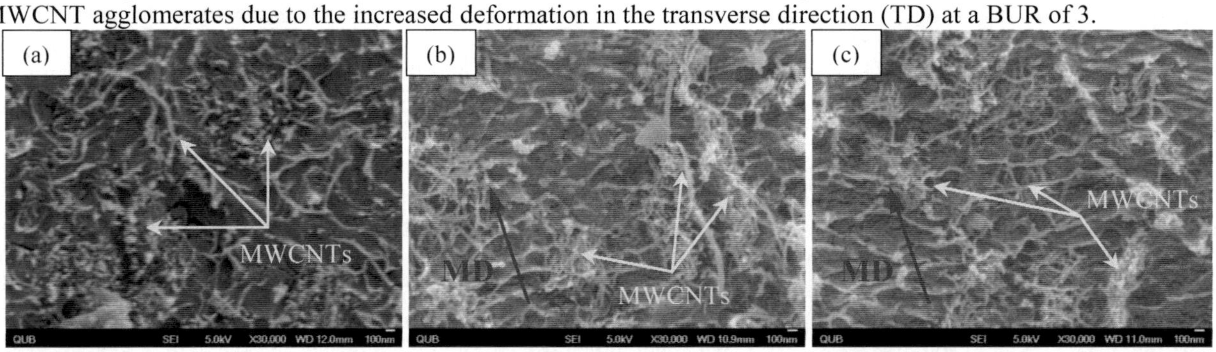

FIGURE 2. SEM images: the compression moulded composites containing 4 wt% MWCNTs (a); the blown films containing 4 wt% MWCNTs with BURs of 2 and 3 respectively (b-c).

Tensile Testing

Tensile testing was carried out to investigate the mechanical properties of the compression moulded sheets and blown films containing 4 wt% MWCNTs. The Young's modulus (E), stress at break (σ_b) and strain at break (ε_b) of the compression moulded sheets and the blown films with increasing BURs are shown in Figure 3. The effect of the addition of 4 wt% MWCNTs on the tensile properties of samples is shown in Table 1. The modulus of compression moulded composite significantly increases by about 112% compared to the pure HDPE. The σ_b and ε_b of the compression moulded composite with 4 wt% MWCNTs decreased by about 40% and 95% respectively due to the presence of nanotube agglomerates (Table 1). An evident anisotropy in modulus can be observed for the composite film with 4 wt% MWCNTs at a BUR of 2, where the modulus in the MD is 51% higher than that in the TD (Figure 3a). At this BUR, the clearly higher modulus in the MD can be attributed to more MWCNTs being oriented along the MD. Then the difference in modulus in the MD and TD is weakened gradually, because some nanotubes may be reoriented along TD with increasing BURs (Figure 2c).

The experimental modulus of the composite blown films is compared with the theoretical values from a Halpin-Tsai composite model. According to the Halpin-Tsai model,[1-4] the longitudinal (E_{11}) and transverse (E_{22}) modulus can be predicted using Equation 1 and Equation 2 respectively. For the composites in which the MWCNTs are randomly oriented, the Halpin-Tsai model can be expressed as Equation (3).

$$E_{11}/E_m = (1 + 2\alpha\eta_L\phi_f)/(1 - \eta_L\phi_f) \quad (1)$$

$$E_{22}/E_m = (1 + 2\eta_T\phi_f)/(1 - \eta_T\phi_f) \quad (2)$$

$$E_r/E_m = 3/8[(1 + 2\alpha\eta_L\phi_f)/(1 - \eta_L\phi_f)] + 5/8[(1 + 2\eta_T\phi_f)/(1 - \eta_T\phi_f)] \quad (3)$$

where $\eta_T = (E_f/E_m - 1)/(E_f/E_m + 2)$, and $\eta_L = (E_f/E_m - 1)/(E_f/E_m + 2\alpha)$. E, E_m and E_f are the modulus of the composites, matrix and filler, respectively. E_f is set as 200 GPa in this case.[10] ϕ_f is the volume fraction of filler. α is the aspect ratio of the filler, which is set as 150 according to the MWCNT dimensional parameters.

The predicted modulus in the MD (E_{11}) and TD (E_{22}) by a fully aligned Halpin-Tsai model (Equation 1 and Equation 2 respectively) shows large errors relative to the experimental values as a result of the biaxial orientation during blown film extrusion. The predicted modulus (Er) by a randomly aligned Halpin-Tsai model (Equation 3) is always between the experimental modulus in the MD and TD, and agrees very well with the experimental values at BURs of 2.5 and 3 due to the enhanced biaxial orientation (Figure 3a). It can be observed in Figure 3b and Figure 3c that the stress at break and strain at break of the blown films of HDPE/MWCNT composite increases steadily in the MD and TD with increasing BURs due to the breakup of MWCNT agglomerates. At a BUR of 3, the modulus and stress at break of the composite film increase by about 114% and 29% respectively in the MD, but the strain at break decreases by about 35% compared to the pure HDPE film (Table 1).

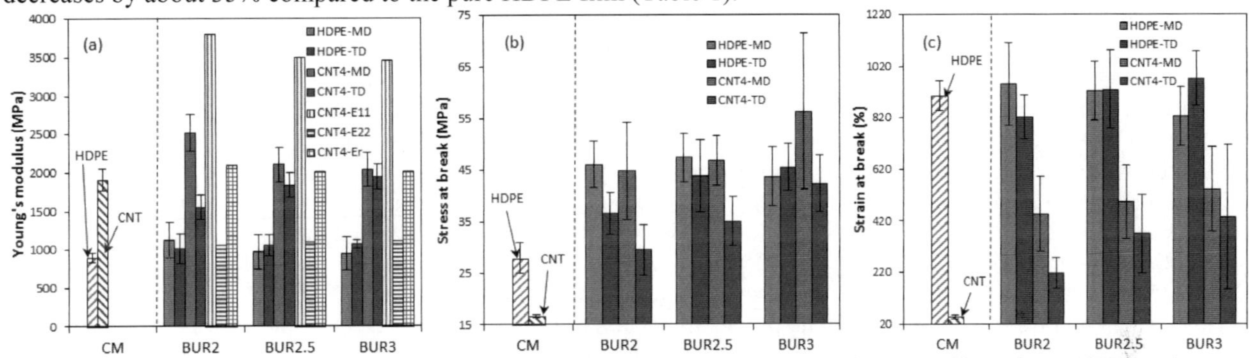

FIGURE 3. Young's modulus (a), stress at break (b), and strain at break (c) of the blown films of pure HDPE and HDPE/MWCNT (4 wt%) composite with increasing BURs.

TABLE 1. Changes in the tensile properties of the compression moulded sheets and blown films of HDPE/MWCNT composites compared to the pure HDPE.

Sample	ΔE(%)	$\Delta\sigma_b$ (%)	$\Delta\varepsilon_b$ (%)
CNT4-CM	+111.6	-40.5	-94.8
CNT4-BUR2-MD	+123.9	-2.6	-53.0
CNT4-BUR2-TD	+52.4	-19.2	-73.7
CNT4-BUR2.5-MD	+116.6	-1.4	-46.6
CNT4-BUR2.5-TD	+74.5	-20.0	-60.3
CNT4-BUR3-MD	+114.1	+28.8	-34.5
CNT4-BUR3-TD	+80.7	-7.1	-55.3

Electrical Resistivity Testing

The electrical properties of the HDPE/MWCNT composites were characterized by measuring the volume resistivity of compression moulded sheets and blown films at different BURs, as shown in Figure 4. The compression moulded composite containing only 2 wt% MWCNTs exhibits a great decrease of 13 orders of magnitude in resistivity relative to the pure HDPE sheet. It indicates that lots of conductive pathways have been formed in the polymer matrix with the incorporation of 2 wt% MWCNTs. At a MWCNT loading of 8 wt%, the compression moulded composite exhibits a low resistivity of ~10 Ωcm as a result of the enhanced conductive networks. It can be observed in Figure 4 that the resistivity of the composite with 2 wt% MWCNTs increases by 10 orders of magnitude at a BUR of 2 before levelling off. For the composite with 4 wt% MWCNTs, the resistivity of blown films increases steadily with increasing BUR, and it also increases by about 10 orders of magnitude when the

BUR increases to 3. However, there is no significant increase in the resistivity of the composite containing 8 wt% MWCNTs after film blowing at increasing BUR. This implies that at this loading there is a sufficient density of nanotubes to form a robust conductive network in the matrix.[1,2] It can be seen in Figure 4 that the resistivity of the composite film with 8 wt% MWCNTs is very close in the MD and TD.

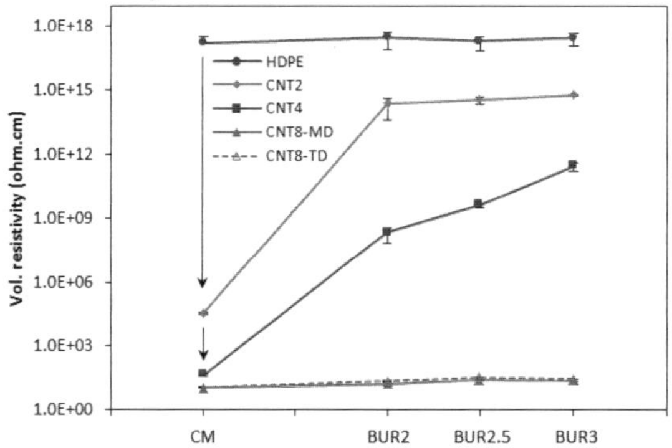

FIGURE 4. Changes in the volume resistivity of the blown films of HDPE/MWCNT composites with increasing BUR.

CONCLUSIONS

In this work, melt mixed HDPE/MWCNT nanocomposites with 2, 4 and 8 wt% MWCNTs were prepared using a twin-screw extruder and then processed by compression moulding and blown film extrusion. SEM results show that the MWCNTs are randomly oriented in the compression moulded composites, while there are more MWCNTs aligned along the MD in the blown composite films. Tensile results show that the modulus of compression moulded composites with 4 wt% MWCNTs significantly increases by 112%. An evident anisotropy in modulus can be observed for the composite with 4 wt% MWCNTs at a BUR of 2, where the modulus in the MD is 51% higher than that in the TD. Then the difference in modulus in the MD and TD is weakened gradually probably due to the reorientation of some MWCNTs along the TD with increasing BUR. The predicted modulus using a randomly aligned Halpin-Tsai model agrees very well with the experimental values at BURs of 2.5 and 3. The stress and strain at break of the composite films increases steadily with increasing BUR due to the breakup of MWCNT agglomerates. Volume resistivity testing indicates that blown film extrusion has a destructive effect on the conductive pathways of nanotubes in the polymer. However, there is no significant increase in the resistivity of the composite containing 8 wt% MWCNTs after film blowing at increasing BURs due to a sufficient density of nanotubes forming a robust conductive network.

ACKNOWLEDGMENTS

The authors would like to thank Nanocyl S.A., Belgium for providing the MWCNTs and the financial support from Sichuan College Key Laboratory of Oil and Gas Field Materials (No. X151516KCL04).

REFERENCES

1. D. Xiang, E. Harkin-Jones and D. Linton, *RSC Adv.* **4**, 44130-44140 (2014).
2. D. Xiang, E. Harkin-Jones and D. Linton, *RSC Adv.* **5**, 47555-44568 (2015).
3. D. Xiang, E. Harkin-Jones and D. Linton et al, *J. Appl. Polym. Sci.* **132**, 42665-42676 (2015).
4. G. Y. H Choong, C. Y. Lew and D. Antonio et al, *J. Appl. Polym. Sci.* **132**, 42277-42287 (2015).
5. G. Kasaliwal, A. Göldel and P. Pötschke et al, *J. appl. polym. sci.* **112**, 3494-3509 (2009).
6. X. Zhang, S. Elkoun and A. Ajji et al, *Polymer* **45**, 217-229 (2004).
7. A. Ghaneh-Fard, P. Carreau, P. Lafleur, *Polym. Eng. Sci.* **37**, 1148-1163 (2004).
8. S. Fatahi, A. Ajji and P. G. Lafleur, *Polym. Eng. Sci.* **47**, 1430-1440 (2007).
9. P. Russo, M. Lavorgna and F. Piscitelli et al, *Eur. Polym. J.* **49**, 379-388 (2013).
10. M. Morcom, K. Atkinson and G. P. Simon, *Polymer* **51**, 3540-3550 (2010).

Development of PLA-PBAT and PLA-PBSA bio-blends: Effects of processing type and PLA crystallinity on mechanical properties

H. Oguz, C. Dogan, D. Kara, Z.T. Ozen, D. Ovali, and M. Nofar*

* *Metallurgical & Materials Engineering Department, Faculty of Chemical and Metallurgical Engineering, Istanbul Technical University, Maslak, Istanbul, 34469, Turkey*

Abstract. In this study, immiscible blends of 75 wt% polylactide (PLA) with 25 wt% of poly[(butylene adipate)-co-terephthalate] (PBAT) were prepared through an injection-molding (IM) process as well as a twin-screw extruder (TSE) followed by IM. An amorphous and a semi-crystalline PLA were also used as the matrix to investigate the matrix crystallization effect on morphology and properties development of the blends. Blend of PLA with 25 wt% of poly[(butylene succinate)-co-adipate] (PBSA) was also prepared through IM process to compare its mechanical properties with those of PLA-PBAT blends. Eventually, the tensile and flexural properties of the prepared blends were explored and their dependency on the evolution of the blend morphology was investigated. The tensile results showed that when IM is used as the sole processing technique, the ductility and toughness were significantly improved only when the semi-crystalline PLA was used as the matrix. Preprocessing through TSE also resulted in enhancement of the PLA blends' strain at break. According to the flexural experimental results, the morphology variation effect on the PLA blends' flexural properties was not as determinative as it appeared in PLA blends' tensile properties.

Keywords: PLA, PBAT, PBSA, blend, morphology, mechanical properties
PACS: 83.30.Tc

INTRODUCTION

Recently, biopolymers have attracted much attention from academia and industries due to environmental concerns and non-composability issues that exist in petroleum-based polymers [1]. Polylactic acid (PLA) is one of the most well-developed commercial bio-based/bio-degradable polymers which is produced from the ring-opening polymerization through the use of renewable resources such as cornstarch or sugarcane [2]. As a linear thermoplastic aliphatic polyester polymer, PLA is considered as a copolymer which contains L-lactide and D-lactide units and by varying the D-lactide content, PLA's crystallization kinetics could be controlled [3].

PLA could be considered as a promising substitute for the petroleum based polymers in various commodity and engineering applications due to the comparable cost and properties of PLA when compared with those of petroleum-based polymers. Furthermore, due to its non-toxic features, PLA is an appropriate candidate for food packaging and other consumer products [1]. It is also a biocompatible polymer which could also be applied in biomedical applications such as bone fixtures, drug delivery and tissue engineering [4].

Despite the noted advantages of PLA, there are several limitations such as brittleness, low toughness, low service temperature, narrow processing window, low melt strength, poor foamability and slow crystallization kinetics that PLA possess [2, 5-8]. In order to overcome some of these limitations, specifically the enhancement of ductility and toughness properties, PLA could be blended with other biodegradable polymers such as poly[(butylene adipate)-co-terephthalate] (PBAT) and poly[(butylene succinate)-co-adipate] (PBSA) that have high ductility, higher melt strength, and better processability. It has been shown by Nofar *et al.* [2, 9, 10] that the viscoelastic properties of PLA are improved when blended with PBAT and PBSA and the interfacial interaction between the matrix and dispersed phase has dominant role on complex viscosity and modulus improvement specifically at low frequencies. In this study, however, we investigated how the mechanical properties of the PLA blends with PBAT could be dependent on the crystallization behavior of PLA matrix and on using various processing techniques. Moreover, we have compared the properties of PLA-PBAT and PLA-PBSA blends.

EXPERIMENTAL

Materials and Preparations

Commercial PLAs of Ingeo 4060D and 2500HP with respective D-lactide contents of 12% and 0.5% were supplied from NatureWorks LLC, USA. In this work, these PLAs are referred to as PLA4060 and PLA2500 which are amorphous and semi-crystalline grades, respectively. The thermograms of neat PLAs from differential scanning calorimetry (DSC, Perkin Elmer DSC400) were investigated and their thermal properties are disclosed in Figure 1. The glass transition temperatures (T_g) of these PLAs were confirmed as 59°C and 63°C, respectively, and the crystal melting temperature of PLA 2500 was appeared at 175.5°C. As shown, the presence of crystalline regions in PLA2500 caused an increase in its T_g when compared to that in PLA4060. PBAT and PBSA, the dispersed phases, were also supplied from Ecoflex® and Showa Denko K.K. (Japan), respectively. The grade of PBAT is F blend C1200 and of PBSA is Bionolle 3001 MD.

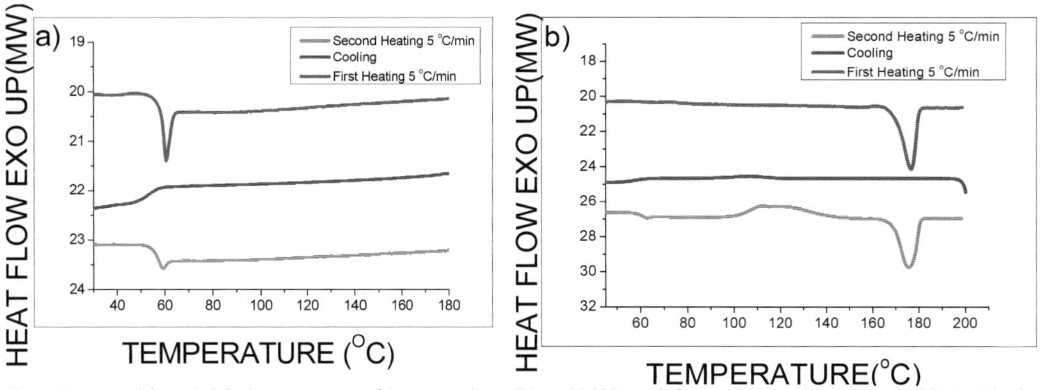

FIGURE 1. Heat-cool-heat DSC thermograms of (a) amorphous PLA 4060D and (b) semi-crystalline PLA 2500HP. The heating and cooling ramp rates was fixed at 5°C/min

All the resins were dried before processing in vacuum oven at 50°C over night to decrease the moisture level of polymers and to prevent their thermal degradation. Blends of 75 wt% PLA as the matrix with 25wt% PBSA or PBAT as the dispersed phases were prepared through injection molding (IM) process by feeding the dry mixed resins of matrix and dispersed phase. The IM process was carried out with a pressure of 1500 bar and with a barrel temperature of 170-190-200-200-200°C in the five zones. Only one mixture with PLA4060-PBAT composition was melt blended using a co-rotating twin-screw extruder (TSE) [Leistritz extruder from Leistritz Advanced Technologies Corp., NJ, USA, with screw diameter (D) of 18 mm and L/D ratio of 40] before running the injection molding (IM). The TSE experiments were conducted at a screw speed of 100 rpm and a set temperature of 160°C for all the heating zones. All of the blend compositions were molded into standard ASTM samples by using injection molding machine [Arburg model, Allrounder 221-K, with L/D ratio of 15]. The samples prepared through only IM process could be summarized as: PLA4060-PBAT, PLA2500-PBAT, PLA4060-PBSA. As noted, PLA4060-PBAT was also premixed in TSE and then processed through IM.

Mechanical Properties

Tensile and flexural properties of the samples were analyzed using a tensile testing apparatus (Schimadzu, ASG-J). Tensile and three-point bending experiments were carried out in order to ASTM638-03 and ASTM-D790 standards with the speed of 5 mm/min, respectively.

RESULTS AND DISCUSSION

Tensile Properties

Figure 2 demonstrates ultimate tensile stress, stiffness and elongation at break values of all samples from tensile testing. As expected from neat PLA samples, PLA4060 and PLA2500 demonstrated brittle behavior, while they have tensile strength values of 62.8 MPa and72.2 MPa, respectively. The tensile strength of PLA2500 seems higher than that of PLA4060, most likely due to the existence of crystallites in PLA2500. Elongation at break values of PLA 4060 and PLA 2500 are in a similar range of 5%. However, the stiffness of PLA4060 shows higher value than PLA2500 which could be due to higher molecular weight of PLA4060 than PLA2500.

The effect of blending polymer type on mechanical properties of PLA was indicated by using PBAT and PBSA as the dispersed phases in the PLA4060 matrix, individually. Ultimate tensile stress of PLA 4060 was diminished by blending with PBAT to 44.6 MPa (29% decrease) and to 42.5 MPa (32% decrease) by blending with PBSA. In both cases, the strain at break values of PLA blends did not show any improvement with the addition of 25wt% PBAT or PBSA. This might have been due to the non-homogeneous blend morphology when using only IM as the processing route and also when an amorphous PLA is used as the matrix. In addition, a decrease in stiffness values of PLA4060-PBAT and PLA4060-PBSA blends comparing to the neat PLA4060 was 28% and 32%, respectively. The reduction of the tensile strength and stiffness (compared to neat PLA) with the addition of PBAT and PBSA to PLA is rational, however, the strain at break did not increase in presence of ductile droplets. This might be due to the non-homogeneous blend morphology when using only IM as the processing route and also when an amorphous PLA is used as the matrix which during the processing can influence the poor dispersion of the dispersed phases.

The results of crystallinity effect on the mechanical properties of PLA were revealed by comparing the PLA2500-PBAT and PLA4060-PBAT blends. Both PLA2500 and PLA 4060 lost their ultimate tensile stress values at the same ratio after blending with PBAT. However, the blend of PLA2500-PBAT revealed superior ductility enhancement; 3700% increase compared to neat PLA2500 (from 5.5% to 205%). This was while the blend of PLA4060-PBAT demonstrated almost no ductility improvement. After blending with PBAT, there has been a decrease in stiffness values of both PLA4060 and PLA2500 with 28% and 27% decrease, respectively. According to these results, one can say that the PLA crystallization during the processing could result in obtaining finer dispersed phase and/or more uniform blend morphology and thereby could profoundly improve the mechanical properties of the blend (specifically the ductility and toughness values which are the main purpose of blending PBAT or PBSA with PLA).

Furthermore, effect of droplet size variation through using different processing methods on mechanical properties of PLA blends were determined by comparing PLA4060-PBAT blends when processed through only IM and when passed through TSE first followed by IM. It can be seen that the blend with TSE preprocessing revealed a much higher elongation at break with the strain value of 121.8% when compared with injection molded case. This was due the much finer and more uniform dispersion of the PBAT droplets when firstly processed through TSE. While injection molded PLA4060-PBAT blend has stiffness of 1522.5 MPa (28% decrease compared with that of neat PLA4060), twin screw extruded PLA4060-PBAT blend has stiffness of 1620.4 MPa (23% decrease from the neat PLA4060). Therefore, PLA4060-PBAT blend preprocessed with TSE also showed less stiffness drop than the injection molded blend.

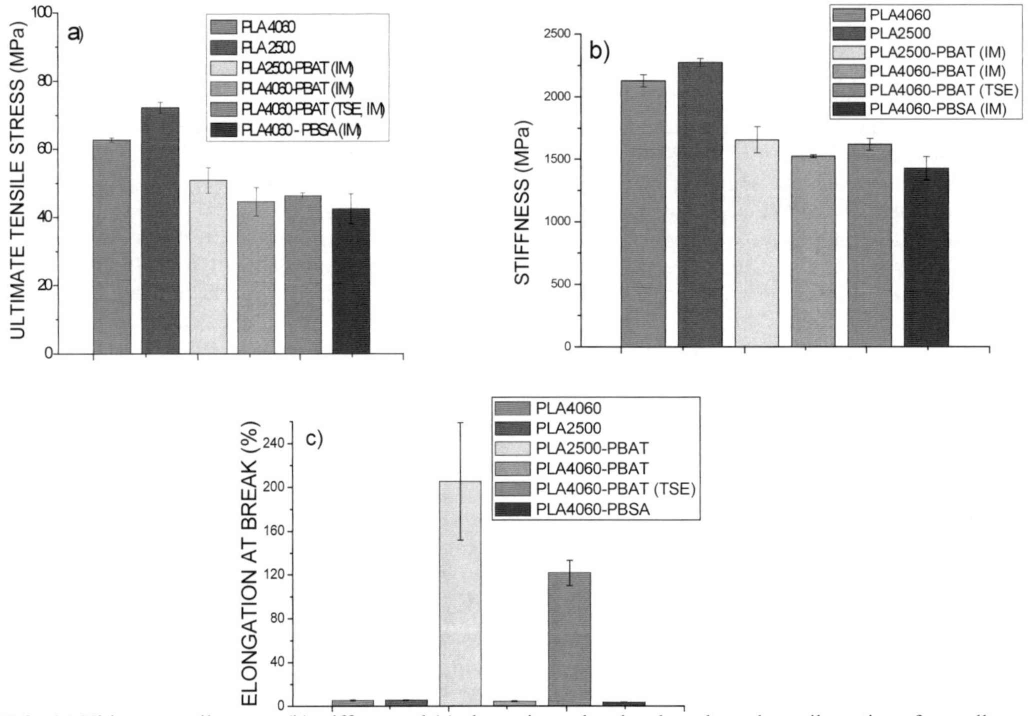

FIGURE 2. (a) Ultimate tensile stress, (b) stiffness and (c) elongation at break values through tensile testing of overall samples

Bending Properties

Figure 3 shows the ultimate flexural stress, flexural modulus and elongation at break values for the neat PLAs and blend samples. Neat PLAs failed just after the elastic region meaning of brittle fracture characteristics. Neat PLA4060 and PLA2500 showed the highest ultimate flexural stress with the values of 100.15 MPa and 112.48 MPa, highest flexural modulus with the values of 2868.7 MPa and 2930 MPa, respectively. Despite the fact that, in tensile testing, the ductility improvement did not appear in all blended samples, the flexural deflection (ductility) of all PLA blend increased. Also, the failure behavior of PLA demonstrated ductile behavior owing to the blending. Elongation at break values obtained from bending test of neat PLA4060 and PLA2500 were 8.27 % and 5.31%, respectively.

Blending PLA with PBAT and PBSA through using different processing or different PLA matrices reduced the flexural strength and stiffness of PLA almost in a similar manner reflecting that the morphology variations in different blends could not vary the flexural properties of PLA however as shown in Figure 2, it significantly influenced the tensile properties. Similar scenario exists in ductility improvement during the flexural testing. In all blend case, the ductility of PLA seems to be improved with the use of any dispersed phase (PBAT or PBSA), any processing technique, or any PLA matrix type. These results reflect that the three-point bending test is more dependent on to the blend composition rather than its morphology.

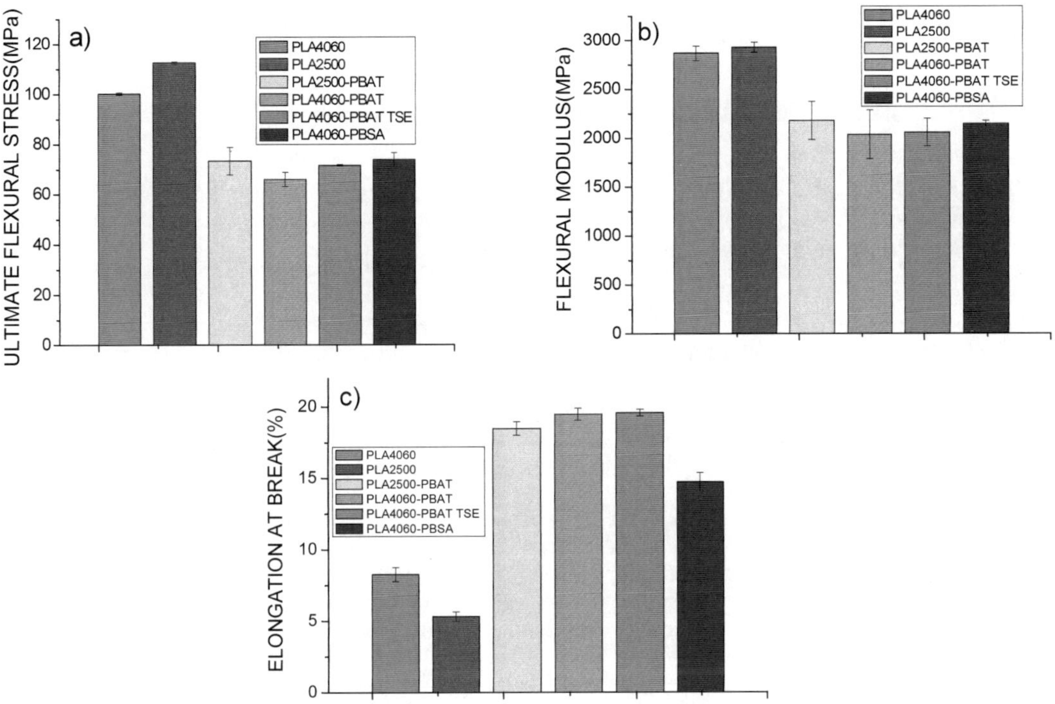

FIGURE 3. (a) Ultimate flexural stress, (b) flexural modulus and (c) elongation at break values through bending test

ACKNOWLEDGMENTS

The authors would like to thank Res. Asst. Dogukan Cetiner and Res. Asst. Onur Tazegul for their supports during mechanical analysis.

REFERENCES

1. Ren J, *Biodegradable Poly(Lactic Acid) Synthesis, Modification, Processing and Applications*, Beijing: Tsinghua University Press, pp. 1-2.
2. Nofar M, Maani A, Sojoudi H, Heuzey M-C, Carreau P.J, *J. of Rheol,* **59 (2)**, 317-333 (2015).
3. Itry R et al., *Polym. Degr. and Stab.*, **97**, 1898-1914 (2012).
4. Domb A.J, Kumar N and Ezra A, *Biodegradable Polymers In Clinical Use and Clinical Development*, New Jersey: John Wiley&Sons, pp. 7.
5. Nofar M and Park C.B, *Prog. Polym. Sci*, **39**, 1721-1741 (2014).
6. Saeidlou S, Huneault M, Li H and Park C.B, *Prog. Polym. Sci*, **37**, 1657-1677 (2012).
7. Rasal R.M, Janorkar A.V and Hirt D.E, *Prog. Polym. Sci*, **35**, 338-356 (2010).
8. Dorgan J and Williams J, *J. of Rheol*, **43**, 1141-1155 (1999).
9. Nofar M, Huzey M-C, Carreau P.J and Kamal M.R, *Polymer*, Accepted (2016).
10. Nofar M, Huzey M-C, Carreau P.J, Kamal M.R and Randall J, *J. of Rheol*, **60 (4)**, 637-648 (1999).

Enhanced performance of polyarylene ether nitriles based composite proton exchange membranes with graphene oxide

Mengna Feng[a], Yumin Huang and Xiaobo Liu[*]

Research Branch of Advanced Functional Materials, School of Microelectronic and Solid-State Electronic, High Temperature Resistant Polymers and Composites Key Laboratory of Sichuan Province, University of Electronic Science and Technology of China, Chengdu 610054, China

Abstract. Sulfonated aromatic polymers such as polyarylene ether nitriles (SPEN) have been extensively investigated to substitute commercial Nafion due to the high proton conductivity, excellent thermal and mechanical properties, and oxidation resistance. Besides, polar nitrile group in sulfonated aromatic polymers could decrease the swelling of membranes and enhance the adhesion between membranes and electrode layers. Meanwhile, graphene oxide (GO), an atomic-layer thick nanosheet containing oxygen-rich functional groups, has been researched extensively for applications in water permeability, supercapacitors, anti-fouling and anti-bacterial properties, attracting more and more researchers to pay attention to it. To our best knowledge, few researchers report the synergistic effects of SPEN and GO applied in the field of proton exchange membrane fuel cell. In this study, we have designed a series of GO/SPEN composite membranes by solution casting technique and investigated the thermal properties, water uptake, swelling ratio and proton conductivity. As a result, the composite membranes possessed better thermal stability, lower water uptake and swelling ratio than that of pure SPEN. Besides, the composite membranes exhibited excellent proton conductivity, which is higher than that of pure SPEN and Nafion 117 membranes. All the data proved that the GO is an effective filler to improve performances of pure SPEN.

Keywords: Sulfonated polyarylene ether nitriles; Graphene oxide; Synergistic effects; Proton conductivity.

PACS: 88. Renewable energy resources and applications

INTRODUCTION

Recently, under the circumstances of resource shortage and serious environment pollution, fuel cells can directly convert chemical energy to electrical energy, attracting more and more researchers to investigate as promising alternative energy conversion devices [1-3]. Due to the low emission, low noise, rich sources of fuel and environment friendly, proton exchange membrane fuel cells (PEMFC) have significant potential for use in the applications of electric vehicles, portable devices, and sensor. Proton exchange membrane is the key component of PEMFC, which plays an important role in provide effective ionic channels for proton transfer and prevent directly contact of anode and cathode [4-5]. Currently, commercial perfluorosulfonic acid membranes such as Nafion have been used extensively as PEM materials, and well known for the superior electrochemical and chemical stability, as well as high proton conductivity. However, the high cost, high methanol crossover and difficulty of preparation prevent the widespread application [6-9]. Therefore, various attempts have been made to overcome, such as developing sulfonated aromatic hydrocarbon polymers.

Polyarylene ether nitrile (SPEN) is considered as one of the potential PEM materials due to the low cost, excellent thermal stability, and low fuel crossover [10]. However, it is heavily influenced by the degree of sulfonation. Organic-inorganic hybridization is the effective methods to overcome the limitation of the membranes as the inorganic particles in the polymer matrix can better improve the dimensional stability, proton conductivity, and resistance against fuel. Graphene oxide (GO), an atomic-layer thick nanosheet containing oxygen-rich functional groups, has been researched extensively for applications in water permeability, supercapacitors, anti-fouling and anti-bacterial properties [11]. To our best knowledge, few researchers report the synergistic effects of SPEN and GO applied in the field of proton exchange membrane fuel cell. So in this study, a series of GO/SPEN composite membranes were prepared by solution casting technique and their properties are investigated in details, including the thermal stability, water uptake, swelling ratio and proton conductivity.

EXPERIMENTAL SECTION

The SPEN was synthesized by DFBN, BP and SHQ via nucleophilic aromatic substitution in N-methylpyrrolidone (NMP) with K_2CO_3 as catalyst, according to the similar procedure in the previous literature [12]. The detail synthesis process is shown in **Scheme 1**.

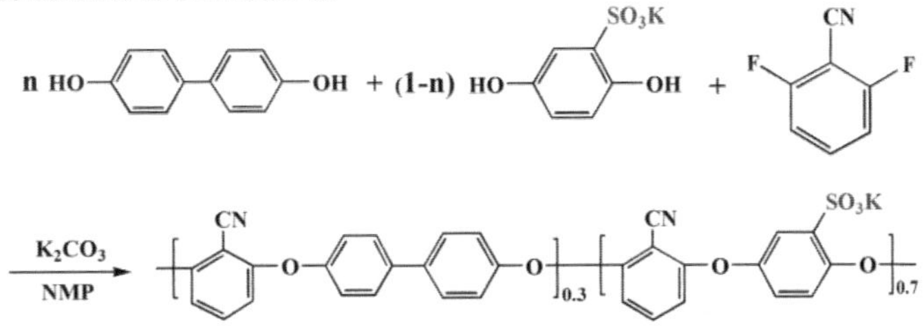

SPEN

SCHEME 1. The schematic of diagrams to synthesize SPEN

The GO was prepared from graphite through the modified Hummers method referred to previously report [13]. The GO/SPPEN composite membranes were fabricated by the facile solution-casting method. A certain amount of GO powder and DMAc solvent were added sequentially into a three necked which was equipped with the mechanic blender and an ultrasonic machine. Meanwhile, the purified SPEN was dissolved in the DMAc solvent, heating to the polymer dissolve completely. After that, the mixture of GO and SPEN solution were kept ultrasounding for 2 h to ensure that the GO was dispersed in the polymer matrix uniformly. Finally, the suspension was cast onto the clean and horizontal glass plate, and dried in an oven at the temperature of 80 °C, 100 °C, 120 °C, 140 °C, and 160 °C each for 2 h to evaporate the solvent completely. The composite membranes with different GO contents (0 wt%, 0.5 wt%, 1 wt%, and 2 wt%) were obtained. The corresponding acid form membranes were obtained by immersing potassium form in 1 M H_2SO_4 for 24 h at room temperature.

RESULTS AND DISCUSSION

SEM characterization

The cross-sectional images of pure SPEN and GO/SPEN composite membranes with 1 wt% and 2 wt% GO content at different magnification time are presented in Fig. 1. The membrane with 1 wt% GO exhibits better interfacial compatibility with SPEN, which can be more clearly observed at the 80,000 magnification time. However, the interfacial interaction of 2 wt% GO/SPEN membrane is bad than that of 1 wt% GO/SPEN membrane. Hydrogen bond and excellent interfacial interaction are the prerequisite for better conductivity performance.

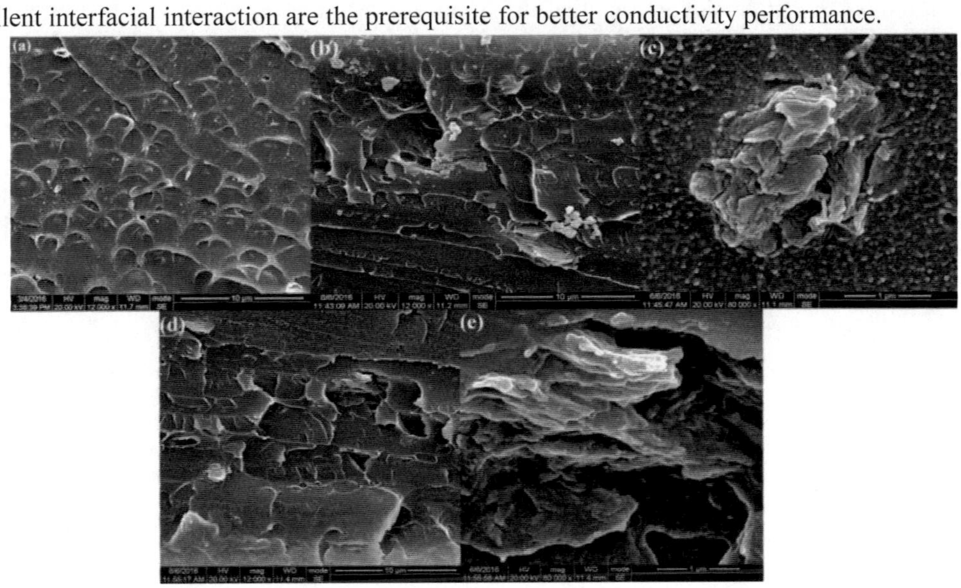

FIGURE 1. The cross-sectional images of GO/SPEN membranes with (a) pure SPEN, (b, c) 1 wt% GO and (d, e) 2 wt% GO at the 12,000 and 80,000 magnification time.

Thermal stability

The thermal stability of pure SPEN and GO/SPEN composite membranes was investigated by the TGA (**Fig. 2**). The 5% weight loss temperature of the membranes is presented in **Table i**. All the membranes exhibited two-step degradation behavior, which is the degradation of sulfonic acid groups and SPEN backbone, respectively. For GO/SPEN, as the filler content increases to 1 wt%, the decomposition temperature increases, while it decreases as the filler content increases to 2.0 wt%. The decrease in the decomposition temperature of the polymer composites with a larger content of graphene oxide is attributed to the phase separation between GO and SPEN matrix. Even so, the decomposition temperature of the GO/SPEN membranes can reach 350 °C, suggesting that the composite membranes possess excellent thermal stability which is high enough for PEMFC applications.

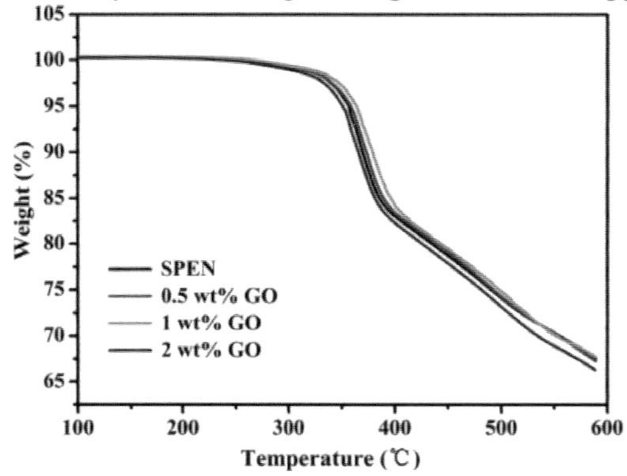

FIGURE 2. The TGA curves of pure SPEN and GO/SPEN composite membranes.

Water uptake and swelling ratio

Fig. 3 and **Table i** exhibits the water uptake and swelling ratio of pure SPEN and GO/SPEN composite membranes. For the SPEN membrane, the acid form membrane swollen excessively in 80 °C and the detailed value cannot be calculated precisely, so the water uptake and swelling ratio of SPEN in this temperature cannot be reported. For the GO/SPEN composite membranes, the water uptake and swelling ratio are lower than that of pure SPEN membrane, which may be caused by the hydrogen bonding interaction between hydroxyl and sulfonic acid groups. From the figure, it can be clearly seen that the composite membrane with 1 wt% GO possesses the lowest water uptake and swelling ratio, indicating that the membrane possesses better interfacial interaction, which can be confirmed by the SEM images. When the filler content increases to 2 wt%, the water uptake and swelling ratio of composite membrane also increase. This is caused by the phase separation and large interface permeation.

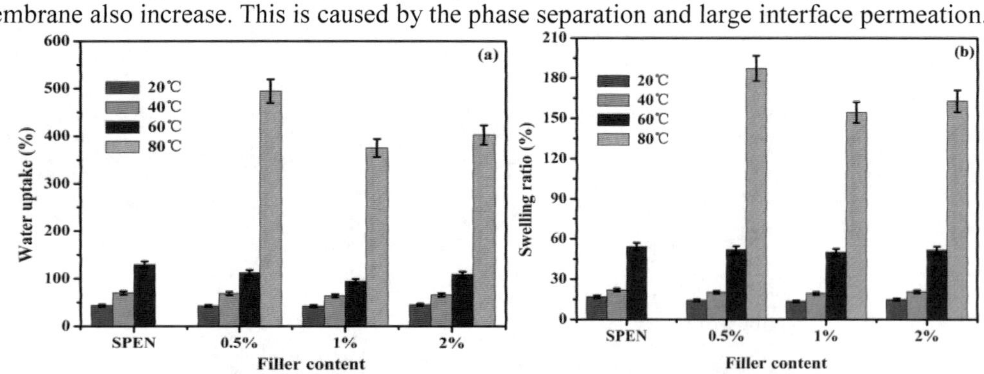

FIGURE 3. The (a) water uptake and (b) swelling ratio of pure SPEN and GO/SPEN composite membranes.

Proton conductivity

Fig. 4 shows the proton conductivity of pure SPEN and GO/SPEN composite membranes at different temperatures in the full wet state and the detail data are also shown in **Table i**. The proton conductivity of composite membranes increases with the increase of GO content until reaching the highest value (0.099 $S \cdot cm^{-1}$) at 1 wt%, and then the proton conductivity begins to descend. The introduction of GO can alter hydrophilic domains and disrupt the original microstructure of pure SPEN. A small amount GO can better enhance the proton conductivity of SPEN, which is caused by the hydrogen bonding interaction between hydroxyl and sulfonic acid groups based on the Grotthus mechanism and vehicle mechanism. However, at relative high filler content, the phase separation and bad interfacial interaction result in the decease of proton conductivity. This is corresponding to the result of SEM images. **Fig. 4** also shows the change of proton conductivity of GO/SPEN membranes with variation of temperatures (20, 40, 60 and 80 °C). In general, the increase of temperature can increase the movement of water molecular and accelerate the proton transfer by the way of hopping or diffusion, causing the increase of proton conductivity. Although the proton conductivity of the GO/SPEN composite membrane with 2 wt% content has a slight decrease, it is still higher than that of Nafion 117 membrane (0.64 $S \cdot cm^{-1}$)) measured in the same condition [14].

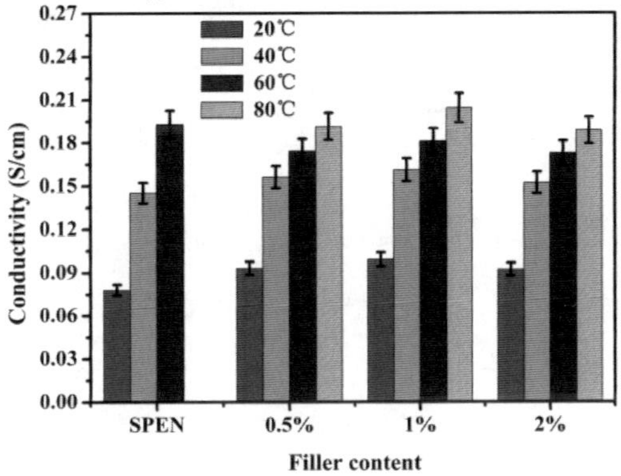

FIGURE 4. The proton conductivity of pure SPEN and GO/SPEN composite membranes at different temperature.

Conclusion

To study the synergistic effects of SPEN and GO applied in the field of proton exchange membrane fuel cell, a series of GO/SPEN composite membranes were prepared by facile solution casting technique. The dispersity of GO in SPEN matrix was observed by SEM. The 1 wt% GO/SPEN membrane possessed better interfacial interaction than that of 2 wt% GO/SPEN membrane, which is the prerequisite for better performances of GO/SPEN membrane. Besides, the GO/SPEN composite membrane exhibited excellent thermal stability, lower water uptake and swelling ratio than that of pure SPEN, suggesting that the GO is an effective filler to enhance the performances of SPEN. The proton conductivity of 1 wt% GO/SPEN membrane reached 0.99 $S \cdot cm^{-1}$, far higher than that of Nafion 117 membrane, indicating that the GO/SPEN composite membrane possesses outstanding ability to proton transfer and has great potential for the application of proton exchange membrane fuel cell. However, the dispersity and compatibility between GO and polymer matrix are still critical and urgent challenge to be resolved.

TABLE i). The thermal stability, water uptake, swelling ratio and proton conductivity of pure SPEN and GO/SPEN composite membranes.

Filler content	Thermal stability (°C)	Water uptake (%)	Swelling ratio (%)	Proton conductivity ($S \cdot cm^{-1}$)
SPEN	356	51	17	0.078
0.5 wt% GO	358	43.02	14.31	0.093
1 wt% GO	355	42.6	13.57	0.099
2 wt% GO	351	45.37	14.92	0.092

ACKNOWLEDGMENTS

The authors wish to thank for financial support of this work from the National Natural Science Foundation (Nos. 51173021, 51373028, 51403029).

REFERENCES

[1] Q. F. Li, J. O. Jensen, R. F. Savinell and N. J. Bjerrum, Prog Polym Sci 34, 449-477 (2009)
[2] N. Radenahmad, A. Afif, P. I. Petra, S. M. H. Rahman, S. G. Eriksson and A. K. Azad, Renewable Sustainable Energy Rev 57, 1347–1358 (2016)
[3] E. Bakangura, L.Wu, L. Ge, Z. J. Yang and T. W. Xu, Prog Polym Sci 57, 103–152 (2016)
[4] Y. Wang, K. S. Chen, J. Mishler, S. C. Cho and X. C. Adroher, Appl Energy 88, 981-1007 (2010)
[5] B. J. Yao, X. L. Yan, Y. Ding, Z. J. Lu, D. X. Dong, H. Ishida, M. Litt and L. Zhu, Macromolecules 47, 1039–1045 (2014)
[6] M. A. Hickner, H. Ghassemi, Y. S. Kim, B. R. Einsla and J. E. McGrath, Chem Rev 104, 4587-4611 (2004)
[7] M. N. Feng, Y. You, P. L. Zheng, J. C. Liu, K. Jia, Y. M. Huang and X. B. Liu, Int J Hydrogen Energy 41, 5113-5122 (2016)
[8] X. H. Yan, R. Z Wu, J. B. Xu, Z. T. Luo and T. S. Zhao, J Power Sources 311, 188-194 (2016)
[9] C. H. Chia, Z. Wu, C. H. Wu, R. H. Cheng and S. Ding, J Mater Chem 22, 22440–22445 (2012)
[10] D. W. Shin, S. Y. Lee, C. H. Lee, K. Soo Lee, C. H. Park, J. E. McGrath, M. Q. Zhang, R. B. Moore, M. D. Lingwood, L. A. Madsen, Y. T. Kim, I. Hwang and Y. M. Lee, Macromolecules 46, 7797−7804 (2013)
[11] C. Y. Tseng, Y. S. Ye, M. Y. Cheng, K. Y. Kao, W. C. Shen, J. Rick, J. C. Chen and B. J. Hwang, Adv Energy Mater 1, 1220–1224 (2011)
[12] Y. Gao, G. P. Robertson, M. D. Guiver, S. D. Mikhailenko, X. Li and S. Kaliaguine, Macromolecules 38, 3237-3245 (2005)
[13] D. C. Marcano, D. V. Kosynkin, J. M. Berlin, A. Sinitskii, Z. Z. Sun, A. Slesarev, L. B. Alemany, W. Lu and J. M. Tour, ACS nano 4, 4806-4814 (2010)
[14] L. Chen, Z. J. Pu, J. Yang, X. L. Yang and X. B. Liu, J Polym Res 20, 45 (2013)

Tribological Properties of Carbon and Aramid Fiber Hybrid Reinforced Polyamide 66 Composites

Tatsuya Shinohara[a], Yosuke Nishitani[b,*] and Takeshi Kitano[c]

[a] *Department of Mechanical Engineering, Graduate School of Engineering, Kogakuin University, 2665-1 Nakano, Hachioji, Tokyo, 192-0015 Japan*
[b] *Department of Mechanical Engineering, Faculty of Engineering, Kogakuin University, 2665-1 Nakano, Hachioji, Tokyo, 192-0015 Japan*
[c] *Polymer Centre, Faculty of Technology, Tomas Bata University in Zlin, T.G.M. 275, Zlin, 767 72 Czech Republic*

Abstract. For the purpose of developing the new engineering materials such as tribomaterials, the tribological properties of carbon fiber (CF) and aramid fiber (KF) reinforced polyamide 66 (PA66) composites were investigated. These hybrid composites (CF/KF/PA66) were melt-mixed by a twin screw extruder and injection-molded. Tribological properties were measured by a ring-on-plate type sliding wear tester under dry condition. The tribological properties such as frictional coefficient and specific wear rate of these hybrid composites showed the different tendencies. The frictional coefficient of these hybrid composites decreased with increasing CF. However, the specific wear rate of these hybrid composites had the minimum peak and showed the curve like as "U" shape. Thus, the filling the combination of CF and KF were effective for the reducing in wear, which is called hybridization effect. This may be attributed to be due to the change in the mode of wear mechanism by the filling combination of CF and KF. Therefore, the difference in the formation of transfer films on the counterface, the shape of wear debris and the worn surface change with the wear mechanisms of CF/KF/PA66 composites.

Keywords: Polyamide 66, Carbon Fiber, Aramid Fiber, Fiber Reinforced Thermoplastic, Tribological Properties
PACS: 81.05.Qk, 81.40.Pq

INTRODUCTION

Over the recent years, fiber reinforced thermoplastic (FRTP) is widely used in mechanical sliding parts such as gears, cams and bearings due to its high mechanical properties, heat resistance, tribological properties and ease of processing [1)-3)]. However, there are still a number of problems that need to be solved. To develop the tribomaterials for mechanical sliding parts, various properties need to be well balanced. In particular, not only are good tribological properties but also suitable mechanical properties and high moldability, which allow flow molding such as compression and injection molding for mass production, are essential [4)-5)]. Several investigations have proved that multi-fiber (hybrid fiber) reinforced thermoplastic composites are effective for realizing good among various properties [6)-7)]. However, the tribological properties of these hybrid fiber reinforced thermoplastic composites are not known enough yet. For the purpose of developing the new tribomaterials, the tribological properties of carbon (CF) and aramid fiber hybrid reinforced polyamide 66 (PA66) composites are investigated experimentally. We discuss the effect of composition (the ratio of CF and KF) on the tribological properties by ring on plate type sliding wear testing against carbon steel ring of these hybrid CF/KF/PA66 composites.

EXPERIMENTAL

The material used in this study was carbon fiber (CF) and aramid fiber (KF) hybrid reinforced polyamide 66 (PA66) composites (CF/KF/PA66). PA66 (Zytel 101L, Du Pont Japan, Ltd., Japan) was used as the matrix polymer. Carbon fiber (CF, HTA-C6-NR, diameter d=6μm and initial length L=6mm, Toho Tenax Co., Ltd., Japan) and aramid fiber (KF, KC40 3WAT-S00-2.3K49ZA, diameter d=12μm and initial length L=3mm, Du Pont-Toray Co.,

Ltd., Japan) were used as the reinforcement. The employed ratios of CF and KF are listed in Table 1. The total amount of fibers was fixed at 20vol.%.

All the components which were dried for 12 hours at 90°C in vacuum oven were dry blended in the small plastic bottle, subsequently melt mixed at 85 rpm and 260°C on a twin screw extruder (TEX-30, Japan Steel Works, Ltd., Japan). After mixing, the extruded strands of CF/KF/PA66 composites were cut in about 5mm length by the pelletizer, and were dried again at 90°C for 12 hours in the vacuum oven. The plate shaped samples for tribological properties testing were injection-molded (NS20-2A, Nissei Plastic Industrial, Japan). The molding conditions were as follows: cylinder temperature of 270°C, mold (cavity) temperature of 80°C and injection rate of $6cm^3/s$. To keep the drying conditions of specimens for all measurements, they were kept in accordance with JIS K 6920-2 for at least 24h at 23°C in desiccators after injection molding.

Tribological properties such as frictional coefficient and specific wear rate were measured using a ring on plate type sliding wear tester (EMF-III-EN, Orientec Co., Ltd., Japan) under dry conditions. A carbon steel (S45C) ring with the surface finished by No.800 polishing paper was used as a counterpart. The tests were conducted using plate specimens with the surface finished by No.800 polishing paper, at the sliding speed of 0.5m/sec, sliding distance 3000m and normal load of 100N. The specific wear rate was calculated by the weight of sample (plate) before and after testing. The counterface after sliding wear testing and wear debris were observed by a scanning electron microscope (SEM, JSM-6360LA, JEOL Ltd., Japan), and the 3D profiles of worn surface of various CF/KF/PA66 composites were measured by a 3D laser scanning confocal microscope (VK-X200, Keyence Co., Japan).

TABLE 1. Composition of various CF/KF/PA66 blends

Code	PA66 (vol.%)	CF (vol.%)	KF (vol.%)
PA66	100		
KF20	80		20
CF2/KF18	80	2	18
CF6/KF14	80	6	14
CF10/KF10	80	10	10
CF14/KF6	80	14	6
CF18/KF2	80	18	2
CF20	80	20	

RESULTS AND DISCUSSION

First, the tribological properties of CF/KF/PA66 hybrid composites by constant normal load and sliding speed testing using a ring-on-plate type sliding wear tester are discussed. Fig. 1(a) shows the relationship between frictional coefficient μ and fiber content (CF contents C_{CF} and KF contents C_{KF}) of various CF/KF/PA66 hybrid composites. μ of these hybrid composites decreases with increasing C_{CF}, and has the tendency of negative deviation which has lower frictional coefficient than simple mixture rule, except of CF18/KF2/PA66 hybrid composites. Fig. 1(b) shows the relationship between specific wear rate V_s and fiber content of CF/KF/PA66 hybrid composites obtained by the same testing method for μ in Fig. 1(a). The V_s of these hybrid composites has the minimum peak and shows the curve like as "U" shape. In short, V_s of these hybrid composites has the tendency of negative deviation (low specific wear rate) in all composition. Thus, the filling the combination of CF and KF were effective for the reducing in wear, which is called hybridization effect. This may attributed to be due to the change in the mode of wear mechanism by the filling combination of CF and KF. Because the tribological properties of polymer composites are strongly influenced by their ability to form a transfer film on the counterface, the shape of wear debris and the morphology of worn surface, it is essential to observe these factors for understanding the mechanisms of tribological behavior [8)].

Fig.2 presents SEM photographs of the surface of metallic counterface (S45C) after sliding wear testing against various CF/KF/PA66 hybrid composites: KF20/PA66 composites (Fig.2 (a)), CF14/KF6/PA66 hybrid ones (Fig.2 (b)) and CF20/PA66 composites (Fig.2 (c)), respectively. The morphology after sliding wear test of KF20/PA66 composites (Fig.2 (a)) shows thin transfer film overall. On the other hand, the morphologies of CF14/KF6/PA66 hybrid composites (Fig.2 (b)) and CF20/PA66 composites (Fig.2 (c)) show thick ones. These morphologies of counterfaces indicate that the presence of CF results in different surface conditions. Fig.3 shows SEM photographs of wear debris after sliding wear testing of various CF/KF/PA66 hybrid composites against S45C: KF20/PA66 composites (Fig.3 (a)), CF14/KF6/PA66 hybrid ones (Fig.3 (b)) and CF20/PA66 composites (Fig.3 (c)), respectively.

The shape and size of wear debris changes with the composition of the composites. First, those of KF20/PA66 composites (Fig.3 (a)) are long flaky and filamentary particles. Second, those of CF14/KF6/PA66 hybrid composites (Fig.3 (b)) are short filamentary and big granular ones. Third, those of CF20/PA66 composites (Fig.3 (c)) are small granular ones. Thus, the size of wear debris of CF/KF/PA66 hybrid composites decrease with increasing CF contents.

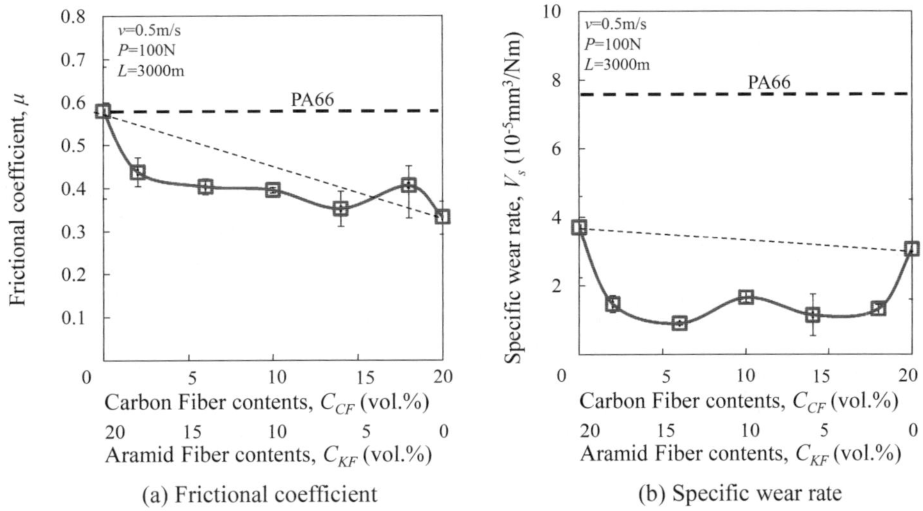

FIGURE 1. Tribological properties of various CF/KF/PA66 hybrid composites

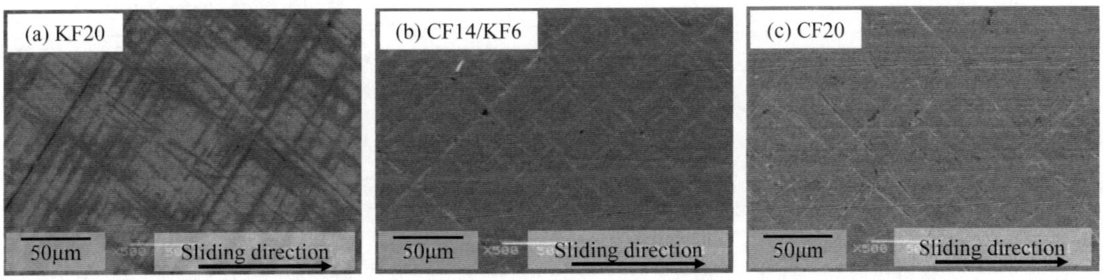

FIGURE 2. SEM photographs of counterface of various CF/KF/PA66 hybrid composites

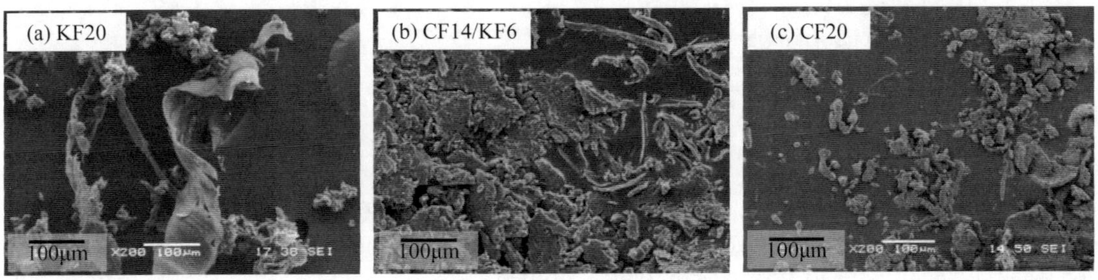

FIGURE 3. SEM photographs of wear debris of various CF/KF/PA66 hybrid composites

Fig.4 presents microscope photographs of worn surfaces after sliding wear testing of various CF/KF/PA66 hybrid composites against S45C: KF20/PA66 composites (Fig.4 (a)), CF14/KF6/PA66 hybrid ones (Fig.4 (b)) and CF20/PA66 composites (Fig.4 (c)), respectively. The worn surface of KF20/PA66 composites (Fig.4 (a)) shows very rough on the overall surface. In contrast, that of CF14/KF6/PA66 hybrid composites (Fig.4 (b)) are relatively smooth surfaces, and the CF and KF clearly are exposed on the surface. On the other hand, that of CF20/PA66 composites (Fig.4 (c)) shows intermediate roughness between KF20/PA66 composites and CF14/KF6/PA66 hybrid composites. Fig.5 shows the profiles of wear track of various CF/KF/PA66 hybrid composites against S45C:

KF20/PA66 composites (Fig.5 (a)), CF14/KF6/PA66 hybrid ones (Fig.5 (b)) and CF20/PA66 composites (Fig.5 (c)), respectively. The wear track of the KF20/PA66 composites (Fig.5 (a)) shows deep and large concave shape, which is almost the same shape as metal counterpart (ring). On the other hand, that of CF14/KF6/PA66 hybrid composites (Fig.5 (b)) and CF/PA66 composites (Fig.5 (c)) is shallower than that of KF20/PA66 composites. The shows surface roughness of wear track of various CF/KF/PA66 hybrid composites measured by 3D laser scanning confocal microscope is listed in Table 2. The surface roughness parameter such as arithmetic mean roughness *Ra* and maximum height roughness *Rz* of various CF/KF/PA66 hybrid composites decrease in the following order: KF20 > CF20 > CF14KF6. This indicates that the surface roughness is closely related to the wear resistance. It follows from these results that it may be possible to develop the new engineering materials with sufficient balance between mechanical and tribological properties.

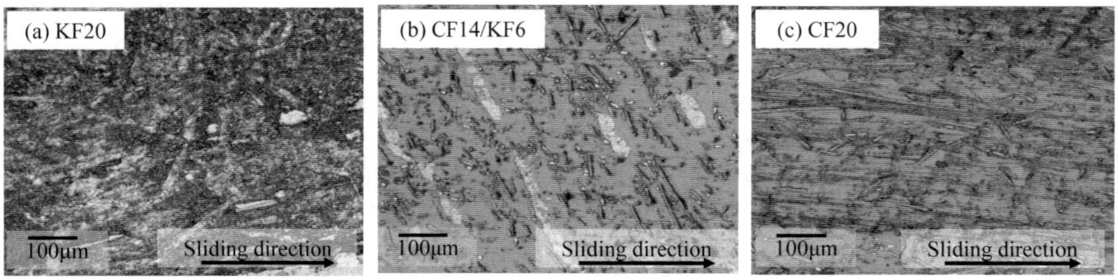

FIGURE 4. Microscope photographs of worn surfaces of various CF/KF/PA66 hybrid composites

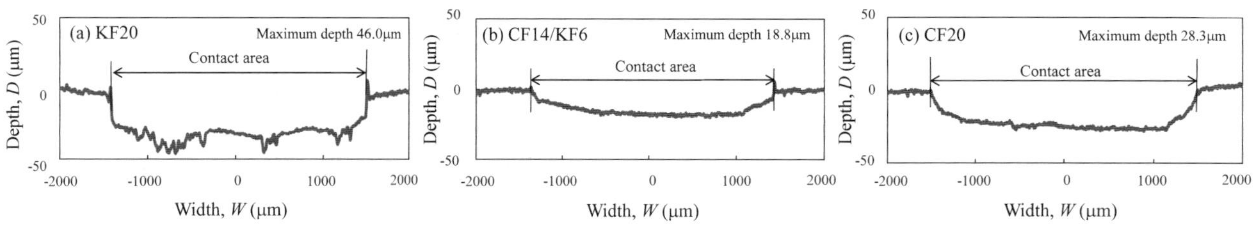

FIGURE 5. Profiles of wear track of various CF/KF/PA66 hybrid composites

TABLE 2. Surface roughness of wear track of various CF/KF/PA66 hybrid composites

	KF20	CF14/KF6	CF20
Arithmetic mean roughness, *Ra* (μm)	1.87	0.56	0.70
Maximum height roughness, *Rz* (μm)	17.3	4.90	6.31

CONCLUSION

For the purpose of developing the new tribomaterials, the tribological properties of carbon (CF) and aramid fiber hybrid reinforced polyamide 66 (PA66) composites were investigated experimentally. We discussed the effect of composition (the ratio of CF and KF) on the tribological properties by ring on plate type sliding wear testing against carbon steel ring of these hybrid CF/KF/PA66 composites. It was found that the tribological properties such as frictional coefficient and specific wear rate of these hybrid composites showed the different tendencies. The frictional coefficient of these hybrid composites decreased with increasing CF. However, the specific wear rate of these hybrid composites had the minimum peak and showed the curve like as "U" shape. Thus, the filling the combination of CF and KF were effective for the reducing in wear, which is called hybridization effect. This may be attributed to due to the change in the mode of wear mechanism by the filling combination of CF and KF. Therefore, the difference in the formation of transfer films on the counterface, the shape of wear debris and the worn surface change with the wear mechanisms of CF/KF/PA66 composites.

ACKNOWLEDGMENTS

We would like to thank the Functional Microstructured Surfaces Research Center (FMS, MEXT, Japan) of Kogakuin University, the Project Research of Research Institute for Science and Technology of Kogakuin University and the Ogasawara Foundation for the Promotion of Science & Engineering for funding this study.

REFERENCES

1. Y. Chen, O. Modi, A. Mhay, A. Chrysanthou and J. O'Sullivan, *Wear*, **255**, 714-721 (2003)
2. D. Mehmet and D. Hayrettin, *Materials and Design*, **57**, 560-567 (2014)
3. T. Shinohara, Y. Nishitani and T. Kitano, *Tribology Conference Spring'16*, Tokyo, B29 (2016) (in Japanese).
4. Y. Nishitani, Y. Yamada, C. Ishii, I. Sekiguchi and T. Kitano, *Polym. Eng. Sci.*, **50**, 100-111 (2010)
5. M. Hasumi, Y. Nishitani and T. Kitano, *The proceedings of the Polymer Processing Society 28th Annual Meeting (PPS-28)*, Pattaya, P-07-324 (2012).
6. Y. Nishitani, I. Sekiguchi, Y. Yoshimitsu, K. Tahira, P. Saha, Y. Nagatsuka and T. Kitano, *Polym.Polym.Composi.*, **7**, 205-215 (1999)
7. T. Kitano, E. Haghani, T. Tanegashima and P. Saha, *Polym. Composi.*, 21, 493-505 (2000)
8. S. Bahadur, *Wear*, **245**, 92-99 (2000)

Gear Strength of Hemp Fiber Reinforced Plants-Derived Polyamide1010 Biomass Composites

Yosuke Nishitani[a,*], Jun Mukaida[a], Yu Yamada[a], Toshiyuki Yamanaka[b], Tetsuto Kajiyama[c] and Takeshi Kitano[d]

[a] *Department of Mechanical Engineering, Faculty of Engineering, Kogakuin University, 2665-1 Nakano, Hachioji, Tokyo 1920015, Japan*
[b] *Tokyo Metropolitan Industrial Technology Research Institute, 2-4-10, Aomi, kotoku, Tokyo 135-0064, Japan*
[c] *Jonan Branch, Tokyo Metropolitan Industrial Technology Research Institute, 1-20-20, Minamikamata, Ota, Tokyo 144-0035, Japan*
[d] *Polymer Centre, Faculty of Technology, Tomas Bata University in Zlin, T. G. M. 275, Zlin, 762 72 Czech Republic*

Abstract. In order to develop the new engineering materials such as mechanical sliding parts (tribomaterials) based on all inedible plants-derived materials, the purpose of this study is to investigate the gear strength of hemp fiber (HF) reinforced plants-derived polyamide 1010 (PA1010) biomass composites. PA1010 was made from sebacic acid and decamethylenediamine, which are obtained from plants-derived castor oil. HF, which is one of the natural fibers, were surface-treated by two types of surface treatment: a) alkali treatment by NaOH solution and b) surface treatment by ureidosilane coupling agent. HF/PA1010 biomass composites were extruded by a twin screw extruder and injection-molded. The test small spur gears (pitch diameter d_0=11mm, module m=0.5mm, number of teeth Z=22, tip diameter d_a=12mm and face width b=1.5mm) were machined from the injection-molded plate and finished by a hobbing machine. Three types of properties of spur gear machined from these HF/PA1010 biomass composites were evaluated: dedendum bending strength of the spur gear by universal tester, gear fatigue strength by using power absorbing typed gear testing apparatus and wear volume of spur gear by same apparatus. It was found that the dedendum bending strength and the wear volume of the spur gear of PA1010 improved with the filling HF although those of HF/PA1010 biomass composites decreased with increasing the volume fraction of fiber. On the other hand, the gear fatigue strength had a maximum peak at 20vol.%. These tendencies may be due to the fundamental mechanical and tribological properties of HF/PA1010 biomass composites.

Keywords: Gear strength, Biomass polymer composites, Natural fiber, Plants-derived Polyamide, Tribology.
PACS: 81.05.Lg, 81.05.Qk, 81.40.Pq, 81.70.Bt

INTRODUCTION

Biomass polymer composites are drawing extensive interest not only as a solution for growing environmental threats but also as a solution for the depletion of petroleum in recent years [1-4]. In addition, the supply of raw materials is becoming increasingly unstable as many biopolymers are made from edible biomass such as corn. In order to solve these problems, new engineering materials made of 100% inedible plants-derived materials are strongly required. In previous studies, we conducted the development of new engineering materials such as structural materials and tribomaterials (for mechanical sliding parts such as bearing, cum, gear, and seal) based on all plants-derived materials. In particular, we investigated the rheological, mechanical, and tribological properties of natural fiber-reinforced biopolymer composites such as hemp fiber (HF)-reinforced plants-derived polyamide 1010 (PA1010) biomass composites [5-10]. PA1010 was made from sebacic acid and decamethylenediamine, which are obtained from plants-derived castor oil [11]. As castor oil is not used for food, there is no competition with human food consumption. It was found that the mechanical and tribological properties of these composites are improved when filled with hemp fibers and surface-treated using silane coupling agent. However, in our knowledge, there are no experimental data on the actual mechanical sliding parts such as gear and so on. In order to develop the new engineering materials such as mechanical sliding parts (tribomaterials) based on all inedible plants-derived materials, the purpose of this study is to investigate the gear strength of hemp fiber (HF) reinforced plants-derived polyamide 1010 (PA1010) biomass composites.

EXPERIMENTAL

The materials used in this study were surface-treated hemp fiber reinforced plants-derived polyamide 1010 biomass composites (HF/PA1010). Plants-derived polyamide 1010 (PA1010, Vestamid Terra DS16, Daicel Evonik Ltd., Japan) was used as the matrix polymer. PA1010 was made from sebacic acid and decamethylenediamine, which are obtained from plants-derived castor oil [11]. Hemp fiber (HF, ϕ50-100μm, Hemp Levo Inc., Japan) was used as reinforcement fiber. The volume fraction of fiber V_f in the composites was 10, 20 and 30vol.%. HFs were pre-cut into 5mm long pieces, and were surface-treated by two methods: a) alkali treatment by sodium hydroxide (NaOH) solution and b) surface treatment using ureidosilane coupling agent (S3, 3-ureidopropyltrimethoxy silane, A-1160, Momentive Performance Material Inc., USA) (NaOH + ureidosilane). Alkali treatment using NaOH was employed as follows: 5% NaOH solution was placed in a stainless beaker. Then chopped HFs were added into the beaker and stirred well. This was kept at room temperature for 4h. The fibers were then washed thoroughly with deionized water to remove excess NaOH sticking to the fibers. The alkali treated HFs by NaOH (HF-A) were dried on air for 12h and in a vacuum oven at 80°C for 5h. The surface treatment of hemp fibers with the concentration of 1wt.% ureidosilane coupling agent (S3) was carried out in 0.5wt.% acetic acid aqueous solution in which the pH of the solution was adjusted to 3.5 and stirred continuously for 15min. Then, the fibers were immersed in the solution for 60min. The surface-treated hemp fibers (HF-S) were removed from the solution and kept in air for 12h and in a vacuum oven at 80°C for 5h. All the components which were dried for 12h at 80°C in a vacuum oven beforehand until the moisture level was below 0.2%, then, dry blended in the small bottle, and subsequently the melt was mixed at 85rpm and 220°C on a twin screw extruder (TEX-30, Japan Steel Works, Ltd., Japan). After mixing, the extruded strands of various HF/PA1010 biomass composites were cut into 5mm long pieces by the pelletizer, and dried again at 12h at 80°C in a vacuum oven. Various shaped samples for various experiments were injection molded (NS20-A, Nissei Plastic Industrial, Japan). The molding conditions were as follows: cylinder temperatures of 220°C, mold (cavity) temperature of 30°C and the injection rate of 13cm^3/s. To maintain the drying conditions of specimens for all measurements, they were kept in accordance with JIS K 6920-2 for at least 24h at 23°C in desiccators after molding.

The gear strength of the CF/PBT composites was investigated by using spur gears. The test spur gears were machined from the injection-molded plate and finished by a hobbing machine. The specification of the test spur gear and the drive gear (SUS304) is shown in Table 1. Two kinds of gear strength were measured: dedendum (gear root) bending strength of the spur gear by universal testing and gear fatigue strength by absorbing typed gear testing. The dedendum bending strength of the gear were measured by an universal testing machine (V-10, Toyo Seiki Seisaku-sho, Ltd., Japan) at the crosshead speed of 2mm/min using test jig. The dedendum bending strength of the gear was obtained as follows: the test spur gear was contacted by the metal drive gear on the pitch circle diameter. The rod fixed with the center of the metal drive gear was extended 50mm in radial direction, and vertical load was applied. The load torque was obtained from the load at the breakage of gear tooth. This load torque was substituted in the equation of Lewis, which regards a pitch point, and the bending stress of dedendum (gear root) was calculated as:

$$\sigma_b = \frac{2000TZ}{by'{d_0}^2} \tag{1}$$

where σ_b is the dedendum bending strength of the gear (MPa), 2000 is the organized coefficient value of equation of Lewis, T is the measured torque (Nm), Z is the number of teeth, b is the face width (mm), y' is the tooth form factor 0.599, which is determined from the gear geometry such as the number of teeth and the pressure angle and d_0 is the pitch circle diameter (mm). The testing in this study was measured by contacting only the tooth of test spur gear, which was cut off the tooth around the testing one in order to eliminate the influence of the interfere of the other tooth.

The gear fatigue strength was measured using an absorbing typed gear testing apparatus which is produced in our laboratory [12, 13] (Fig.1). The measuring conditions for this testing were as follows: the applied load T_a is 10-150Nmm, the constant rotating speed n is 1740rpm (conversion peripheral speed v is 1m/s), at room temperature and under dry (no lubrication) condition. The lifetime at breaking of the gear was defined by the integrated number of revolution N when the power transmission was stopped. The stress amplitude (the repeated stress) S was calculated beforehand the testing from Equation (1). The gear fatigue strength σ_u was defined by the stress amplitude at N=10^7 from S-N curves. In addition, the wear loss of spur gear were measured by the same apparatus. The measuring conditions for wear loss were as follows: the stress amplitude σ is 20MPa, the constant rotating speed n is 1740rpm (v=1m/s), the integrated number of revolution N is $1\text{x}10^4$, $2\text{x}10^4$, $5\text{x}10^4$, $10\text{x}10^4$ and $20\text{x}10^4$, at room temperature and under dry condition. The wear loss was calculated by the weight of sample (spur gear) before and after testing.

TABLE 1 Specification of test gear and drive gear

	Test gear	Drive gear
Materials	HF/PA1010 composites	SUS304
Module, m (mm)	0.5	0.5
Number of teeth, Z	22	44
Pitch diameter, d_0 (mm)	11	22
Tip diameter, d_a (mm)	12	23
Face width, b (mm)	1.5	2

FIGURE 1. Image of absorbing typed gear testing apparatus.

RESULT AND DISCUSSION

First, we discuss about the dedendum bending strength of the spur gear σ_b as one of the gear strength. σ_b of hemp fiber reinforced plants-derived polyamide 1010 biomass composites (HF/PA1010) is shown in Fig.2. σ_b increases with filling hemp fiber (HF) and decreases with increasing with the volume fraction of fiber V_f. These tendencies may be due to the fundamental mechanical properties and tribological properties of HF/PA1010 biomass composite [14]. In general, the mechanical properties of the polymer composites such as strength and modulus increase with increasing the fiber content. However, by filling the fiber, the material property changes from ductile to brittle.

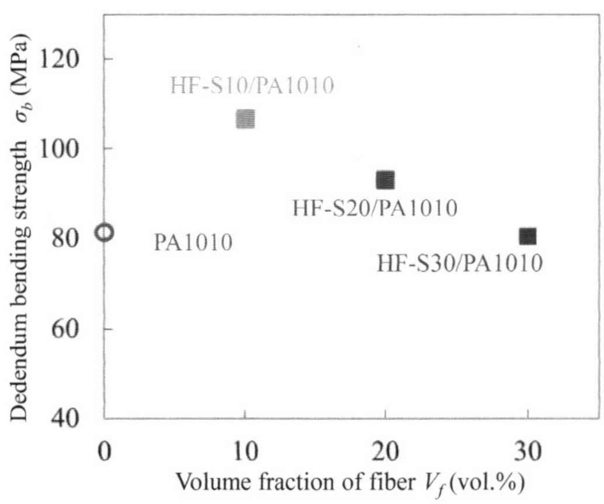

FIGURE 2. Influence of volume fraction of fiber on the dedendum bending strength of HF/PA1010 biomass composites.

The strength of HF/PA1010 biomass composites increases with V_f. Moreover, the elongation at break of HF/PA1010 biomass composites was remarkably decreased with V_f. In order to improve the dedendum bending strength of the spur gear, not only the strength but also the elongation should be improved. These behaviors have been reported for other polymer composites such as various CF/PBT composites [13].

Next, the gear fatigue strength measured by using absorbing typed gear testing apparatus is discussed. Relationship between the stress amplitude *S* and the integrated number of revolution *N*, called as "*S*-*N* curves", is shown in Fig.3 for HF/PA1010 biomass composites. *S*-*N* curves show different behavior according to the volume fraction of fiber and change with the stress amplitude *S*. In high *S* region, *N* increases with the following order: HF-S30/PA1010 < HF-S10/PA1010 < HF-S20/PA1010 < PA1010. On the other hand, in low *S* region, *N* increases with the following order: HF-S30/PA1010 < PA1010 < HF-S10/PA1010 < HF-S20/PA1010. In short, the gear fatigue strength had a maximum peak at 20vol.%. These behavior may be attributed to be corresponding to the mechanical properties such as modulus, hardness, toughness and fatigue properties. However the order of improving the gear fatigue strength is different from those of the mechanical properties.

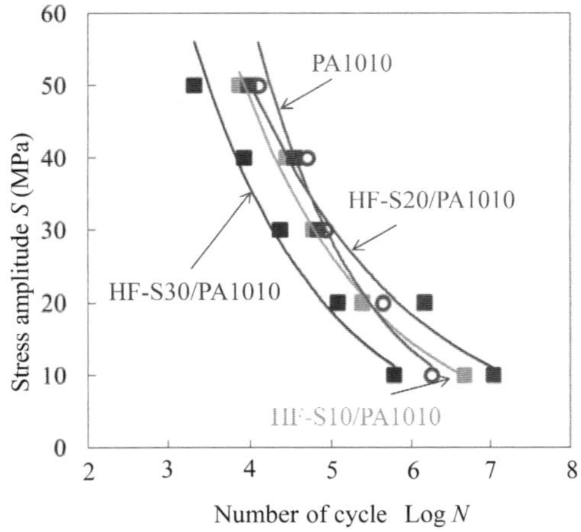

FIGURE 3. *S*-*N* curves of HF/PA1010 biomass composites using an absorbing typed gear testing apparatus.

The wear loss (volume) of HF/PA1010 biomass composites measured by using absorbing typed gear testing apparatus is discussed. Relationship between the wear volume *V* and the integrated number of revolution *N* is shown in Fig.4 for HF/PA1010 biomass composites. It is found that the wear volume of the spur gear of PA1010 improved with the filling HF although those of HF/PA1010 biomass composites decreased with increasing the volume fraction of fiber. This has the same tendency as the dedendum strength. In short, the wear volume are closely related to the fundamental mechanical and tribological properties of HF/PA1010 biomass composites.

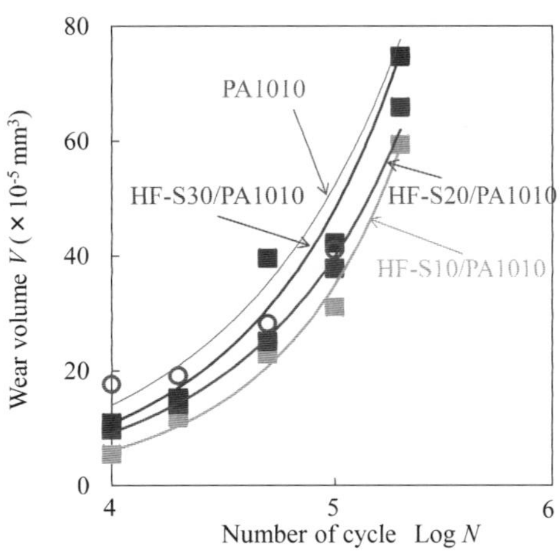

FIGURE 4. Relationship between wear volume and number of cycle for HF/PA1010 biomass composites using an absorbing typed gear testing apparatus.

CONCLUSION

In order to develop the new engineering materials such as mechanical sliding parts (tribomaterials) based on all inedible plants-derived materials, the purpose of this study is to investigate the gear strength of hemp fiber (HF) reinforced plants-derived polyamide 1010 (PA1010) biomass composites. It was found that the dedendum bending strength and the wear volume of the spur gear of PA1010 improved with the filling HF although those of HF/PA1010 biomass composites decreased with increasing the volume fraction of fiber. On the other hand, the gear fatigue strength had a maximum peak at 20vol.%. These tendencies may be due to the fundamental mechanical and tribological properties of HF/PA1010 biomass composites.

ACKNOWLEDGMENTS

This work was supported by JSPS KAKENHI Grant Number 16K06750 and 25420735. We would like to thank the Functional Microstructured Surfaces Reserch Center (FMS, MEXT, Japan) of Kogakuin University and the Project Rresearch of Research Institute for Science and Technology of Kogakuin University for funding this study, and partial support by national budget of Czech Republic within the framework of the Centre of Polymer Systems project (reg. number: CZ. 1.05/2.1.00/ 03.0111).

REFERENCES

1. A. C. Wibowo, A. C. Mohanty, M. Misra, T. Lawrence, L. T. Drzal, *Ind. Eng. Chem. Res.* **43**, 4883–4888 (2004).
2. L. Petersson, K. Oksman, *Composi. Sci. Technol.* **66**, 2187–2196 (2006).
3. A. K. Bledzki, J. Gassan, *Prog. Polym. Sci.* **24**, 221–274 (1999).
4. F. Omar, A. K. Bledzki, H.-P. Fink, M. Sain, *Prog. Polym. Sci.* **37**, 1552–1596 (2012).
5. M. Hasumi, Y. Nishitani, T. Kitano, The proceedings of Seikei-kako symposia'11, Akita, 491-492 (2011) (in Japanese).
6. M. Hasumi, Y. Nishitani, T. Kitano, The proceedings of the Polymer Processing Society 28th Annual Meeting (PPS-28), Pattaya, P-07-324 (2012).
7. Y. Nishitani, M. Hasumi, T. Kitano, *AIP Conference Proceedings* **1664**, 060007 (2015).
8. J. Mukaida, Y. Nishitani, T. Kitano, *AIP Conference Proceedings* **1664**, 060008 (2015).
9. J. Mukaida, Y. Nishitani, T. Yamanaka, T. Kajiyama, T. Kitano, The Proceedings of Regional Conference of Polymer Processing Society (PPS-2105), Graz, Nat-104 (2015).
10. Y. Nishitani, J. Mukaida, T. Yamanaka, T. Kajiyama, T. Kitano, *AIP Conference Proceedings* **1713**, 120007 (2016).
11. M. Naganuma, *JETI* **59**, 88–90 (2011).
12. S. Sato, N. Saeki, M. Horiguchi, T. Kataoka, Y. Nishitani, *Research Report of Kogakuin University*, **111**, 7-13 (2011) (in Japanese).
13. Y. Nishitani, I. Sekiguchi, T. Kitano, The Proceedings of the Polymer Processing Society 27th Annual Meeting (PPS-27), Marrakech, P-3-521 (2011).
14. J. Mukaida, Y. Nishitani, T. Yamanaka, T. Kajiyama, T. Kitano, The Proceedings of International Tribology Conference, Tokyo 2015 (ITC2015), Tokyo, 904-905 (2015).

Preparation of dynamically vulcanized SEBS/PP TPVs with high solvent-resistance and good tensile properties

Yijian Wu, Baoqing Shentu* and Zhixue Weng
State Key Lab of Chemical Engineering, College of Chemical and biological Engineering, Zhejiang University, 38 Zheda Road, Hangzhou 310027, China

Abstract. Dynamically vulcanized blends of polypropylene (PP) and styrene-ethylene-butadiene-styrene block copolymer (SEBS) with crosslinking coagent-assisted peroxide were prepared by melt blending to obtain SEBS/PP thermoplastic vulcanizates (TPVs). In addition to di-tert-butyl peroxide (DTBP), trimethylopropane trimethacrylate (TMPTMA) and poly(styrene-b- butadiene-bstyrene) (SBS) were used as the crosslinking coagents. TMPTMA and SBS showed synergistic effect on the properties of the TPVs. With the same dosage of DTBP and coagents, simultaneous loading of TMPTMA and SBS could generate the TPVs with shorter reaction time, higher solvent-resistance and better tensile properties than solely addition of TMPTMA or SBS. The synergistic effect mechanism of TMPTMA and SBS was also studied.

Keywords: blends, crosslinking, elastomers, properties and characterization
PACS:

Torque-time curing characteristics of the TPVs

The oil-filled elastomer, PP, DTBP and TMPTMA were simultaneously added into the internal mixer at 180 °C with a rotor speed of 120 rpm. The mixing time for each batch was 10 min. The compositions of SEBS/PP blend and TPVs for this study are shown in Table 1. The torque-time curve allows for dynamic evaluation of the change in viscosity which was closely related to the reactions in the blends[1-2]. Torque-time curves of the SEBS/PP TPVs were shown in Figure 1. In figure 1 B (sample T_1S_4), point "a" represented the loading peak of the crude materials, the valley "b" indicated that the crude materials were fully melted, and point "c" was considered as the termination of the crosslinking reaction[2]. Finally the crosslinked melts were mechanically crushed by the mixer blade (region "d"). The time span of b-c was regarded as the crosslinking reaction time[2].

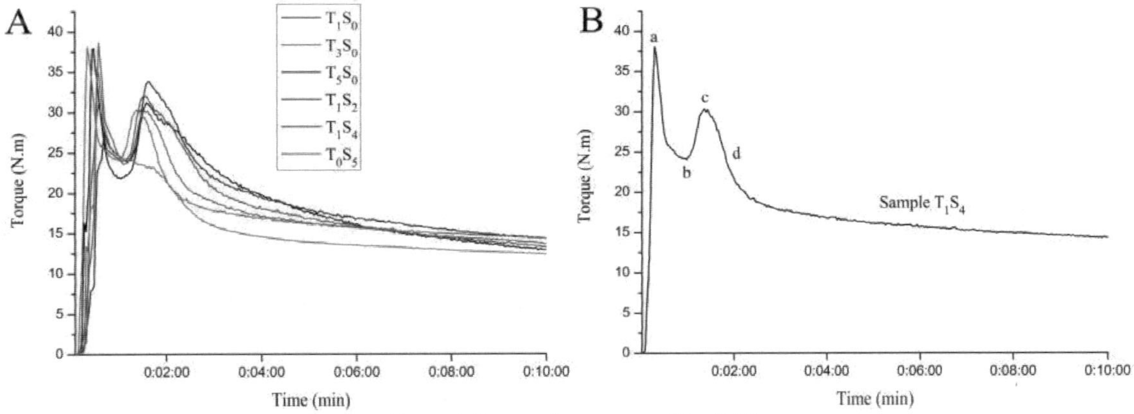

Figure 1. Torque-time curves of the TPVs.

Most of the samples (exclusive of T_0S_5) showed apparent crosslinking peaks (Figure 1A). Crosslinking reaction time and final torque of these samples were displayed in Table 2. Reaction time of sample T_1S_0, T_3S_0, T_5S_0, T_1S_2, T_1S_4 were 36 s, 32 s, 28 s, 26 s and 22 s, respectively. It could be concluded that crosslinking reaction time decreased with the increasing loading of TMPTMA. Besides, simultaneous addition of TMPTMA and SBS further shortened the crosslinking reaction time. In other words, TMPTMA could accelerate the crosslinking reaction of the SEBS/PP blend, and acceleration effect became more obvious in the present of SBS.

Table 1. Weight composition of SEBS/PP TPVs

Sample name	SEBS	PP	naphthenic oil	DTBP	TMPTMA	SBS

T_0S_0	60	20	20	0	0	0
T_1S_0	60	20	20	1	1	0
T_3S_0	60	20	20	1	3	0
T_5S_0	60	20	20	1	5	0
T_1S_2	58	20	20	1	1	2
T_1S_4	56	20	20	1	1	4
T_0S_5	55	20	20	1	0	5

Table 2. Cross-linking reaction time of the TPVs

Sample	T_1S_0	T_3S_0	T_5S_0	T_1S_2	T_1S_4
Crosslinking reaction time (s)	36	32	28	26	22

Solvent-resistance of the TPVs

In order to get a material with good solvent-resistance, the continuity of the noncrosslinked SEBS, the gel content, the swelling degree and v_e of different samples were determined, and the results were shown in Table 3.

For the noncrosslinked SEBS/PP blend (T_0S_0), the continuity of SEBS reached 98.6%, which indicated that the SEBS phase was continuous in the blend. Meanwhile, as the SEBS was extracted, the shape of the sample remained, which proved that the PP phase was also continuous. The extraction experiment proved the co-continuous structure of the sample T_0S_0, which was in accordance with the SEM result. For the crosslinked TPVs, the continuities of all the samples were less than 40%, and ranked in this order: $T_0S_5 > T_1S_0 > T_3S_0 > T_1S_2 > T_5S_0 > T_1S_4$. As both SEBS and PP were soluble in boiled xylene, all the noncrosslinked SEBS and PP could be extracted. The gel contents of the TPVs were between 40% and 70%, and ranked in this order: $T_1S_4 > T_5S_0 > T_1S_2 > T_3S_0 > T_1S_0 > T_0S_5$, which were contrary to the continuity results. The swelling degree was used to evaluate the average molecular weight between crosslinks. The swelling degree of the TPVs ranked in the same order as the continuity results, while the v_e ranked in the reverse order. Specially, the value of v_e of sample T_0S_5 was approximately half of other TPVs, which indicated that SBS was not an efficient crosslinking coagent in the absence of TMPTMA and the crosslinking peak (Figure 1) was not as recognizable as other samples.

The results of continuity, gel content and v_e closely related to each other. Take the sample T_1S_4 for example, it owned the maximal gel content and v_e among all the TPVs, which meant it had the maximal crosslinking degree. Considering the results of continuity, gel content and v_e of all the samples, it could be concluded that the solvent-resistance ranked in this order: $T_1S_4 > T_5S_0 > T_1S_2 > T_3S_0 > T_1S_0 > T_0S_5 > T_0S_0$. The results indicated that TPVs showed much better solvent-resistance than noncrosslinked SEBS/PP blend. Besides, TMPTMA and SBS showed synergistic effect on improving the solvent-resistance, and the simultaneous addition of TMPTMA and SBS supplied the TPVs better solvent-resistance.

Table 3. Continuity, gel content (GC), swelling degree (SD) and overall crosslinking density (v_e) of the samples

Sample	Continuity (%)	GC (%)[1]	SD	v_e (10^{-4} mol/mL)
T_0S_0	98.6	0	-	-
T_1S_0	33.6	48.1	4.19	1.01
T_3S_0	23.0	50.4	4.09	1.09
T_5S_0	21.8	59.9	3.87	1.29
T_1S_2	22.6	59.6	4.06	1.11
T_1S_4	18.0	66.4	3.83	1.34
T_0S_5	38.8	40.6	5.15	0.54

[1]Naphthenic oil was excluded

Crosslinking mechanism of the TPVs

The mechanism of coagent assisted peroxide crosslinking had been a topic of debate for many years. It was well known that the dynamic vulcanization included the radical generation, the transfer of free radicals and the crosslinking reaction[3]. It was clear that the reactions occur in a very complex way, and the key issue was the mechanism of the crosslinking reaction. To further simplify the dynamic vulcanization, we proposed a mechanism of the crosslinking reaction (Scheme 1). To highlight the motion of the polymer chains during the crosslinking reaction, the polymer chains of PP/SEBS blends before crosslinking were expressed as straight lines. As TMPTMA was exclusively used as crosslinking coagent, the polymeric radical with TMPTMA formed, and rapidly coupled with another radical. However, as the crosslinking reaction proceeded and the network formed, the mobility of polymeric chains reduced and the reaction gradually slowed down. Thus the sample T_1S_0, T_3S_0 and T_5S_0 took much longer time to accomplish the crosslinking reaction than the sample T_1S_2 and T_3S_0 (Figure 1A and Table 2). Meanwhile, for the TPVs with solely SBS as crosslinking coagent, SBS chains within the network could still move, and the vinyl groups in SBS further coupled with the nearby chains. However, the molecular number of SBS was far less than that of TMPTMA under the same dosage, which resulted in much less formation of polymeric radical with SBS in the radical transfer reaction. Thus, the crosslinking density of sample T_0S_5 was lower than that of other TPVs (Table 3). To overcome the shortage of TMPTMA and SBS, TMPTMA and SBS were added simultaneously as the coagent. On the one hand, TMPTMA could form plentiful polymeric radical with coagent in the radical transfer reaction. On the other hand, the movable SBS chains could keep the reactivity as the network formed, consequently promote the crosslinking reaction. Eventually, simultaneous addition of TMPTMA and SBS as crosslinking coagent could accelerate (Table 2) and deepen (Table 3) the crosslinking reaction than sole loading of TMPTMA or SBS.

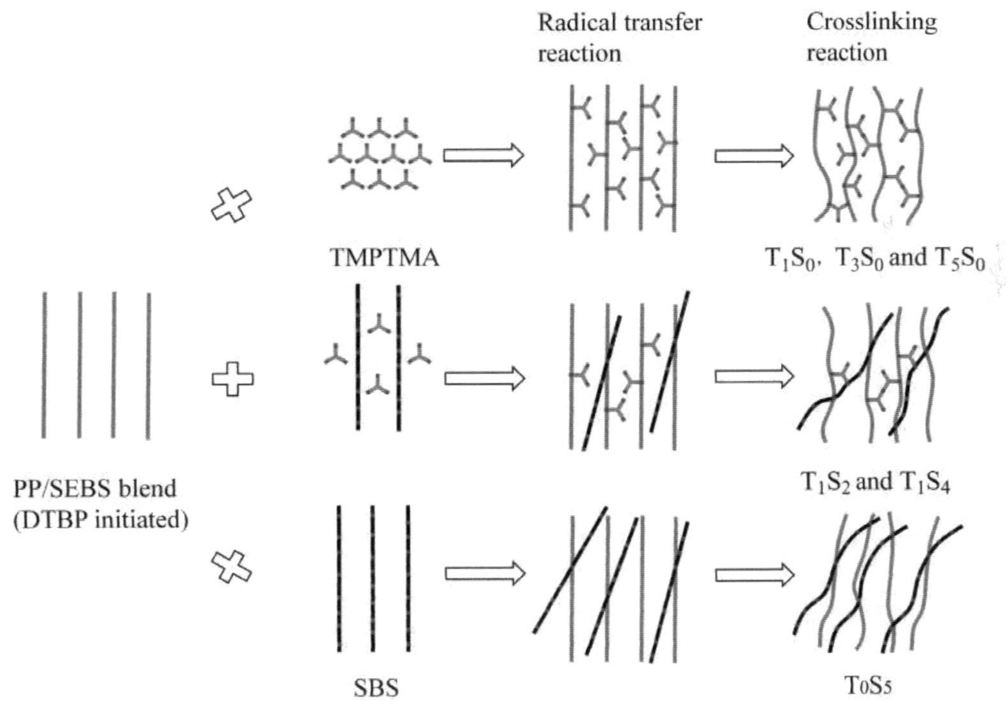

Scheme 1. Mechanism of the crosslinking reaction in the presence of TMPTMA and (or) SBS (red spot represented the unsaturated bond).

Mechanical properties of the TPVs

The tensile properties and hardness of these TPVs were listed in Table 4. For the SEBS/PP blends with the sole addition of TMPTMA (T_1S_0, T_3S_0, and T_5S_0), modulus at 100% ($E_{100\%}$) and hardness increased gradually with the increase of TMPTMA loading, while the elongation at break (ε_b) and compression set (CS) had the opposite tendency, which was ascribed to the increased crosslinking degree (Table 3). For other SEBS/PP blends (T_1S_2, T_1S_4, and T_0S_5), $E_{100\%}$ and hardness also increased with the increased crosslinking degree. Besides, ε_b of all the samples with SBS were more than 330%, and were larger than those of the samples with the sole addition of TMPTMA. As described in the mechanism part, the SBS chains had much better extensibility than TMPTMA, accordingly the samples with more SBS loading had bigger ε_b (T_1S_2 and T_1S_4). Due to the low crosslinking degree of sample T_0S_5, PP degradation still existed[3], thus the sample T_0S_5 had smaller ε_b than sample T_1S_4.

It was found from Table 4 that sample T_1S_4 had higher tensile strength (15.9 MPa) than samples T_5S_0 and T_0S_5 under the same loading of crosslinking coagent. Based on these results, TMPTMA and SBS also showed synergistic effect on tensile properties, and the simultaneous addition of TMPTMA and SBS could get TPVs with high $E_{100\%}$ and large ε_b.

Table 4. Tensile properties, hardness and compression set of the TPVs.

Sample	σ_t (MPa)	ε_b (%)	$E_{100\%}$ (MPa)	Hardness	CS (%)
T_1S_0	9.2	298	2.7	72 A	21.2
T_3S_0	11.0	286	3.3	75 A	19.4
T_5S_0	11.4	265	3.6	77 A	18.3
T_1S_2	12.1	332	3.0	75 A	19.2
T_1S_4	15.9	366	3.6	77 A	18.1
T_0S_5	10.0	338	2.5	70 A	22.3

.ACKNOWLEDGMENTS

We acknowledge the financial supports of the Natural Science Foundation of China (20974100 and 20674075) and Natural Science Foundation of Zhejiang Provence (Y404299).

REFERENCES

1. L. Yang, J. Huang, X. Lu, S. Jia, H. Zhang, G. Jin and J. Qu, *J. Appl. Polym. Sci.* **132**, 41543-41554 (2014).
2. TK. Kang and CS. Ha, *Polym. Test.* **19**, 773-783 (2000).
3. R. R. Babu, N. K. Singha and A. S. Sirqueira, *J. Appl. Polym. Sci.* **117**, 1578-1590 (2010).

Effects of coupling agents on the thermal and mechanical performance of glass fiber reinforced PA6T/66 composites

Xiaoling Luo[1], Qi Zhang[1], Meiling Zhang[2], Gang Zhang[2], Xiaojun Wang[2], Shengru Long[2] and Jie Yang[2,3*]

[1] *College of Polymer Science and Engineering, Sichuan University, Chengdu China;*

[2] *Analytical and testing Center, Sichuan University, Chengdu China;*

[3] *State Key Laboratory of Polymer Materials Engineering of China, Sichuan University Chengdu, China*

Abstract. Polyhexamethylene terephthalamide (PA6T) is a kind of semi-aromatic polyamide with excellent performance. Glass fiber (GF) reinforced composites of PA6T's random copolymer with polyhexamethylene adipamide (PA66) were prepared via melt blending method. In order to improve the inferior mechanical properties of the composites, silane coupling agent KH550 was also employed to modify the interfacial adhesion between resins and glass fibers. Differential scanning calorimetry (DSC) test results showed that with the addition of KH550 in composites, the melting temperature changed little while the melting enthalpy decreased obviously, besides, 30wt%GF caused a further decrease, which suggested a decline of the crystallization ability of the matrix. Thermogravimetric analysis (TGA) test revealed that the decomposition temperature of the composite materials had no obvious changes with different content of the GF. The initial decomposition temperature of composites contained various content of GF were about 406°C, keeping its initial excellent thermal performance well. Mechanical properties test showed when the content of GF was 30wt%, the tensile strength, flexural modulus and unnotched impact strength of composites was 96.1 MPa, 4944MPa and 18.0 kJ/m^2 respectively, increased by 123%, 171% and 100% to the resin matrix correspondingly. On the other hand, those composites with 2wt% KH550 get additional increase by 53.5%, 102% and 224% in tensile strength, flexural modulus and unnotched impact strength respectively, comparing to the PA6T/66-30wt%GF composites.

Keywords: coupling agents; glass fiber; PA6T/66; composites; performance

PACS: S06-159

1. Introduction

The polyamided have attracted a great deal of interest over the last few years due to superior properties such as high specific strength and stiffness, wear resistance, dimensional stability, low weight, and directional properties compared to conventional metallic materials 1. Especially, a new class of polymer has been successfully developed and commercialized. These are known as polyphthalamides (PPA), which are polyamides (PA) containing aromatic rings in their backbones, giving them high mechanical strength and thermal resistance. This leads to polymer materials that have a high commercial potential, competing with metals, especially in the automotive industry. The semicrystalline PPA are based on PA6T, obtained through condensation of hexamethylene diamine (HMDA) and terephthalic acid. These polyamides are generally characterized by higher melting points, higher glass transition temperatures, better chemical resistance 2 lower moisture absorption and better thermal stability compared to common aliphatic polyamides such as nylon 6 and nylon 66.

Glass fibre reinforced polymers (GFRPs) are growing steadily because these materials offer opportunities of cost and weight saving in many industrial sectors, such as automotive, furniture and home appliances. GFRPs are employed for manufacturing mass production components, which may undergo significant mechanical cyclic loading, in the framework of metal replacement 3. Glass fibre-reinforced aliphatic crystalline polyamide, such as nylon 6 or nylon 66, is an excellent reinforced thermoplastic in terms of exhibiting toughness, good heat resistance and oil resistance 4.

The properties of the fiber/matrix interphase in glass fiber-reinforced composites can play a dominant role in governing overall composite performance. As is known, the interfacial interaction can be improved through using silane coupling agents, resulting in better interfacial strength as well as the toughness of the composites 5. This work was aimed to involve silane coupling agents in the glass fiber reinforced PA6T/66, The change of the interfacial interaction, thermal and mechanical performance are realized.

2. Results and discussion

2.1. Thermal analytical studies

2.1.1 Crystallization and melting behaviour

FIGURE 1. shows the variation of the melting temperature (T_m) of PA6T/66 under different loadings of the glass fiber. With the increase of carbon fiber content, the Tm of PA6T/66-CF composites were 304.3°C, 303.3°C, 302.8°C, 299.8°C, 295.0°C and 304.4°C, respectively, the T_m values of composites were not changed significantly with increasing CF content.

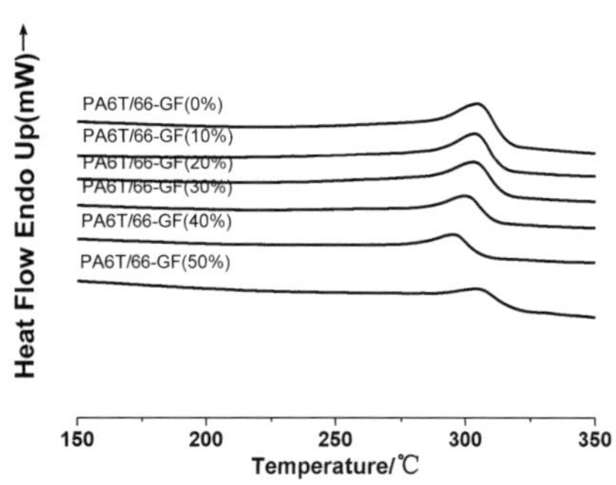

FIGURE 1. DSC curves of PA6T/66-GF composites.

The crystallization and melting behaviour of the PA6T/66 resin with different content of coupling agent KH550 were tested. The melting peak shifted to lower temperature with higher concentration of KH550. The value of T_m tabulated in TABLE 1. It demonstrated that the crystalline morphologies of the resin were slightly altered. On the other hand, the crystallization temperature (T_c) and melting enthalpy (ΔH_c) decreased obviously, the T_c and ΔH_c of PA6T/66 were 268.1°C and 26.08J/g, when the content of KH550 was 4%, the T_c and ΔH_c was 254.8°C and 19.38J/g, which indicated the blending of KH550 with polymer resin reduced the crystallizability of PA6T/66. This might be ascribed to the inhibition of the molecule chains by the small molecules of KH550.

The crystallization and melting behaviour of the PA6T/66-GF composite with different content of coupling agent KH550 were also investegated. All of the composites contained 30wt% glass fiber. TABLE 2 showed the similar phenomenon with PA6T/66/KH550. However, the PA6T/66-GF/KH550 had lower ΔH_c than that without glass fiber, which indicated the decrease of crystallizability. At this fiber content, fibers prevented the mobility of polymer matrix chains, obstructing the crystal growth in the matrix.

TABLE 1. Crystallization and melting data of PA6T/66 with different content of KH550

Sample number	T_c/ °C	T_m/°C	ΔH/ (J/g)	T_m-T_c/ °C
PA6T/66-KH550 0%	268.1	300.6	26.08	32.5
PA6T/66-KH550 0.5%	269.5	303.5	24.57	34.0
PA6T/66-KH550 1%	257.5	297.5	24.40	40.0
PA6T/66-KH550 2%	255.2	295.3	22.06	40.1
PA6T/66-KH550 3%	255.6	295.9	20.81	40.3
PA6T/66-KH550 4%	254.8	293.5	19.38	38.7

TABLE 2 Crystallization and melting data of PA6T/66-GF composites with different content of KH550

Sample number	T_c/ °C	T_m/°C	ΔH/ (J/g)	T_m-T_c/ °C
PA6T/66-GF/KH550 0%	270.8	299.8	16.92	29.0
PA6T/66-GF/KH550 0.5%	268.8	303.7	15.80	34.9
PA6T/66-GF/KH550 1%	265.3	300.0	16.32	34.7
PA6T/66-GF/KH550 2%	265.6	298.8	11.93	33.2
PA6T/66-GF/KH550 3%	269.1	301.5	12.63	32.4
PA6T/66-GF/KH550 4%	261.3	296.2	10.39	34.9

2.1.2 Thermogravimetric analysis

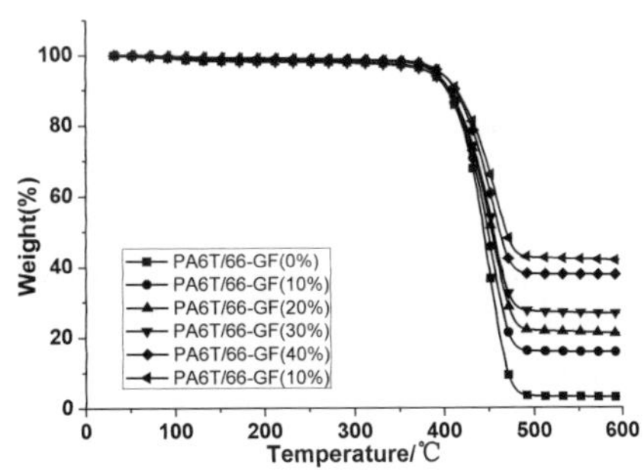

FIGURE 2. TGA thermograms of PA6T/66-GF composites (initial decomposition temperature was determined for 5% mass-loss in nitrogen).

All polymer composites demonstrated excellent thermal stability with no significant weight loss up to temperatures of approximately 300°C, as shown in FIGURE 2. The 5% weight loss temperatures and maximum decomposing temperature of PA6T/66-GF composites under nitrogen atmosphere determined by TGA are almost 406 °C and 450 °C, respectively. Combining the TGA results and above analysis of melt behavior through the DSC measurement, we can conclude the PA6T/66-GF composites held the initial excellent thermal performance of PA6T/66 well.

2.2 Mechanical properties

2.2.1 Mechanical properties of PA6T/66-GF composites without KH550

The mechanical properties of PA6T/66-GF with different content of glass fiber and the values were presented in TABLE 3. The tensile strength, tensile modulus, flexural strength, flexural mudulus and impact strength of PA6T/66 were 43.1MPa, 1834MPa, 70.9MPa, 1822MPa and 9.0kJ/m^2, respectively. When the content of GF was 30%, the tensile strength, tensile modulus, flexural strength, flexural mudulus and impact strength of PA6T/66-GF were 96.1MPa, 3663MPa, 125.3MPa, 4944MPa and 18.0kJ/m2, respectively. The tensile, flexural and impact strength increased by 123%, 77%, 100%. When the glass fiber weight was equal to the resin, the tensile strength slightly decreased reversely. But other mechanical properties increased sequentially.

2.2.2 Mechanical properties of PA6T/66-GF composites with KH550

TABLE 3 Mechanical property values of PA6T/66-CF composites.

Glass fiber content	Tensile strength (MPa)	Elastic modulus (MPa)	Flexural Strength (MPa)	Flexural modulus (MPa)	Notched impact strength (KJ/m^2)
0	43.1	1834	70.9	1822	9.0
10%	56.5	2465	89.6	3313	9.3
20%	39.3	1991	79.2	3487	11.8
30%	96.1	3663	125.3	4944	18.0
40%	72.2	3586	86.7	3417	9.9
50%	74.6	3963	189.8	8397	26.8

The effect of KH550 on the mechanical properties of PA6T/66-GF composites was shown in FIGURE 3. From TABLE 3. one can get the tensile strength and modulus of PA6T/66-GF without KH550 was 96.1MPa and 3663MPa, but with the addition of KH550, obvious and continuous improvement were obtained. The tensile strength and modulus of comosites were 147.5MPa，increased by 53.5% when the KH550 was 2%. The effect of KH550 on the flexural and impact properties of composites were similar with that of tensile properties. This was attributed to the improvement of interfacial bonding between matrix and fiber involved by KH550.

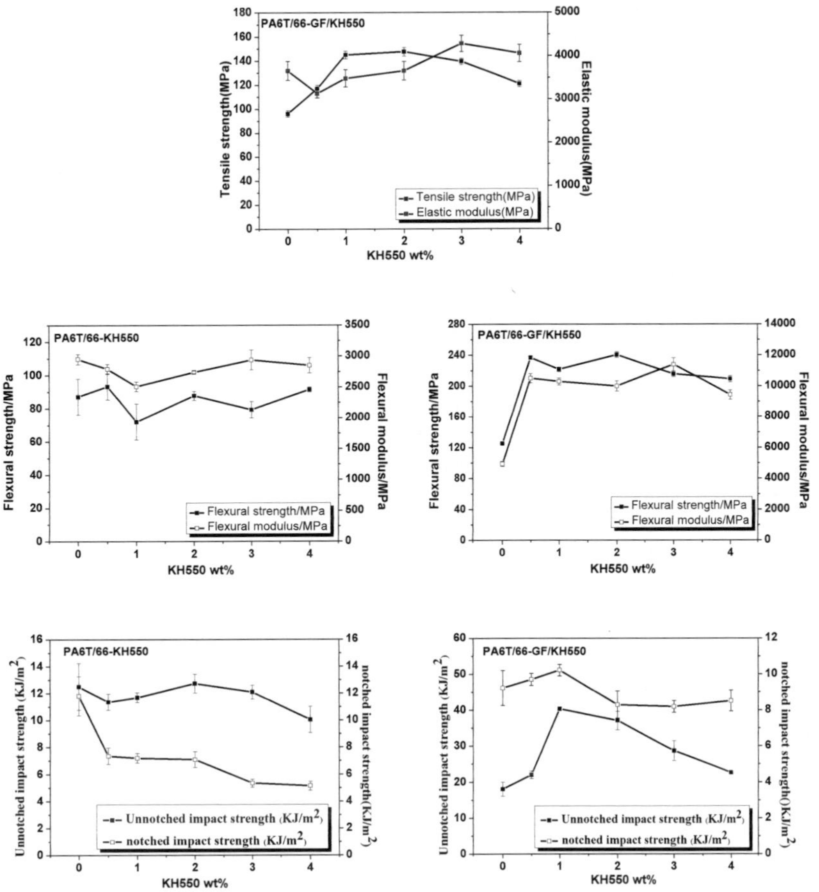

FIGURE 3. Mechanical properties of PA6T/66-GF composites with different content of KH550

REFERENCES

[1] Gaitonde VN, Karnik SR, Mata F, Davim JP. Modeling and Analysis of Machinability Characteristics in PA6 and PA66 GF30 Polyamides through Artificial Neural Network. Journal of Thermoplastic Composite Materials. 2009;23(3):313-336.

[2] Cousin T, Galy J, Dupuy J. Molecular modelling of polyphthalamides thermal properties: Comparison between modelling and experimental results. Polymer. 2012;53(15):3203-3210.

[3] Novitsky TF, Lange CA, Mathias LJ, Osborn S, Ayotte R, Manning S. Eutectic melting behavior of polyamide 10,T-co-6,T and 12,T-co-6,T copolyterephthalamides. Polymer. 2010;51(11):2417-2425.

[4] Bernasconi A, Conrado E, Hine P. An experimental investigation of the combined influence of notch size and fibre orientation on the fatigue strength of a short glass fibre reinforced polyamide 6. Polymer Testing. 2015;47:12-21.

[5] Jing Shi, Yong Wang, Yao Gao, Hongwei Bai. Effects of coupling agents on the impact fracture behaviors of T-ZnOw/PA6 composites. Composites Science and Technology. 2008;68: 1338–1347.

Antioxidative Modification for Semi-aromatic Polyamide PA6T/66

Weiming Peng[a], Xiaoling Luo[a], Meilin Zhang[b], Shengru Long[b], Xiaojun Wang[b,*] and Jie Yang[b,c,*]

[a] *College of Polymer Science and Engineering, Sichuan University, Chengdu , China*
[b] *Analytical & Testing Center, Sichuan University, Chengdu , China*
[c] *State Key Laboratory of Polymer Materials Engineering (Sichuan University),Chengdu, China*

Abstract. In the paper, five different antioxidants were adopted to improve the antioxidant capacity and to prevent color change of the semi-aromatic polyamide PA6T/66 caused by oxidation in the application. Thermal gravimetric analysis (TGA) showed that the initial decomposition temperatures of the modified plastics in air atmosphere were elevated effectively. Thermal oxidation of the suppressed films and mechanical testing samples was performed at 170 ℃ for 8h. The differential scanning calorimetry(DSC) results suggested that the melting points of the samples were all at about 310 ℃, but the crystallization temperature increased after thermal oxidation. Decrease after initial growth of the intrinsic viscosity suggested that there existed a competition between crosslinking and degradation in PA6T/66. The antioxidants inhibited the degradation, and result in a maintained high intrinsic viscosity of PA6T/66. UV-visible (UV-Vis) spectroscopy showed the chromophore absorbance at 325nm increased with increasing of oxidation time, which slowed down after adding antioxidants. No obvious change in mechanical properties was detected before and after oxidation since oxidation only happened on the surface of the samples. Comprehensive tests results suggested that the antioxidation of PA6T/66 was successfully improved with the present of antioxidants. The antioxidant effect of the organic compound was superior to the organic antioxidants alone, the effect of inorganic and organic antioxidants compound effect showed little different.

Keywords: semi-aromatic polyamide; PA6T/66; antioxidant; properties
PACS: S06-168

EXPERIMENTAL

Four groups of antioxidants, groups B to E, were dissolved in acetone, and antioxidant F was dissolved in water. Then the solution would be mixed with PA6T/66 pellets respectively, the mixture was extruded by a twin-screw extruder after fully drying, and finally moulded into films and standard mechanical samples for following test.

	Pure PA6T/66	1010	1010+S-9228	PW-9225	H161	$CuCl_2$+KI
	A	B	C	D	E	F
Antioxidant content (wt%)	0	0.5%	0.25%+0.25%	0.5%	0.5%	0.02%+0.1 %
Mass (g)	0	5	2.5+2.5	5	5	0.2+1

RESULTS AND DISCUSSION

Effect of Antioxidants on Intrinsic Viscosity

Figure 1 shows the relationship between intrinsic viscosity of PA6T/66 with different antioxidants addition and thermal oxidation time. It can be seen that intrinsic viscosity decreases with the increase of thermal oxidation time for PA6T/66 with no antioxidants addition. But for PA6T/66 with different antioxidants addition, intrinsic viscosity increases at the first stage and then it decreases. We can infer that degradation and crosslinking both exist during the thermal oxidation process. At the beginning of thermal oxidation, antioxidant can effectively suppress the degration so that crosslinking plays a main role. With thermal oxidation time increases, antioxidant is expended and degration turns to play a main role. Figure 1 also shows that intrinsic viscosity of PA6T/66 with antioxidant is larger than that of pure PA6T/66. And for the antioxidant effect, the effect of phenolic antioxidant used alone is weaker than that of antioxidants used together, at the same time, inorganic antioxidant also has a good effect on oxidation resistance.

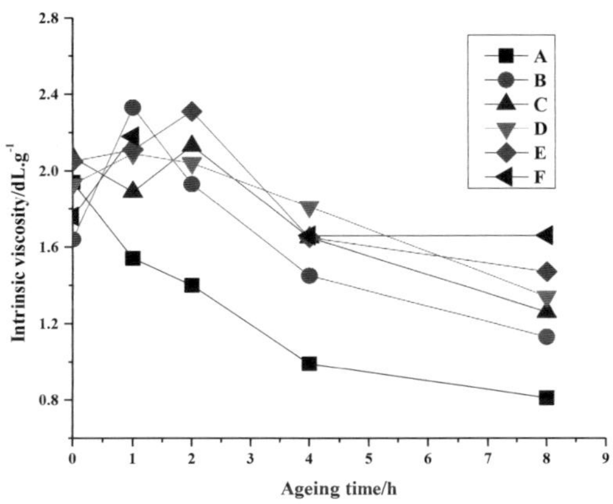

FIGURE1. Intrinsic viscosity with different treatment time at 170℃

Effect of Antioxidants on Thermal Properties

The differential scanning calorimetry(DSC) results show that the melting points of the samples were all at about 310 ℃, there is no obvious change with the increase of thermal oxidation time, as is shown in figure 2. But the crystallization temperature increased after thermal oxidation, the maximum crystallization temperature is 255 ℃ at the beginning and it increases to 270℃ after thermal oxidation for 8h, as is shown in figure 3. Thermal gravimetric analysis (TGA) showed that the initial decomposition temperatures of the modified plastics in air atmosphere were elevated effectively, as is shown in table 2, the initial decomposition temperatures in air atmosphere is almost 50℃ lower than that in N2 atmosphere. What' s more, little change of the initial decomposition temperatures for different samples occurs in N2 atmosphere. But in air atmosphere, the initial decomposition temperatures of the modified plastics in air atmosphere were elevated effectively, in which group D,E and F have relatively better effect for the increase of 11.6℃,12.1℃ and 15.3℃.

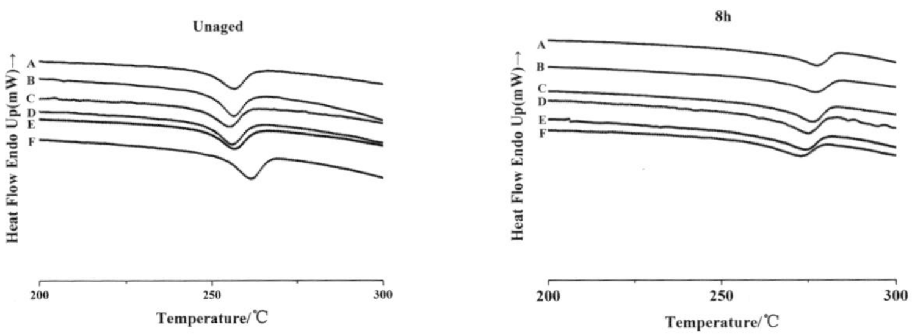

FIGURE2. Crystallization temperature of samples before and after oxidation

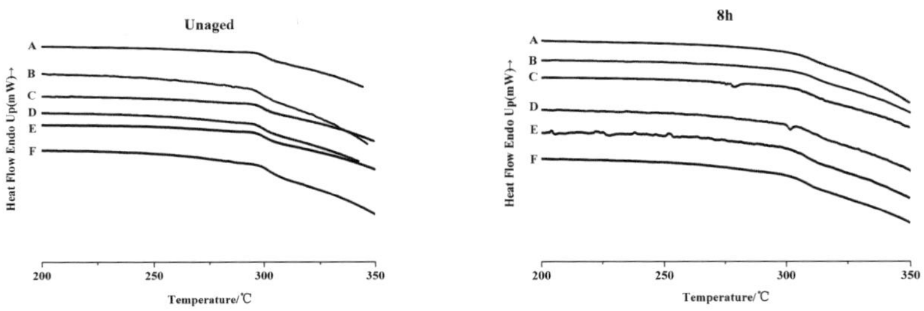

FIGURE3. Melting point of samples before and after oxidation

TABLE(2). Decomposition temperature of samples

Atmosphere	**N_2**			**Air**		
	Initial decompositi-on temperature/℃	Maximum decompositi-on temperature/℃	Residual mass%	Initial decompositi-on temperature /℃	Maximum decompositi-on temperature /℃	Residual mass %
A	392.4	460.6	5.16	341.7	452.9	7.51
B	391.4	461.2	5.7	341.6	455.6	6.95
C	387.3	451.6	4.37	347.3	454.2	5.27
D	390.8	450.6	5.81	353.1	449.1	7.7
E	383.8	451.5	6.05	353.8	444.9	7.44
F	388.5	448.5	5.11	357	447.7	6.14

Effect of Antioxidants on Mechanical Properties

From figure 7 and figure 8, it can be seen that no obvious change in stretch and flexure properties was detected before and after oxidation for 8h since oxidation only happened on the surface of the samples. It is different from films which will be obviously embrittled after oxidation, there is only color change on the surface for samples of mechanical test. Figure 9 shows that unnotched impact strength after treatment for 8h decreases, but the opposite result occurs in notched impact strength. We can infer that the notch sensitivety decreases after oxidation. Besides, the mechanical properties of pure PA6T/66 is close to that of antioxidative PA6T/66, so we infer that mechanical properties will not be influenced within short thermal oxidation time with the addition of antioxidant.

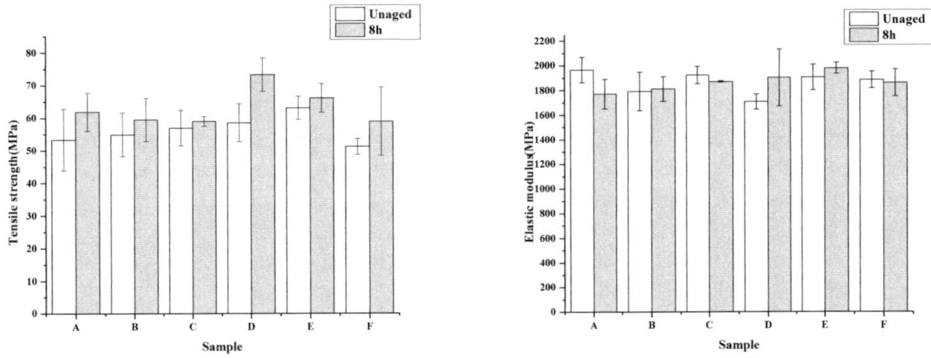

(a) (b)

FIGURE4. Effect of antioxidants on the tensile properties of samples before and after oxidation (a)stretch strength, (b)stretch modulus

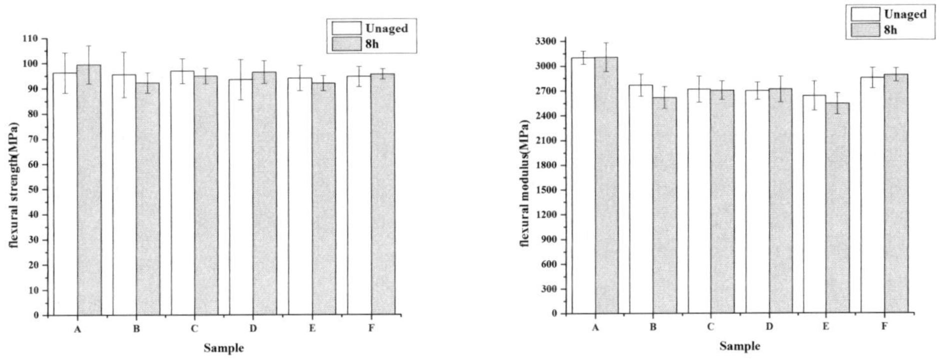

(a) (b)

FIGURE5. Effect of antioxidants on the flexural properties of samples before and after oxidation (a)flexural strength, (b)flexural modulus

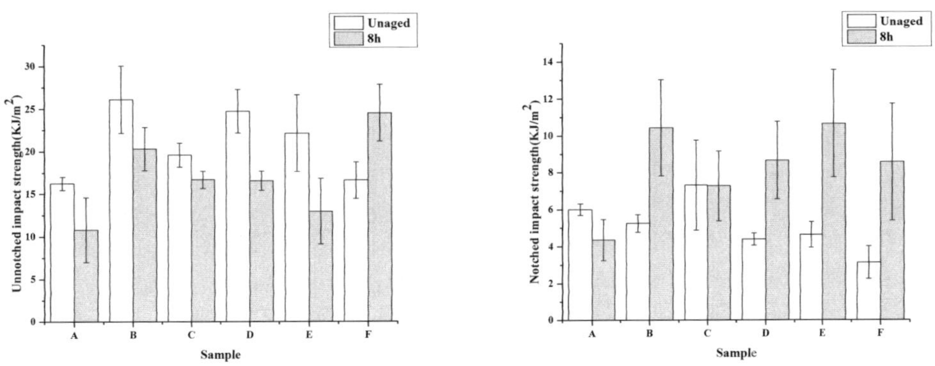

(a) (b)

FIGURE5. Effect of antioxidants on the impact properties of samples before and after oxidation (a)unnotched impact strength, (b) notched impact strength

CONCLUSION

In the paper, five different antioxidants were adopted to improve the antioxidant capacity and to prevent color change of the semi-aromatic polyamide PA6T/66 caused by oxidation in the application. Thermal gravimetric analysis (TGA), The differential scanning calorimetry (DSC), UV-visible (UV-Vis) spectroscopy and mechanical properties test were applied to analyse antioxidative PA6T/66, the following conclusions were obtained. Samples would be degraded in air atmosphere, and the antioxidation of PA6T/66 was successfully improved with the present of antioxidants. The antioxidant effect of the organic compound was superior to the organic antioxidants alone, the effect of inorganic and organic antioxidants compound effect showed little different. Mechanical properties will not be influenced within short thermal oxidation time with the addition of antioxidant.

REFERENCES

1. Klaus-juergen. Lead-free soldered MID assembly made from high-temperature polyamide[J]. Engineering Plastics, 2005, 9,1-4
2. Watanabe K， et al. Production of crystalline polyamide molded product： JP， 5104650A ［P］.1993-04-27.
3. Janssen,K,Gijsman,P,Tummers,D. Polymer Stability andDegradation,1995,(49):127
4. McIntosh, Stanler A,et al. Spinning and stability of solution-dyed nylon fibers [P]. US, US 6 358 458, 2002
5. La Mantia F P, Scaffaro R. Polymer Degradation and Stabili-ty, 2002,(75):473～477
6. Beer Ludwig,Matthies Paul,Mell Karlheinz, et al. Inherentlylight-and heat-stabilized polyamides [P]. WO, WO 9 528443,2002
7. B1Lanska, et al Eru1Polym1J11986, 22 (8):687~689

Effects of Injection Molding Process Parameters on the Mechanical Performance of Long Glass Fiber Reinforced PPS Composites

XiangZhang[a], Gang Zhang[b], XiaojunWang[b], ShengruLong[b] and JieYang[b,c*]

[a] *College of Polymer Science and Engineering, Sichuan University, Chengdu China;*
[b] *Analytical and testing Center, Sichuan University, Chengdu China;*
[c]*State Key Laboratory of Polymer Materials Engineering, Sichuan University Chengdu, China*

Abstract. In this paper, the injection processing parameter of long glass fiber reinforced polyphenylene sulfide(PPS/LGF) composite was optimized. Then composite with different fiber content was prepared and the effect of glass fiber on the mechanical properties was evaluated. It was found that when the temperature of the first heating section was higher than the melting point of PPS, the mechanical performance of the composite was significantly improved. When the temperatures of the heating sections were 315°C, 320°C and 315°C, optimum comprehensive mechanical property of the composite was obtained. Under this situation, tensile strength, flexural strength and flexural modulus nearly linearly increased with the increase of fiber content. The tensile, flexural and impact strength of the composite with 40 wt% fiber were 154MPa, 250MPa, 13KJ/m^2 respectively. In addition, the results of Dynamic Mechanical Analysis(DMA)showed that storage modulus and loss modulus of the composites obviously increased with the increase of fiber content under the optimized injection molding temperatures.

Keywords: Polyphenylene Sulfide; Long glass fiber; Injection molding; Composite.

INTRODUCTION

The ever-growing use of reinforced thermoplastics in various industrial applications has led to a demand of always-higher mechanical performances for injection-moulded parts. However because of the low residual fiber length after processing, mechanical properties are often limited when short fiber reinforced thermoplastic pellets are used. In order to overcome this limitation, long fiber thermoplastic (LFT) pellets have been developed so as to answer these new market requirements leading to higher mechanical properties. The improvement of mechanical properties however also depends on the homogeneity and isotropy of the injection-moulded plastic parts, which are governed by the fiber distribution and orientation mechanisms, and on the capacity of the processing technologies to limit fiber breakage.

In the case of long fiber reinforced thermoplastic injection moulding, even if long fibers sometimes behave as short fibers, the fiber interactions are more complicated due to the higher fiber length and the reinforcement tangling. Even if a multilayered structure is kept, an increase of the core layer thickness with increasing fiber length is observed, thus suggesting a different mechanism. Bijserbosch and Gaymans have noticed for example a long fiber orientation mechanism similar to that observed for short fibers, which depend on the flow behavior of the polymer melt during the filling and packing stages on the injection process with a great influence on the gate geometry.

Some authors have investigated the effects of injection moulding processing conditions on the long fiber degradation. All processing parameters that decrease the shear stresses in the melt generally decrease the fiber degradation (high melt temperature, high flow rate, low back pressure etc). However, Hafellner et al. have noticed an improvement of the residual fiber length with a short plastication time for a long fiber reinforced polypropylene.

Most of the studies carried out up to now don't describe the influence of the temperature of feeding section.so it is important to investigate the process parameters.

EXPERIMENT AND RESULT

First, Long glass fiber reinforced PPS composites were produced and the contents of fiber were 40%. Then, the injection processing parameters of composites were investigated. The temperatures of the heating sections used are described in table 1.

TABLE (1). Heating temperatures

Group	Third section/°C	Second section/°C	First setion/°C
1	325	325	260
2	315	320	315
3	320	325	320
4	325	330	325

The mechanical properties of the parts with different heating temperatures were investigated. The mechanical properties are shown in fig 1. It can be seen that the mechanical property of group 1 was lower than the mechanical properties of group 2, 3 and 4. The reason of the situation is that the interaction of fiber is more low due to the resin melt early when the temperature of feed section is higher than the melting point of resin, so fiber can retain greater length and the parts have higher mechanical property.

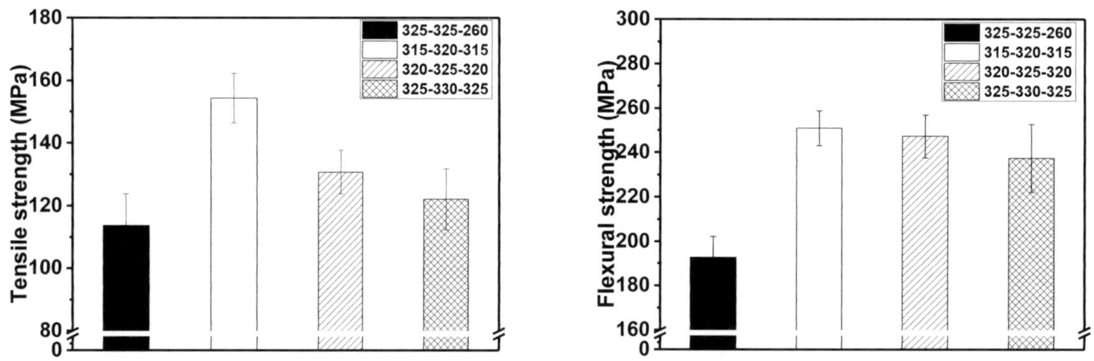

(a) (b)

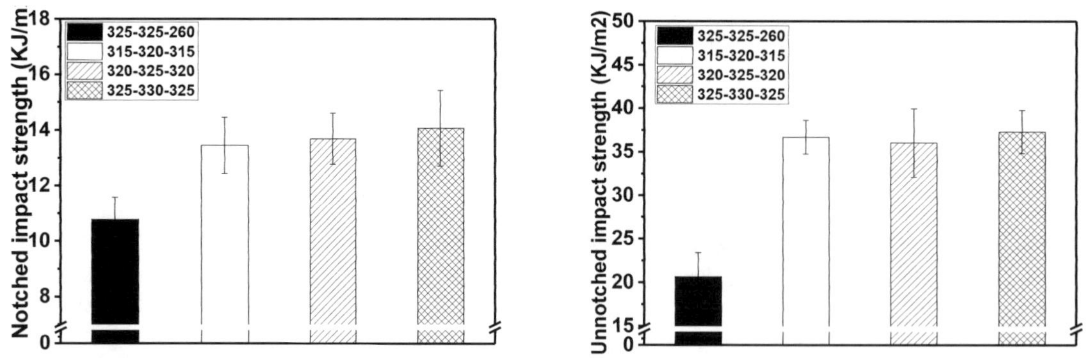

(c) (d)

FIGURE 1. Mechanical property of the composite in different temperatures: (a) Tensile strength; (b) Flexural strength; (c) Notched strength; (d) Unnotched impact strength

Comparing group 2, 3 and 4 separately, impact strength have no obvious change and the strength of tensile and flexural decrease with increase of the temperature. In conclusion, the comprehensive mechanical property of group 3 was best and tensile strength, flexural strength, notched impact strength and unnotched impact strength of the composite were 154MPa, 250MPa, 13KJ/m2, 36.68KJ/m2 respectively.

Injection moulding of composites with different fiber contents were prepared under the optimized injection molding temperatures. The mechanical property of the composites can be seen in fig 2. It can be seen that the tensile strength, flexural strength and flexural modulus nearly increased with the increase of fiber content linearly.

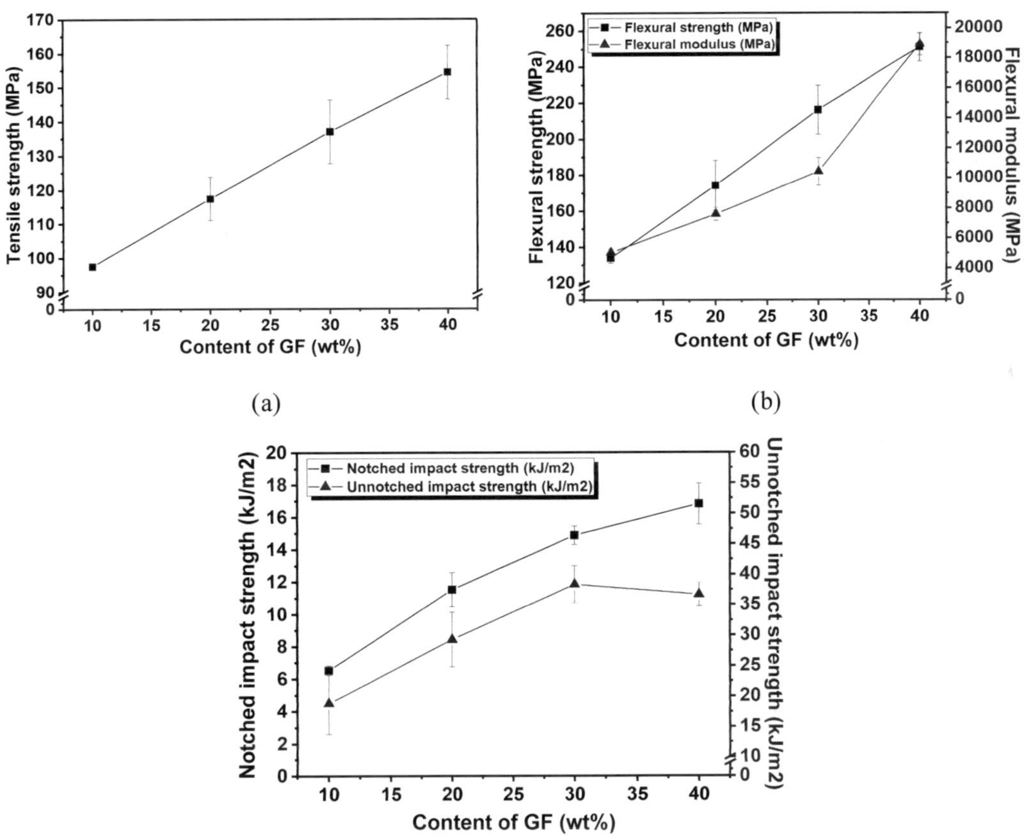

(a) (b)

(c)

FIGURE 2. Mechanical property of composites with different fiber content under optimized injection molding temperatures: (a) Tensile strength; (b) Flexural strength and modulus; (c) Notched and Unnotched impact strength

The results of Dynamic Mechanical Analysis(DMA) in fig 3 showed that storage modulus and loss modulus of the composites obviously increased with the increase of fiber content under the optimized injection molding temperatures.

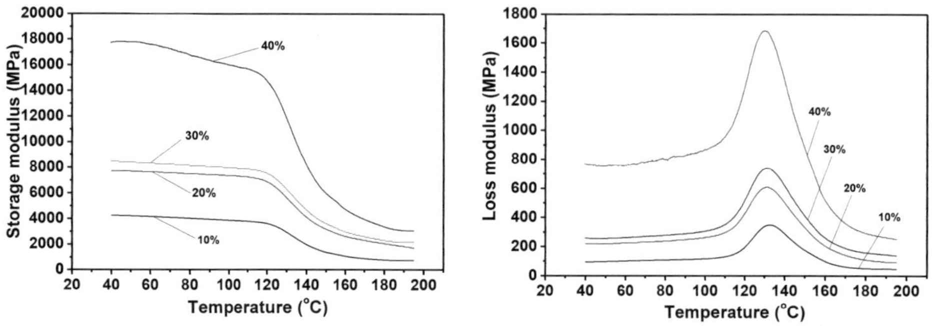

(a) (d)

FIGURE 3. Dynamic Mechanical Analysis of the composites with different fiber content under optimized injection molding temperatures: (a) Storage modulus; (b) Loss modulus

CONCLUSION

In conclusion, increasing the temperature of feeding section can efficiently improve the mechanical property of long glass fiber reinforced PPS composites. The optimizing temperatures were $315^{O}C$, $320^{O}C$ and $315^{O}C$.

REFERENCES

1. G´erard, P. Ph.D. Thesis, University of Lille I - Ecole des Minesde Douai (F), 1998.
2. Bailey, R.; Rzepka, B. Int Polym Process 1991, 6(1), 35–41.
3. Bijsterbosch, H.; Gaymans, R. J. Polym Compos 1995, 16(5),363–369.
4. Hafellner, R.; Picher, M.; W¨orndle, R. Kunststoffe 2000, 90(3),44–48.
5. Metten, M.; Cremer, M. Kunststoffe 2000, 90(1), 80–83.

Preparation and mechanical properties of glass fiber cloth reinforced poly (Arylene Sulfide Suofone) laminate via solution impregnation.

Yang Jiacao[a], Long shengru[b], Zhang Gang[b], Wang Xiaojun[b], Yang Jie[b,c,*]

a College of Polymer Science and Engineering, Sichuan University, Chengdu, China

b Analytical and Testing Center, Sichuan University, Chengdu, China

c State Key Laboratory of Polymer Materials Engineering, Sichuan University, Chengdu, China

Abstract. Poly (arylene sulfide sulfone) (PASS) is the thermoplastic engineering resin with excellent integrated performance. However, PASS have high melt viscosity and is prone to be oxidized at high temperature. It means that the melt processing is difficult for PASS. In this paper, glass fiber reinforced PASS (PASS/GFC) laminates were prepared via solution impregnation. The impregnation condition was evaluated via SEM observation of prepregs of different impregnation parameter. The results show that the optimum processing parameter may be 130 °C, 5min, 576 Pa for temperature, time and pressure, respectively. The prepregs of optimum parameter were molten and laminated, and the mechanical properties of laminate with different glass fiber content were tested. The tensile strength and impact strength increase with the increasing content of glass fiber, reach 221 MPa and 74.3 kJ/m^2, at 75 wt% glass fiber content, respectively. The flexural strength reaches its peak 348 MPa, at 65 wt% glass fiber content.

Keywords: Poly (arylene sulfide sulfone); glass fiber cloth; solution impregnation;

PACS: S06-160

Poly (arylene sulfide sulfone) (PASS) is one of the modified species of Poly (phenyl sulfide) (PPS), which have the good integrated performance of PPS. Moreover, the introduction of sulfone group in the backbone endue PASS with higher thermal resistance and toughness. Yielding the great properties listed above, PASS and its fiber reinforced composites have great potential application in area of aerospace, vehicle, military industry and machine [1]. However, PASS is prone to be oxidized at high temperature and have very high melt viscosity [2]. It is difficult for PASS melt to impregnate the fiber yarn. Studies show that the impregnation condition of PASS/GFC laminates is poor and the mechanical performance is weak. Fortunately, the sulfone group guarantee the solubility by particular solvents, such as N-methyl-2-pyrrolidone (NMP), 1,3-Dimethyl-2-imidazolidinone (DMI), which makes it possible to prepare the laminate via solution impregnation [3]. In this study, the optimum processing parameter were investigated and the mechanical properties of laminates were studied.

THE INVESTIGATION OF OPTIMUM IMPREGNATION PARAMETER

In this study, the optimum processing parameter were investigated and the mechanical properties of laminates were studied.

The PASS/NMP solution with 25 wt% PASS resin were prepared, and cooled to 90 °C, 110 °C, 130 °C, 150 °C and 170 °C , respectively. The temperature of hot plate of hydraulic machine was set at corresponding temperature during impregnation. Considering that the viscosity of PASS/NMP solution is very low, the impregnation was completed in 5 min with no extra pressure except the gravity of mould, about 576 Pa. The prepregs were dipped into boiled water 4 hour to remove solvent. The fracture surface of prepregs of different impregnation temperature was shown in Fig. 1.

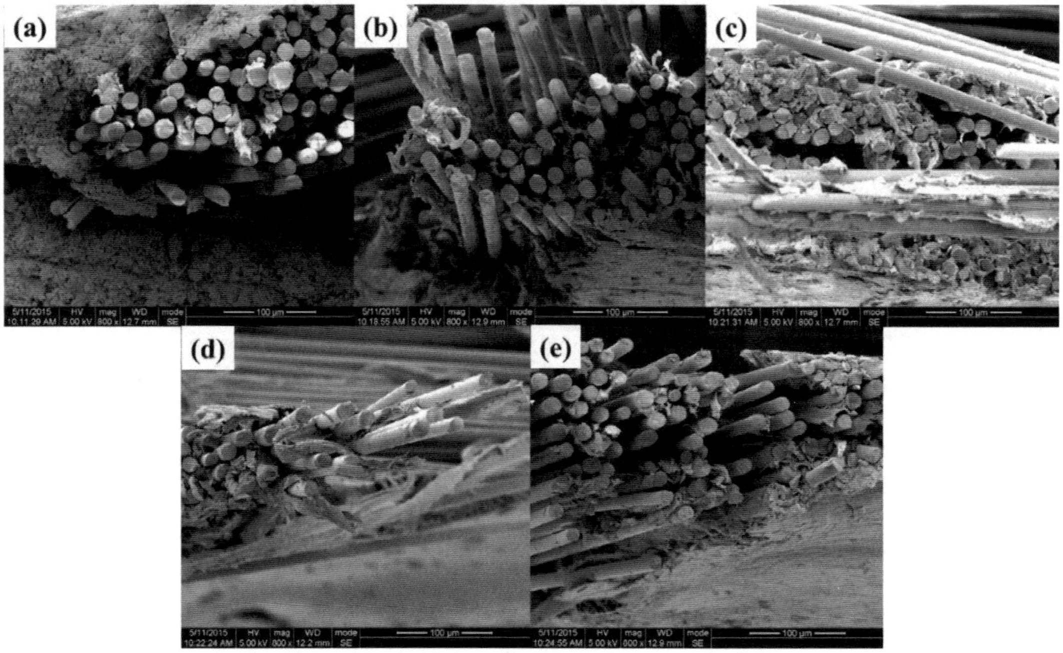

FIGURE 1. Impregnation condition of prepregs of different impregnation temperature under SEM

(a) 90 °C; (b) 110 °C; (c) 130 °C; (d) 150 °C; (e) 170 °C

As displayed in Fig. 1, the impregnation condition of prepregs show a strong dependence on impregnation temperature. When the temperature is below 110 °C, the viscosity of solution is high and of poor fluidity, which limits the impregnation of resin solution. When the temperature is above 150 °C. The viscosity is so low that the thickness of solution is small which cannot cover the thickness of glass fiber cloth. 130 °C is moderate, thus the viscosity of solution can cover the glass fiber and penetrate through the fiber bundle.

The effect of impregnation was also studied. The temperature was set at 130 °C, and the pressure 576 Pa. The impregnation lasted for 3 min, 5 min and 7 min respectively. The fracture surface of prepregs of different impregnation time was observed under SEM, and the images were shown in Fig. 2.

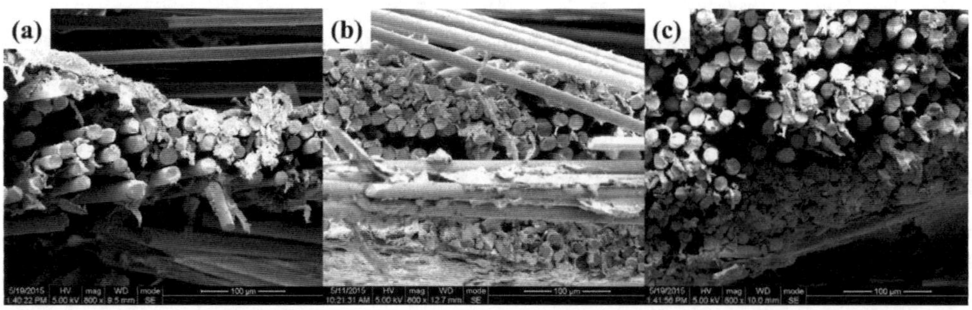

FIGURE 2. SEM image of fracture surface of prepregs with different impregnation time

(a) 3 min; (b) 5 min; (c) 7 min

As shown in SEM image, the impregnation comdition of 5 min and 7 min is almost equal, and better than 3 min. the reason may be that the flow of resin in glass fiber bundle is not complete in 3 min.

Taking efficiency into account, the 5min was selected as the optimum impregnation time.

The studies above reveal that the optimum impregnation parameter may be 130 °C, 5 min, 576 Pa for impregnation temperature, time and pressure, respectively.

MECHANICAL PROPERTIES OF PASS/GFC LAMINATE

The mechanical properties of laminates were also studied. The prepregs were prepared with optimum processing parameter and molten and consolidated into laminate. The PASS films were sandwiched between the prepregs to adjust the content of PASS resin in laminate via controlling the amount of PASS film and laminate with 45 wt%, 55 wt %, 65 wt% and 75 wt% glass fiber content were prepared with 325 °C, 5MPa, 15min, following a pre-heat with no pressure of 5min.

The tensile, flexural and impact properties of laminates were tested according to GB/T 1040-1992, GB/T 9341-2000, 1451-2006, respectively. And the results are displayed in Fig. 3, 4 and 5.

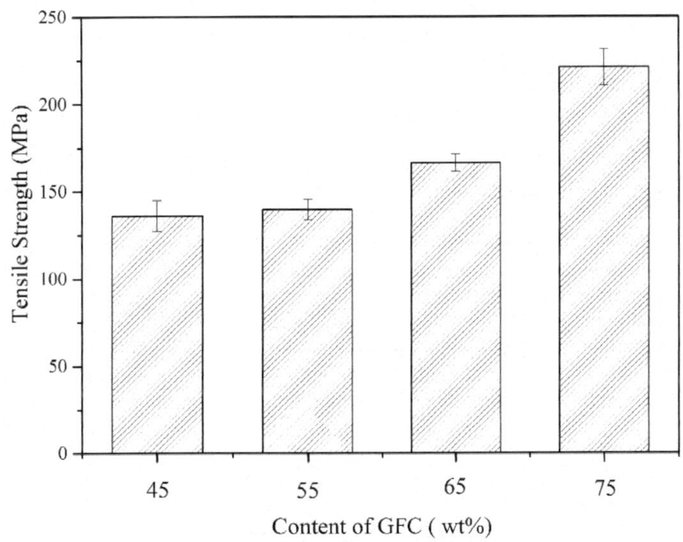

FIGURE 3. Tensile strength of laminate with different glass fiber content

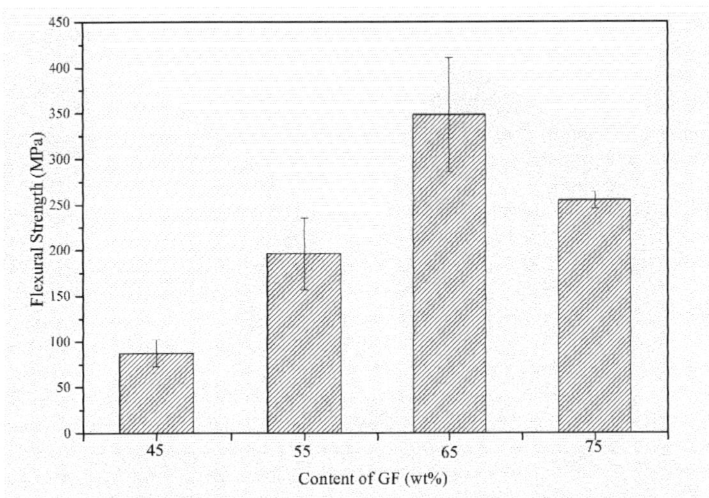

FIGURE 4. Flexural strength of laminate with different glass fiber content

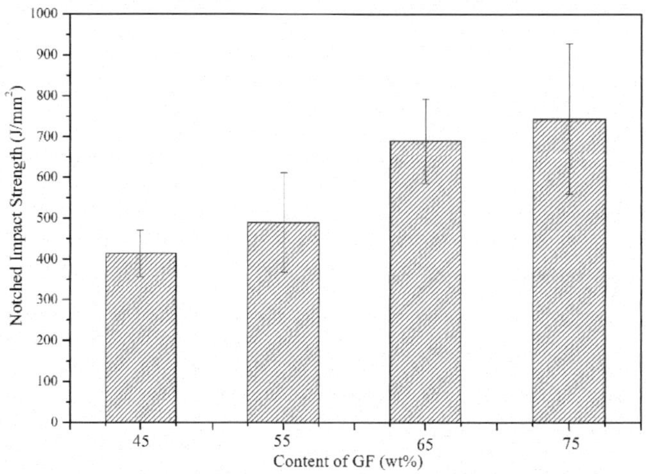

FIGURE 5. Notched impact strength of laminate with different glass fiber content

The test results show that the tensile and impact strength increase with the increasing of glass fiber content. The laminate with 75 wt% glass fiber cloth yield the biggest tensile strength and impact strength, 221 MPa and 74.3 kJ/m^2, respectively. And the flexural strength increase with glss fiber content under 65 wt% glass fiber content and reduces when over 65 wt%.

The main effect of resin in continuous fiber reinforced composite is to fix the fibers and transmit the stress among fibers. The fiber bear the majority of load. The different role of glass fiber and matrix resin lead to the different trend between tensile and impact strength and flexural strength. The fiber can bear the remarkable load in the fiber direction, as in tensile condition. The more fiber content, the higher tensile strength. However, it is the matrix resin that bear the flexural load. A proper amount of glass fiber can reinforce the matrix resin so increase the flexural strength, when above the proper amount, the amount of resin is not enough to transmit the stress and the flexural strength is deteriorated. As to impact strength, the increase of fiber content can increase the interface between matrix and fiber, which would consume a large amount of energy when damaged, thus the impact strength is increased.

REFERENCES

1. J. Yang, The high performance structure aterial-polyarylene sulfide [J]. Engineering Plastics Application 4 (2003): 020
2. G. Huang and S. Zhang, Facile synthesis of processable aromatic polyamides containing thi-oether units. Polymer International. 2013, 62: 411-418
3. J. Yang, Study on the synthesis, structure and performance of high molecular weight polyarylene sulfide sulfone Ph.D. Thesis, Sichuan University, 2005

An Universal Method to Fabricate Polymer Composites with High Thermal Conductivity at Low Filler Loading

Jinrui Huang[a*], Yutian Zhu[b], Wei Jiang[b], and Xiaoan Nie[a]

[a] *Institute of Chemical Industry of Forest Products, Chinese Academy of Forestry, 16 Suojin Wucun, Nanjing 210042, Jiangsu Province, China*
[b] *Changchun Institute of Applied Chemistry, Chinese Academy of Sciences, 5625 Renmin Street, Changchun 130022, Jilin Province, China*

Abstract. For conventional thermally conductive polymer composites (TCPCs), very high loading of conventional thermally conductive fillers (> 30 vol%) is usually needed to obtain appropriate thermal conductivity, which reduces product quality through diminished mechanical properties, increased weight, reduced flexibility, increased fragility, and increased the difficulty and cost of production. Thus, a method of TCPC fabrication with high thermal conductivity at low filler loading is needed. In the current study, adsorption-desorption of polymer chains on the graphene (GE) surface was used to trap GE at the interface of an immiscible poly(ε-caprolactone) (PCL)/poly(lactic acid) (PLA) blend. The resulting composite exhibits remarkably high thermal conductivity due to the graphene sheets being controlled at the interface of the polymer blend. At 0.53 vol%, when a good co-continuous structure is formed, the thermal conductivity of the graphene composite is nearly 4 times higher than that of the pure PCL/PLA blend with co-continuous structure. Moreover, the blend achieves an extremely low thermal percolation threshold (0.11 vol%), the lowest to date, because most of the graphene sheets are selectively located at the interface of the blend with co-continuous structure.

Keywords: Graphene, Polymer Blend, Interface, Thermal conductivity.
PACS: 65

In recent years, thermally conductive polymer composites (TCPCs) used for heat removal have received a lot of attention. However, for conventional TCPCs, a very high loading of thermally conductive fillers (generally > 30 vol%) is needed to obtain an appropriate thermal conductivity, which reduces product quality through diminished mechanical properties, increased weight, reduced flexibility, and increased fragility [1]. Thus, a method of TCPC fabrication with high thermal conductivity at low filler loading is needed.

In the current study, adsorption-desorption of polymer chains on the grapheme (GE) surface was used to trap GE at the interface of an immiscible poly(ε-caprolactone) (PCL)/poly(lactic acid) (PLA) blend with the expectation of fabricating polymer composites with high thermal conductivity at low GE loading [2-5], as shown in Figure 1.

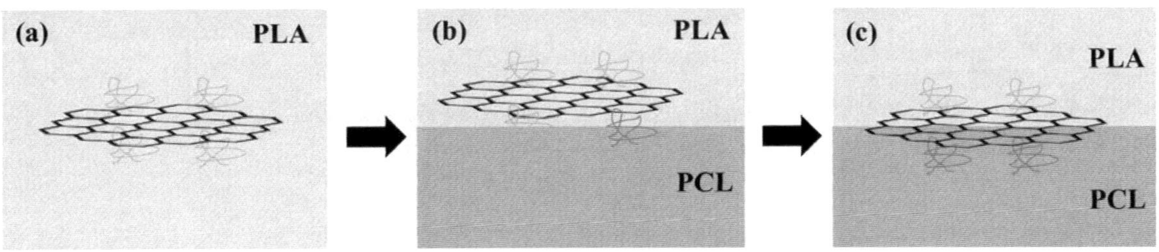

FIGURE 1. Scheme of the adsorption-desorption process of polymer chains on the GE surface for tapping GE at the interface of the PCL/PLA blend.

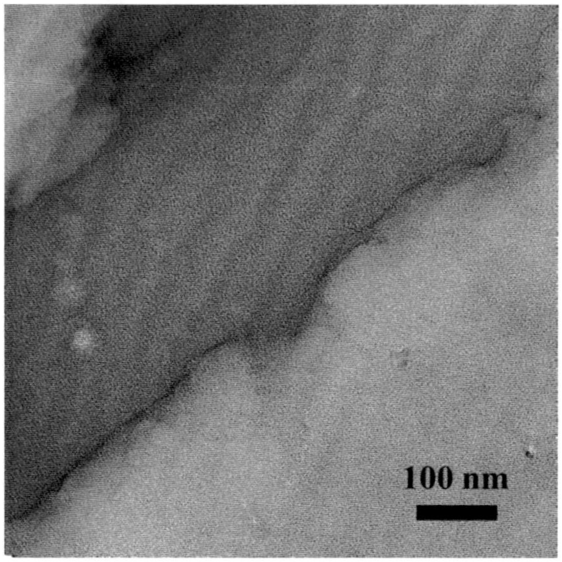

FIGURE 2. TEM image of PCL/PLA/0.053 vol% GE composite (PCL/PLA = 50w/50w). PLA phase: bright; PCL phase: gray region. The black lines are GE sheets.

The selective distribution of GE in the composites was visualized using TEM (Figure 2). From the figure, it can be seen that the GE sheets were aligned along the co-continuous interface between the PCL and PLA phases [5]. Thus, GE can be trapped at the interface through adsorption-desorption of polymer chains on the GE surface.

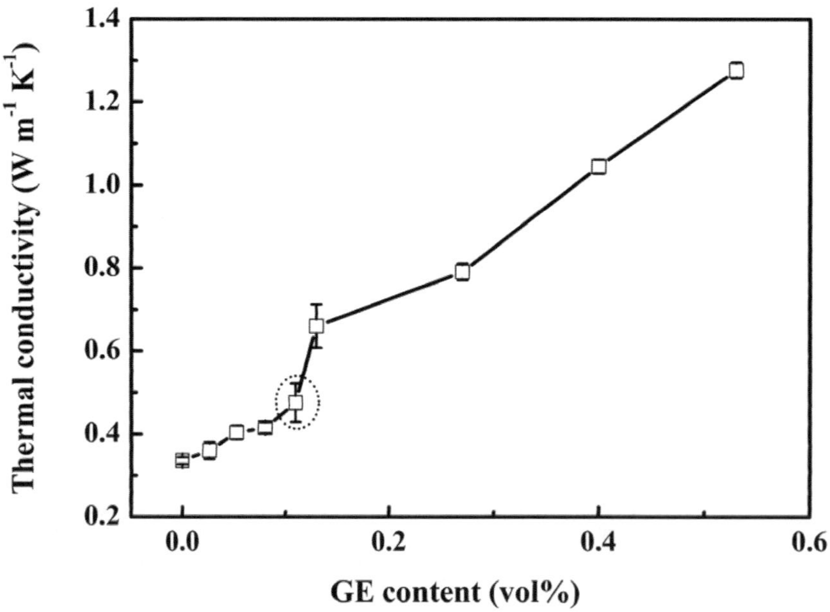

FIGURE 3. Thermal conductivities of the PCL/PLA/GE (PCL/PLA=50w/50w) composite as a function of the GE content. The dashed circle indicates the thermal percolation threshold.

Clearly, controlling the GE sheets at the interface of an immiscible polymer blend can significantly improve the composite's thermal conductivity. The improvement is similar to the improvement of electrical properties in electrically conductive polymer composites [5-7]. The thermal conductivities of the PCL/PLA/GE composites with different GE contents are shown in Figure 3. Significant improvement was observed in the thermal conductivities of PCL/PLA/GE composites as GE content increased. Figure 3 shows that 0.11 vol% is a critical level for PCL/PLA/GE composites. When the GE content is below 0.11 vol%, thermal conductivity increased slightly with increased GE; this is because the GE sheets are unable to contact one another to form a percolated, thermally conductive pathway. With further increases in GE content, thermal conductivity in PCL/PLA/GE composites increases remarkably. This is attributed to the formation of a three-dimensional heat conduction network. Thus, the thermal percolation threshold of PCL/PLA/GE composites is ~0.11 vol%. Obviously, this value is much lower than reported in the literature for TCPCs where fillers are selectively located in one phase of an immiscible polymer blend [8-10]. This is because only a very small amount of GE is needed to form the percolated thermal conduction network at the continuous interface of the PCL/PLA blend. To further confirm formation of a conducting network, the electrical properties of PCL/PLA/GE composites were measured, as shown in Figure 4. The electrical percolation threshold of the composite was estimated to be 0.105 vol%. Thus, a percolated conductive network can be formed when GE is 0.11 vol% in the PCL/PLA blend with co-continuous structure.

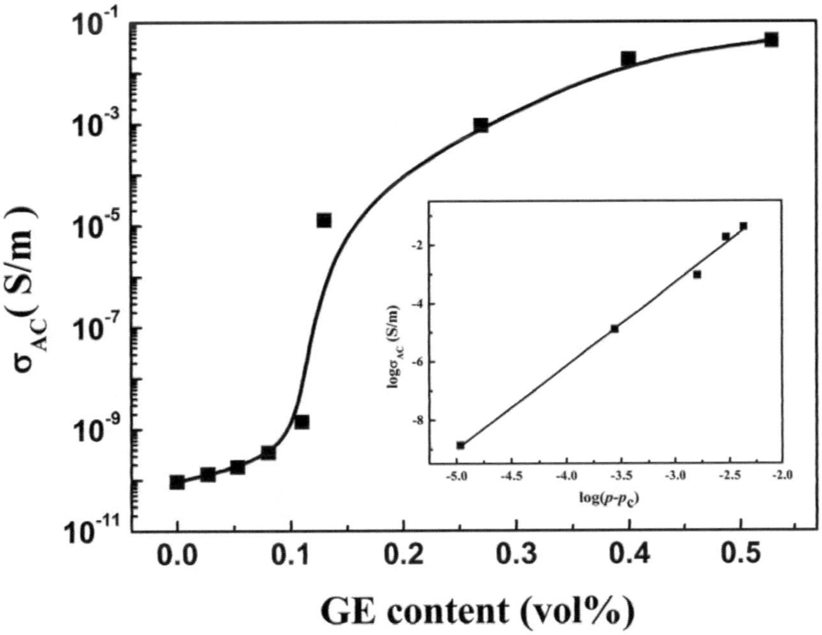

FIGURE 4. AC conductivities (at 1 Hz) of the PCL/PLA/GE (PCL/PLA = 50w/50w) composites as a function of GE content. The inset is A log–log plot of σ_{AC} (at 1 Hz) vs. $(p - p_c)$. The linear line is the best fit of data according to the classical percolation theory $\sigma \propto (p - p_c)^t$ for $p > p_c$, with an exponent t of 2.86 and critical weight concentration of $p_c = 0.105$ vol%.

We demonstrate a simple, effective strategy to trap GE at the interface of an immiscible PCL/PLA blend with co-continuous structure utilizing adsorption-desorption of polymer chains on the GE surface. This PCL/PLA/GE composite with co-continuous structure has a very low thermal percolation threshold (0.11 vol%). To the best of our knowledge, it is the lowest TCPC thermal percolation threshold yet. The current study demonstrates that TCPCs with high thermal conductivity and low filler loading can be produced by controlling filler at the interface of an immiscible polymer blend with a good co-continuous phase structure.

ACKNOWLEDGMENTS

This work was financially supported by Natural Science Foundation of Jiangsu Province for Youth Science Funds (BK20150073), the Fundamental Research Funds for the Central Non-profit Research Institution of CAF (CAFYBB2016QA012), National Natural Science Foundation of China for General Program (51373172), National Natural Science Foundation of China for Major Program (51433009).

REFERENCES

1. Z. D. Han and A. Fina, *Prog. Polym. Sci.* **36**, 914-944 (2011).
2. A. C. Baudouin, C. Bailly and J. Devaux, *Polym. Degrad. Stab.* **95**, 389-398 (2010).
3. A. E. Zaikin, R. R. Karimov and V. P. Arkhireev, *Colloid J.* **63**, 53-59 (2001).
4. A. C. Baudouin, J. Devaux and C. Bailly, *Polymer* **51**, 1341-1354 (2010).
5. J. R. Huang, C. Mao, Y. T. Zhu, W. Jiang and X. D. Yang, *Carbon* **73**, 267-274 (2014).
6. J. Chen, Y. Y. Shi, J. H. Yang, N. Zhang, T. Huang, C. Chen, Y. Wang and Z. W. Zhou, *J. Mater. Chem.* **22**, 22398-22404 (2012).
7. Y. Q. Tan, L. J. Fang, J. L. Xiao, Y. H. Song and Q. Zheng, *Polym. Chem.* **4**, 2939-2944 (2013).
8. D. Yorifuji and S. Ando, *J. Mater. Chem.* **21**, 4402-4407 (2011).
9. J. P. Cao, J. Zhao, X. D. Zhao, F. You, H. Z. Yu, G. H. Hu and Z. M. Dang, *Compos. Sci. Technol.* **89**, 142-148 (2013).
10. J. P. Cao, X. D. Zhao, J. Zhao, J. W. Zha, G. H. Hu and Z. M. Dang, *ACS Appl. Mater. Interfaces* **5**, 6915-6924 (2013).

The intumescent flame retardant (IFR) content-gradient distribution in poly (butylene terephthalate) was prepared in Co-rotating twin-screw extruder and plate vulcanizing technology. The PBT/IFR particles were fabricated by Co-rotating twin-screw extruder. The structure of IFR content-gradient distribution in PBT was fabricated by plate vulcanizing technology through melting blended two different IFR content in the PBT/IFR particles of uniform size and the same volume. The flame retardant, mechanical properties and morphological features of IFR content-gradient distribution in PBT, which were respectively obtained by the limited oxygen index (LOI), UL-94, thermogravimetry analysis (TGA), scanning electron microscope (SEM), the polarizing microscope. LOI and UL-94 tests results showed that with a little change of particle content-gradient distribution in PBT/IFR, the flame retardant property of PBT/IFR was better. the optimal LOI value of 20wt% IFR particle content-gradient distribution in PBT is equal to 25wt% IFR uniform distribution in PBT ， which the LOI reached to 23.7%. When a little change of IFR content-gradient distribution in PBT, on the one hand , the flame retardant property and the thermal stability of IFR content-gradient distribution in PBT are superior to the IFR evenly distribution in PBT, the T_{max} reaches to 405.8℃. on the other hand, it had no effect on the mechanical properties. Tensile and impact tests presented that a little change of particle content-gradient distribution in PBT/IFR had no effect on its mechanical properties compare to uniform PBT/IFR composites of the same IFR content. The tensile strength increased to 50.5Mpa from 45.6Mpa， the impact strength increased to 2.39 KJ/M^2 from 2.28 KJ/M^2.

Controlling the crystallization and crosslinking of phthalonitrile end-capped polyarylene ether nitrile

Lifen Tong, Renbo Wei*, Xiaobo Liu*

Research Branch of Functional Materials, Institute of Microelectronic & Solid State Electronic, High-Temperature Resistant Polymers and Composites Key Laboratory of Sichuan Province, University of Electronic Science and Technology of China, Chengdu 610054, P. R. China

Abstract. A series of phthalonitrile end-capped polyarylene ether nitrile (PEN-Ph) films which possesses crosslinking reaction combined with crystallization behavior had been prepared successfully, and their crystallization and crosslinking degree can be controlled by altering the heat-treating temperature. With heat-treating temperature increasing from 200 °C to 280 °C, the crystallization of the PEN-Ph films increased greatly, which could be proved by the DSC and XRD analysis results. Besides, because of the crosslinking reaction of the phthalonitrile groups at high temperature to form phthalocyanine ring, the crosslinking degree calculated by Soxhlet extraction method also increased as the heat-treating temperature increased from 200 °C to 260 °C. Moreover, when the heat-treating temperature reaches 300 °C, although the crystallization has little change compared with the sample with heat-treating temperature at 280 °C, the crosslinking degree increased dramatically. Most importantly, the phthalonitrile groups on the PEN-Ph crystals can react with the phthalonitrile groups in the matrix resin to form phthalocyanine ring, so that the PEN-Ph crystals can be embedded in the crosslinked matrix, resulting in the improvement of the crystals thermal stability. Therefore, a novel PEN-Ph film with high crystallization and crosslinking degree can be obtained by appropriate temperature treatment.

Keywords: Polyarylene ether nitrile; crystallization; crosslinking
PACS: 61, 65

INTRODUCTION

High performance thermoplastic matrix materials are rapidly becoming a popular alternative to epoxy matrix materials in carbon fiber reinforced composite systems, due to the increased toughness, ease in processing, and longer shelf life offered by thermoplastics. Polyarylene ether nitrile (PEN), as a high-performance thermoplastic engineering resin, have attracted considerable attention in the fields of the scientific community and accompanying commercial interest due to its outstanding properties such as excellent thermal and thermo-oxidative stability, good mechanical properties, superior chemical inertia and radiation resistance.[1-5] In recent years, Liu X. research team synthesize the crosslinkable PEN through nucleophilic substitution to introduce the phthalonitrile at the end of the PEN main chain.[6] Due to the crosslinkable property of the phthalonitrile, PEN-Ph integrates the advantages of the thermosetting resin and the thermoplastic resin, making it available to be fabricated by thermoplastic processing while used in thermosetting fields. Compared with the conventional PEN, the crosslinkable PEN possesses more excellent thermal and mechanical properties.[6, 7] Besides, owing to the regular structure of the part of the molecular chain, the phthalonitrile end-capped polyarylene ether nitrile (PEN-Ph) also possesses partially crystallinity. Thus, the PEN-Ph system possesses crosslinking reaction combined with crystallization behavior.

Nowadays, most researchers have devoted to single-component composites which with lightweight and recyclable properties. Compared to the conventional composites, the single-component composites based on semi-crystalline thermoplastics polymers such as polyethylene [8-10], polyester [11,12], isotactic polypropylene [13, 14], have attracted the increasing interesting in commercial and academic fields because of the absence of any filler or dense reinforcing components [15-20]. Moreover, in the single-component composites, the interfaces between composite components are potentially strong due to that they are chemically identical/compatible, resulting in the superior mechanical properties of the single-component composites [21-23]. However, the study about the single-component composites based on phthalonitrile end-capped polyarylene ether nitrile has little report.

In present work, a novel series of single-component composites based on PEN-Ph which possesses crosslinking reaction combined with crystallization behavior had been prepared successfully, and their crystallization and crosslinking degree can be controlled by altering the heat-treating temperature. Moreover, the thermal properties and mechanical properties had been studied in detail.

EXPERIMENTAL

Materials

N-Methyl-2-Pyrrolidone (NMP) was supplied by Tianjin BODI chemicals, Tianjin, China. Polyarylene ether nitrile terminated with phthalonitrile (PEN-Ph) was synthesized via nucleophilic aromatic substitution polymerization in our laboratory according to the previous reports [6].

Preparation of PEN-Ph films

The preparation of the PEN-Ph crosslinking films was carried out as follows. Firstly, we prepared the PEN-Ph film by solution casting method with the procedure of 80 °C, 100 °C, 120 °C, 140 °C, 160 °C, 180 °C and 200 °C for 1 h, respectively. Then, the obtained PEN-Ph films were heat-treated at different temperatures (200 °C, 220 °C, 240 °C, 260 °C, 280 °C, and 300 °C) for 4 h to prepare the PEN-Ph crosslinking films.

Characterizations

Differential scanning calorimetry (DSC) was performed on TA Instruments DSC-Q100 at a heating rate of 10 °C/min and a nitrogen flow rate of 50 mL/min. The phase analysis of PEN-Ph sheet was carried on through X-ray powder diffraction (XRD, Rigaku RINT2400 with Cu Kα radiation).

RESULTS AND DISCUSSIONS

Thermally induced phase transition behavior of the PEN-t-Ph/TPh system was investigated by DSC under a nitrogen atmosphere. Figure 1 shows the DSC curves of the crystalline PEN-Ph films, which depicts the crystal melting status, and the detailed data were listed in table 1. It can be seen that the melting enthalpy increases with the heat-treating temperature increasing from 200 °C to 260 °C, demonstrating that the crystalline degree of the PEN-Ph films increased. Moreover, when the heat-treating temperature is above 260 °C, the melting peak becomes more and more obvious and shaper, demonstrating that the PEN-Ph crystals becomes more and more stable. The crosslinking degree of the PEN-Ph films was reflected by the gel fractions which were calculated by the Soxhlet extraction, and the results were shown in figure 2. It is obviously obtained that the gel fraction increases with the heat-treating temperature increasing. When the PEN-Ph films heat-treated at lower temperature, the gel fraction is almost as zero, indicating that there is little crosslinking reaction in the PEN-Ph system at low temperature. However, when the heat-treating temperature reaches 300 °C, the gel fraction increases fiercely. It is due to the crosslinking reaction among the phthalonitrile at the end of the PEN main chain to form the stable phthalocyanine ring. Therefore, it draws a conclusion that when the heat-treating temperature reached to 300 °C, although the crystallization has little change compared with the sample with heat-treating temperature at 280 °C, the crosslinking degree increased dramatically.

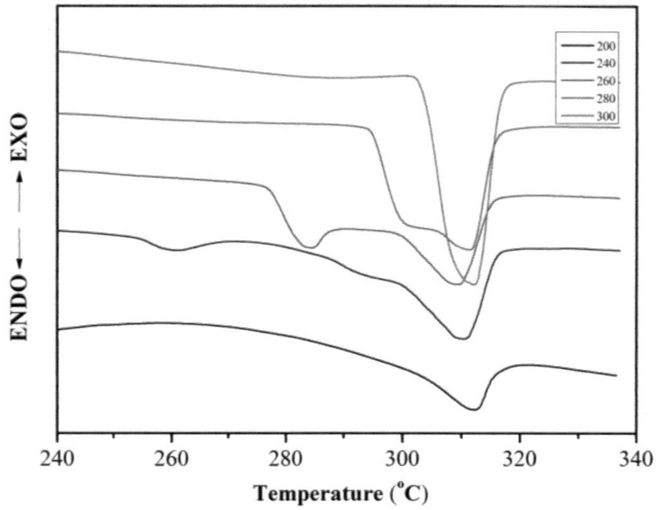

FIGURE 1. DSC curves of the crystalline PEN-Ph films with different treating temperatures (200 °C, 240 °C, 260 °C, 280 °C, and 300 °C)

TABLE 1. Thermal properties of the the crystalline PEN-Ph films with different treating temperatures (200 °C, 240 °C, 260 °C, 280 °C, and 300 °C)

Samples	Melting point (°C)	Melting enthalpy(J/g)	Glass transition temperature (°C)
200 °C	311.4	10.8	165.1
240 °C	260.5, 310.0	21.6	183.0
260 °C	284.0, 309.1	23.3	180.5
280 °C	302.7, 311.4	21.9	181.6
300 °C	312.1	21.5	180.6

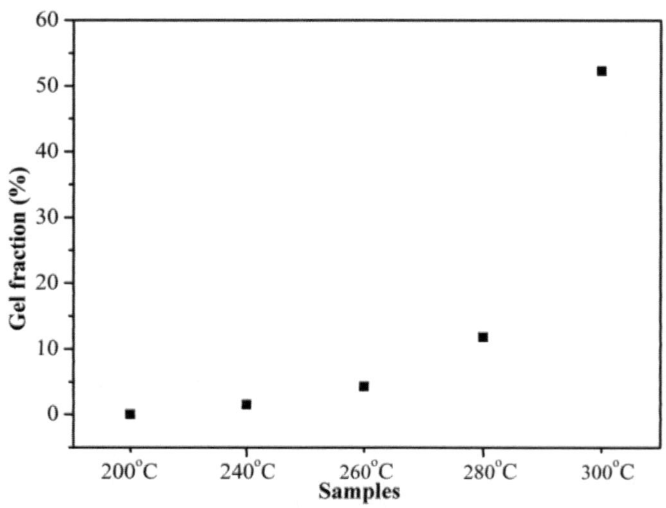

FIGURE 2. Gel fraction of the crosslinked PEN-Ph films with different treating temperatures (200 °C, 240 °C, 260 °C, 280 °C, and 300 °C)

Figure 3 shows the XRD results of the PEN-Ph films with different treating temperatures. It is obvious that all of the PEN-Ph sheets have crystallization behaviour. Because of the same peak position, it can be believed that the PEN-Ph sheets with different treating temperature possess the similar crystallization behavior, and the crystal structure is almost the same. Furthermore, the area of crystallization peak which reflects the degree of crystallinity trends to increase firstly and then decrease. This result is in well agreement with the analysis result of the DSC.

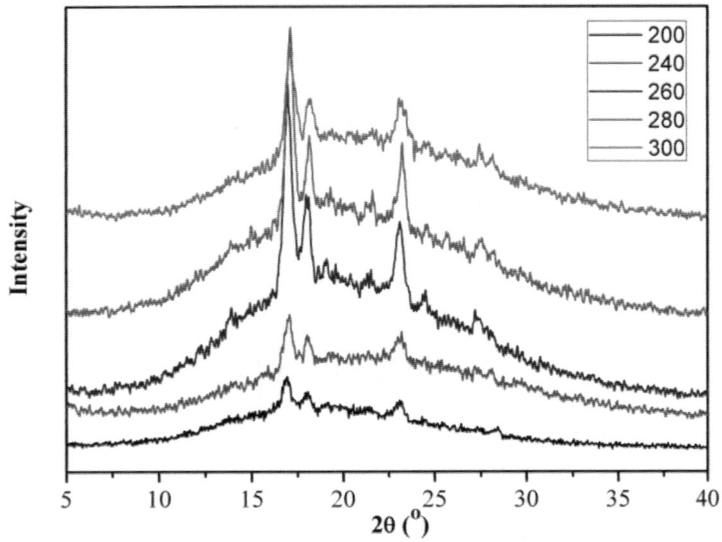

FIGURE 3. Mechanical properties of the crosslinked PEN-Ph films with different treating temperatures (200 °C, 240 °C, 260 °C, 280 °C, and 300 °C)

The mechanical properties of the crystalline PEN-Ph films with different treating temperatures were shown in figure 4. It can be seen that the tensile strength decreases firstly and then increases with the heat-treating temperature increasing from 200 °C to 300 °C. Because of the formation of a large number of crystals when the heat-treating temperature is at 260 °C, the tensile strength decrease greatly. The increase of the tensile strength when the heat-treating temperature increasing from 260 °C to 300 °C is mainly due to the crosslinking reaction between the PEN-Ph to form the stable phthalocyanine ring. Moreover, the elastic modulus shows the similar tendency to the tensile strength.

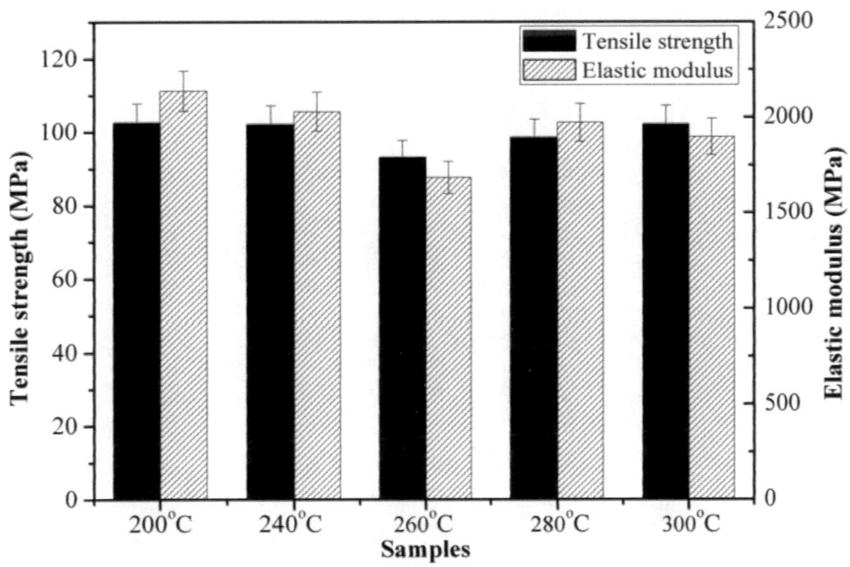

FIGURE 4. Mechanical properties of the crosslinked PEN-Ph films with different treating temperatures (200 °C, 240 °C, 260 °C, 280 °C, and 300 °C)

CONCLUSIONS

A series of PEN-Ph films which possesses crosslinking reaction combined with crystallization behavior had been prepared successfully, and their crystallization and crosslinking degree can be controlled by altering the heat-treating temperature. DSC analysis demonstrates that as the heat-treating temperature increases, the crystallization increases firstly and then decreases slightly. With the heat-treating temperature increasing, the crosslinking degree increases. Compared with the sample with heat-treating temperature at 280 °C, although the crystallization of the sample with heat-treating temperature at 300 °C has little change, the crosslinking degree increased dramatically. The mechanical testing indicates that the increase of the tensile strength when the heat-treating temperature increasing from 260 °C to 300 °C is mainly due to the crosslinking reaction between the PEN-Ph to form the stable phthalocyanine ring. Most importantly, the phthalonitrile groups on the PEN-Ph crystals can react with the phthalonitrile groups in the matrix resin to form phthalocyanine ring, so that the PEN-Ph crystals can be embedded in the crosslinked matrix, resulting in the improvement of the crystals thermal stability. Therefore, a novel PEN-Ph film with high crystallization and crosslinking degree can be obtained by appropriate temperature treatment.

ACKNOWLEDGMENTS

The authors wish to thank for the financial support of this work from the National Natural Science Foundation (No. 51373028, No. 51403029).

REFERENCES

1. F. Meng, R. Zhao, Y. Zhan, et al. *J. Mater. Chem.* **21**, 16385-16390 (2011).
2. Y. Zhan, X. Yang, H. Guo, et al. *J. Mater. Chem.* **22**, 5602-5608 (2012).
3. A. Saxena, R. Sadhana, V. L. Rao, et al. *Polym. Bull.* **50**, 219-226 (2003).
4. J. Verborgt, C. S. Marvel, *J. Polym. Sci., Part A: Polym. Chem.* **11**, 261-273 (1973).
5. A. Saxena, V. L. Rao, K. N. Ninan, *Eur. Polym. J.* **39**, 57-61 (2003).
6. L. Tong, Z. Pu, X. Huang, et al. *J. Appl. Polym. Sci.* 130, 1363-1368 (2013).
7. L. Tong, K. Jia, X. Liu, *Mater. Lett.* **128**, 267-270 (2014).
8. F. S. Lai, S. P. McCarthy, D. Chiu, et al. *Polymer* **38**: 5251-5253 (1997).
9. H. X. Huang, *Polym. Eng. Sci.* **38**, 1805-1811 (1998).
10. X. Zhuang, X. Yan, *Compo. Sci. Technol.* **66**, 444-449 (2006).
11. P. Rojanapitayakorn, P. T. Mather, A. J. Goldberg, et al. *Polymer* **46**, 761-773 (2005).
12. J. M. Zhang, C. T. Reynolds, T. Peijs, *Compo. Part A* **40**, 1747-1755 (2009).
13. J. Loos, T. Schimanski, J. Hofman, et al. *Polymer* **42**, 3827-3834 (2001).
14. H. Li, S. Jiang, J. Wang, et al. *Macromolecules* **36**, 2802-2807 (2003).
15. C. Gao, L. Yu, H. Liu, et al. *Prog. Poly. Sci.* **37**, 767-780 (2012).
16. G. Zhang, L. Jiang, K. Shen, et al. *J. Appl Polym. Sci.* **71**, 799-804 (1999).
17. G. Zhang, Q. Fu, K. Shen, et al. *J. Appl. Polym. Sci.* **86**, 58-63 (2002).
18. A. Y. Zhang, E. Jisheng, P. S. Allan, et al. *J. Mater. Sci.* **37**, 3189-3198 (2002).
19. O. A. Khondker, T. Fukui, M. Inoda, et al. *Compo. Part A* **35**, 1195-1205 (2004).
20. J. C. Chen, C. M. Wu, F. C. Pu, et al. *Express Polym. Lett.* **5**, 228-237 (2011).
21. O. A. Khondker, X. Yang, N. Usui, et al. *Compo. Part A* **37**, 2285-2299 (2006).
22. Á. Kmetty, T. Bárány, J. Karger-Kocsis, *Prog. Polym. Sci.* **35**, 1288-1310 (2010).
23. A. Gurarslan, J. Shen, A. E. Tonelli, *Polymer* **54**, 5747-5753 (2013).

Facile fabrication of 3D ordered layer-by-layer graphene oxide/copper phthalocyanine films with high dielectric properties

Zicheng Wang, Renbo Wei and Xiaobo Liu*

Research Branch of Advanced Functional Materials, High-Temperature Resistant Polymers and Composites Key Laboratory of Sichuan Province, Institute of Microelectronic & Solid State Electronic, University of Electronic Science & Technology of China, Chengdu 610054, P.R. China

Abstract. A novel three-dimensional ordered layer-by-layer graphene oxide/copper phthalocyanine polymer (GO/CuPc) films were fabricated by the orientational ordering in liquid crystal (LC) dispersions and immobilization of ordered structure upon casting and drying. The results of POM, UV-vis and EIS indicated that the prolongation of standing days and the introduction of GO effectively promote the self-assembly of CuPc and the formation of LC GO/CuPc dispersions. And the GO/CuPc dispersions exhibited strongly self-assembly behavior of CuPc by using GO as template and remains excellent texture of liquid crystal. By simple casting and drying, the long range orientation and alignment of GO/CuPc films can be successfully immobilized, which drastically enhanced the dielectric constant of GO/CuPc. Compared with the disordered GO/CuPc, the dielectric constants at 1 KHz and 100 KHz reach 164.82 and 31.95, leading to an increment of 319% and 275%, respectively.

Keywords: copper phthalocyanine; graphene oxide; liquid crystal; orientational ordering; dielectrics properties
PACS: 73, 77

INTRODUCTION

GO has a large 2D planar structure, which is consisted of two parts: oxygen-functional groups and graphitic domain. The presence of versatile oxygen functional groups (hydroxyl, epoxide and carbonyl groups) on the basal planes and edges endows GO with excellent dispersibility in water and polar organic solvents. The two major aspects of structural asymmetry and stabilized dispersibility are considered to determine the possible formation of nematic and lamellar liquid crystals (LCs) in water and polar organic solvents [1]. The novel LCs self-organization behavior of GO sheets in the fluid phase opens up a way to fabricate 3D macroscopically ordered materials. In addition, the graphitic domain is mainly composed of sp^2 carbon. The local delocalized π-electron system of GO effectively facilitates interaction with CuPc through the π-π stacking interaction. Hence, in this paper, we report a facile and effective approach to prepare GO/CuPc nanocomposites, which use LCs GO as an ordered template for the self-assembly of CuPc through π-π interactions between the aromatic sheets of GO and Pc rings of CuPc. And the LCs formulation is utilized for fabrication of self-assembled 3D macroscopically ordered layer-by-layer GO/CuPc nanocomposites upon simple casting and drying. And the effect of long range orientation and alignment on dielectric properties GO/CuPc nanocomposites will also be investigated.

EXPERIMENTAL

A typical procedure for preparing 3D macroscopically ordered layer-by-layer GO/CuPc nanocomposites is presented. The three-step process includes (1) LC GO/CuPc fabrication by the means of vigorous mechanical stirring for achieving intercalation of CuPc onto/into GO layers. (2) The local delocalized π-electron system of GO effectively facilitates interaction with CuPc through the π-π interaction. And GO is used as an ordered template for the self-assembly of CuPc through π-π interactions between CuPc molecules with standing for long time. (3) 3D macroscopically ordered layer-by-layer GO/CuPc preparation by the method of simple casting and drying for immobilization of ordering texture of LCs GO/CuPc.

RESULTS AND DISCUSSION

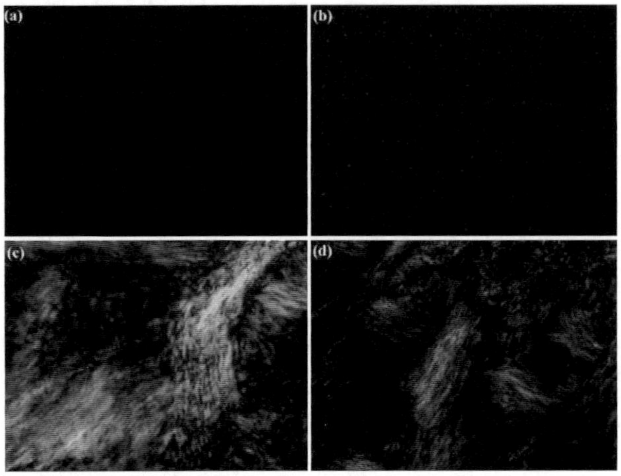

Figure 1. POM images of pure CuPc (a) and GO/CuPc dispersions with various weight ratios of (b)1:2 ; (c) 1:1 ; (d) 2:1 with standing for 0 day.

In order to fabrication of 3D macroscopically ordered GO/CuPc nanocomposites, fluid assembly is the only viable approach to realize ordering of GO/CuPc nanocomposites, due to the impossibility of melting processing for GO and CuPc. The formation of LCs GO in water and polar organic solvents paves the way to fabricate highly ordered, macroscopic assembled graphene materials. Therefore, the preparation of liquid crystal of GO/CuPc becomes the primary challenge. To meeting the requirement, the water/NMP (1:9) dispersions of GO/CuPc at various ratios are prepared and observed by POM as shown in Figure 1. As shown in Figure 1a, no birefringence is observed for the solution of pure CuPc. Isolsted birefringence domains appear as GO is added into CuPc solution as shown in Figure 1b. As the weight ratio of GO/CuPc is continuously elevated to 1:1, the stable birefringence appears and displays vivid aligned banded texture, which is a typical texture of lamellar phases. When the ratio increases to 2:1, the birefringence extends to the whole region with parallel-banded structures, implying the LCs GO/CuPc composites evolved into more ordered lamellar mesophase, which confirms the formation of the highly uniform orientational ordering LCs GO/CuPc nanocomposites.

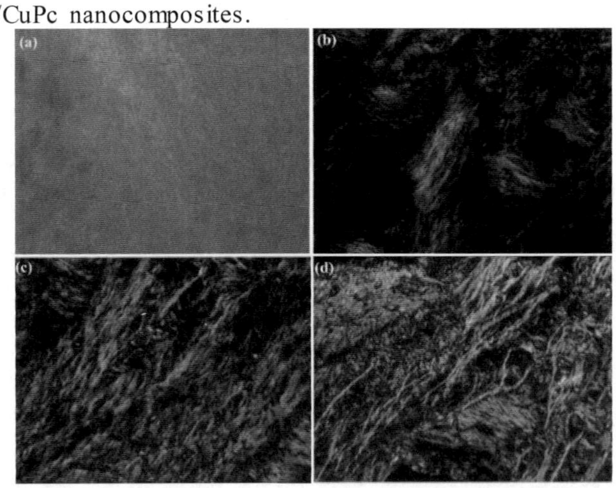

Figure 2. POM images of pure GO (a) and GO/CuPc (2:1) suspensions with standing for (b) 0; (c) 2; (d) 4 day.

For the GO/CuPc (2:1) dispersions, it exhibits strongly self-assembly behavior of CuPc by using GO as template [2, 3] and remains excellent texture of liquid crystal, which paves way to fabricate of 3D macroscopically ordered GO/CuPc nanocomposites. In order to further investigate the transform in self-assembly behavior of CuPc, polarized optical microscopy (POM) is utilized to observe the evolution in birefringence of GO/CuPc dispersions with prolongation of standing days as shown in Figure 2. As shown in Figure 2a, the GO sheets are highly dispersible in mixed solvents of water/NMP, forming typical nematic LCs. As the CuPc is added into GO dispersions, GO/CuPc evolves into lamellar LC phase as shown in Figure 2b. And it can be obviously seen that the orientation and alignment of CuPc is successfully achieved by using GO LCs as templates. Relative to the free CuPc, the CuPc closed to GO sheets are tightly adsorbed on the surface of GO through the strongly π-π interaction between

GO and CuPc, which exhibits different birefringent effect in different colour (blue). As the standing day increases, the blue area gradually extends to the whole region. It can be ascribed to the self-assembly of CuPc at high concentration through π-π stacking interaction between free CuPc and CuPc adsorbed on the surface of GO sheets, which confirms the above results of UV-vis and EIS.

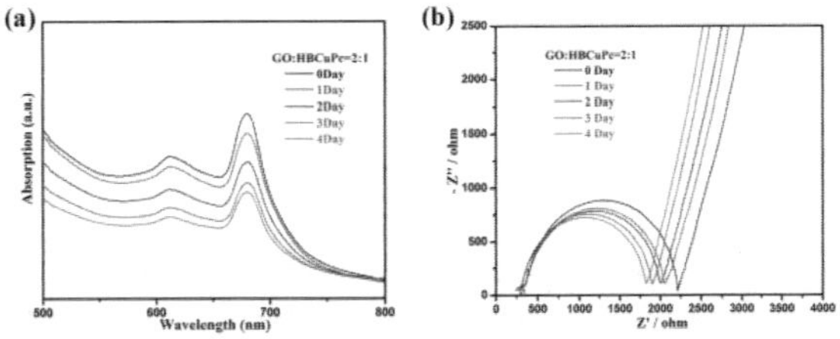

Figure 3. UV-vis curves (a) and Electrochemical impedance spectroscopy (EIS) (b) of GO/CuPc 2:1 with standing for several days.

In order to further investigate the self-assembling properties of GO and CuPc, UV-vis spectrophotometer is used to monitor the progress. The results show that the absorption of Q-band and shoulder peak on the UV-Vis curve decreases gradually with the increase of standing days (Figure 3a) [4, 5]. However, the ratio of the absorption of the shoulder peak and Q-band (I_S/I_Q) increases with the increase of standing days, indicating more CuPc aggregated. The obvious evolution on UV-vis spectra supports the formation of assembled structures through π-π interaction between CuPc and GO. Meanwhile, the electrochemical impedance spectroscopy (EIS) also is employed to record the changes in alternating-current impedance (AC Impedance) of GO/CuPc dispersions as shown in Figure 3b. At low frequency, the impedance plot exhibits gradient line, implying excellent capacitor behavior. At higher frequency, the plot has a semicircle and the intersecting point with the real axis corresponds to internal resistance of GO/CuPc dispersions. With the prolongation of standing days, the radius of semicircles gradually decreases and the intersecting point shifts to low resistance. It means that the internal resistance of GO/CuPc dispersions decreases with the prolongation of standing days, which may be due to π-π stacking between CuPc molecules [6], corresponding to self-assembly of CuPc in solvent. The strong π-π stacking interaction between CuPc molecules make it possible that distance of molecules reduces and the stacking of CuPc becomes tighter, as a result, the internal resistance decreases.

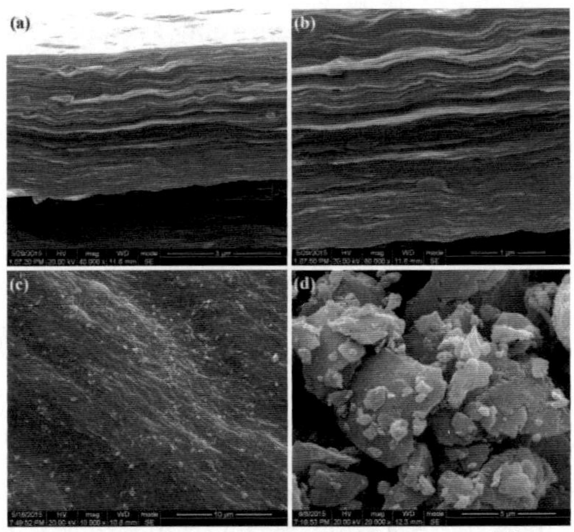

Figure 4. SEM images of cross section morphology of GO/CuPc multilayer film (a) and (b); SEM image of surface morphology of GO/CuPc multilayer film (c); SEM image of cross section morphology of disordered GO/CuPc composite (d).

The formation of long-range orientation in the lamellar phase of LCs GO/CuPc (2:1) dispersions offers the opportunity for facile fabrication of 3D macroscopically ordered materials. After self-assembling in the solvent of NMP and H_2O for 4 days at room temperature, the homogeneous dispersion of GO/CuPc was cast on a preheated ITO plate and laxly dried to obtain the multilayer films. From the SEM images of the cross section and surface of as-cast dried GO/CuPc paper as shown in Figure 4a, 4b and 4c, the LC properties of as-prepared LCs GO/CuPc dispersions in solvent induce a spontaneous self-assembly into tightly stacked layer-by-layer 3D ordered structures. Combined the results of surface and cross section morphology of as-cast dried GO/CuPc, it can be obviously observed that some of CuPc are laid on the surface of the paper while others are placed between layers of GO sheets, which confirms the fabricating structure of layer-by-layer GO/CuPc 3D assemblies. In order to compare with macroscopic assembled materials, the disordered GO/CuPc composite are prepared by dropping into methylbenzene for rapidly precipitating. As shown in Figure 4d, the disordered GO/CuPc composite shows an incompact and randomly aggregated morphology as CuPc is adsorbed on the surface of GO sheets.

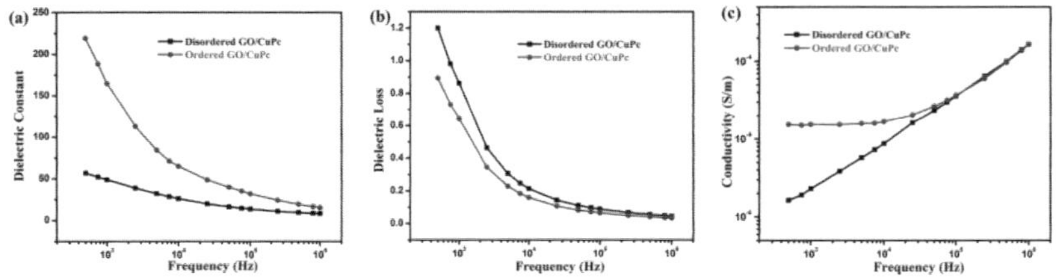

Figure 5. Dielectric constant (a), loss (b) and alternating current conductivity (c) of GO/CuPc multilayer film and disordered GO/CuPc composite measured at room temperature.

In order to investigate the effect of the long range orientation on the dielectric properties of GO/CuPc nanocomposites, the dielectric constant, dielectric loss and alternating current (AC) conductivity as a function of applied frequency for the ordered GO/CuPc paper and disordered GO/CuPc are measured at room temperature. A simple parallel-plate capacitor was fabricated by drop-coating GO/CuPc dispersions onto indium-tin-oxide (ITO) glass substrate and subsequent drying, while the top electrode was prepared by evaporation of Au. For comparing, similar capacitor of the disordered GO/CuPc composite was also prepared. As shown in Fig.5, the dielectric constant and loss of the disordered GO/CuPc decreases form 49.0 to 7.02 and 0.78 to 0.03 with frequency increasing from 500 Hz to 1 MHz, respectively. As can be seen, the dielectric constant of the ordered GO/CuPc paper increases obviously. Compared with the disordered GO/CuPc composite, the dielectric constants at 1 KHz and 100 KHz reach 164.82 and 31.95, leading to an increment of 319% and 275%, respectively. While the AC conductivity of the GO/CuPc multilayer films increases from 2.29×10-6 S/m to 1.54×10-5 S/m at 1 KHz. The substantial improvement in dielectric constant is attributed to the alignment and orientation of GO/CuPc composites [7, 8]. High aspect ratios and large surface area of flake-like graphene oxide make it possible to form a nanocapacitor combined with the adsorbed CuPc through strong π-π interaction. The accumulation of charges at the interfaces of GO and CuPc results in a high dielectric properties of the disordered GO/CuPc composite [9]. However, the incompact, randomly disordered aggregation of the GO/CuPc nanocapacitors hinders further enhancement in the dielectric properties. Hence, in this paper, LCs GO/CuPc formulation is utilized for a facile fabrication of ordered layer-by-layer GO/CuPc 3D assemblies. The long range orientation and alignment of the GO/CuPc nanocapacitors make it possible that the parallel connection of the nanocapacitors in the in-layer direction and series connection of the nanocapacitors in the perpendicular-layer direction can be formed. The formation of the more compact and highly ordered structures promotes dramatic improvement in the electrical properties of GO/CuPc nanocomposites.

CONCLUSIONS

In summary, the novel 3D macroscopically ordered layer-by-layer GO/CuPc nanocomposites was fabricated by LC formulation, involving the orientational ordering in LC GO/CuPc dispersions and ordering phase-transformation from ordered LC phase into 3D ordered solid materials. The results of POM, UV-vis and EIS indicated that the prolongation of standing days and the introduction of GO effectively promote the self-assembly of CuPc and the formation of LC GO/CuPc dispersions. By simple casting and drying, the long range orientation and alignment of GO/CuPc can be successfully immobilized, which drastically enhanced the dielectric constant of GO/CuPc. Compared with the disordered GO/CuPc composite, the dielectric constants at 1 KHz and 100 KHz reach 164.82 and 31.95, leading to an increment of 319% and 275%, respectively. Therefore, such kind of GO/CuPc nanocomposites will have great potential for application as organic dielectric materials.

ACKNOWLEDGMENTS

The financial support from National Natural Science Foundation of China (Project No. 51373028, No. 51403029), "863" National Major Program of High Technology (2012AA03A212), South Wisdom Valley Innovative Research Team Program and Ningbo Major (key) Science and Technology Research Plan (2013B06011) are gratefully acknowledged.

REFERENCES

1. Z. Xu, C. Gao. *ACS Nano* **5**, 2908-2915 (2011)
2. H. K. Bisoyi and S. Kumar, *Chem. Soc. Review.* **40**, 306-319 (2011)
3. T. M. Dellinger and P. V. Braun, *Chem. Mater.* **16**, 2201-2207 (2004)
4. A.W. Snow and N.L. Jarvis, *J. Am. Chem. Soc.* **106**, 4706-4711 (1984)
5. V. Iliev, V. Alexiev and L. Bilyarska, *J. Mol. Catal. A: Chem. 137*, 15-22 (1999)
6. H. L. Wang, Q. L. Hao, X. J. Yang, L. D. Lu and X. Wang, *ACS Appl. Mater. Interf.* **2**, 821-828 (2010)
7. X. Huang, K. Wang, K. Jia and X. B. Liu, *RSC Adv.* **5**, 51975-51982 (2015)
8. S. N. Song, J. W. Zhai, L. N. Gao and X. Yao, *J. Phys. Chem. Solids* **70**, 1213-1217 (2009)
9. T. Ohta, A. Bostwick, T. Seyller, K. Horn and E. Rotenberg, *Science* **313**, 951-954 (2006)

Crystallization coexisting with crosslinking in phthalonitrile end-capped polyarylene ether nitrile / copper phthalocyanine nanocomposite

Renbo Wei[1], Lifen Tong, Yong You and Xiaobo Liu[2]

Research Branch of Advanced Functional Materials, School of Microelectronics and Solid-State Electronics, High Temperature Resistant Polymer and Composites Key Laboratory of Sichuan Province, University of Electronic Science and Technology of China, Chengdu, 610054.

Abstract:

Crystallization coexisting with crosslinking was obtained in novel phthalonitrile end-capped polyarylene ether nitrile (PEN-Ph) / copper phthalocyanine (CuPc) nanocomposites by in-situ crosslinking reaction. The PEN-Ph/CuPc nanocomposite films were prepared by pre-polymerization of PEN-Ph and 1,3,5-tri-(3,4-dicyanophenoxy) benzene (TPh) with the present of copper ion, followed by solution-casting and hot-compression methods. The influence of the crosslinking reaction and crystallization behaviour with different content of copper phthalocyanine as well as the changing of crystallization behaviour before and after hot compressing were investigated. The results show that the crystallization and crosslinking coexist in the system at the same time. Differential scanning calorimetry (DSC) analysis indicates that the degree of crosslinking increases with the increasing amount of copper ion, indicating that the copper ion can be used as a crosslinking agent in this system. The X-ray diffraction (XRD) result shows that crystallinity decreases slightly as the copper ion increases, owing to the effect of the restriction of the crystallization behavior by the crosslinking reaction. Scanning electron microscope (SEM) images and X-ray diffraction (XRD) patterns show that the crystals of the PEN-Ph matrix grow after the hot-compressing procedure. All of the results show that the crystallization and crosslinking can coexist in this system through in-situ compositing, and the hot-compressing is a useful method to improve the crystallization of the composites.

[1] Corresponding author. Tel: +86-28-83207326; Fax: +86-28-83207326; E-mail address: weirb10@uestc.edu.cn
[2] Corresponding author. Tel: +86-28-83207326; Fax: +86-28-83207326; E-mail address: liuxb@uestc.edu.cn

The Capacitive Performance Research based on Polypyrrole/ Carbon Nanotubes/Graphene Ternary Composites

Xi Zhou, Weigang Wang and Shishan Wu*

School of Chemistry and Chemical Engineering, Nanjing University, 163 Xianlin Avenue, Qixia District Nanjing 210023, China

Abstract. In response to the worldwide energy shortage and environmental pollution, the development of energy storage devices with higher efficiency and sustainability has become a common goal among scientists in various countries. Supercapacitors promise to be a new class of electrochemical capacitors that are lightweight, eco-friendly and superior performance. Meanwhile, graphene and carbon nanotubes (CNTs) as the most widespread carbon nanomaterials are an attractive solution for energy storage devices due to their unique properties such as excellent electronic conductivity and activated surface area, high mechanical strength and chemical stability. The agglomeration appears among carbon nanomaterials, which lead to the lower surface area and slower electron transport rate and therefore unsatisfactory capacitive performance. Much effort has gone into preventing the accumulation of carbon based materials, such as compositing conductive polymers with them. Polypyrrole (PPy) is one of the important conducting polymers and has been extensively studied because of its ease of fabrication, high conductivity, good biocompatibility, and low cost. In this work, a kind of novel highly dispersed PPy is firstly designed for decoration onto the surface of pristine multiwalled CNTs, which facilely synthesize a CNTs/PPy core/shell composite via π-π interaction. The decorative PPy shell could prevent the aggregation of CNTs by electrostatic repulsive interaction between them and lead to a high dispersibility of the prepared composite, which could effectively insert into the lamellar structure of graphene, resulting in a high active surface area and capacitive performance of the prepared composite.

Keywords: Polypyrrole, Carbon materials, Dispersion, Capacitors.
PACS: 68

INTRODUCTION

With the increasing demand of clean and sustainable energy, due to the energy crisis and environmental disruption, scientists from various countries are committed to developing a novel kind of energy storage device to alleviate the current situation. Supercapacitors are considered as a promising choice among many kinds of candidates because of their superior features such as higher power density than batteries, longer cycling life, higher energy density than traditional capacitors and excellent reversibility [1]. According to charge-storage mechanism, supercapacitors can be ranked into two types which are known as electric double layer (EDL) capacitors with carbon electrodes and pseudocapacitors with metal oxide or conducting polymer electrodes [2]. Compared with the latter, EDL capacitors commonly use carbon materials like carbon nanotubes (CNTs) and graphene, which have high surface area to improve capacitance performance. Meanwhile, conducting polymers with high conductivity are usually used for supercapacitors electrode materials to enhance the capacitive ability [3].

Carbon nanotubes and graphene, as the novel carbon materials with high specific surface area, high chemical stability, strong mechanical strength and excellent electrochemical performance, are widely used for energy storage devices. Nevertheless, carbon nanotubes and graphene tend to form irreversible agglomerates or even stack due to strong π-π stacking and van der Waals interaction, leading a low specific surface area and electroactivity [4]. Thus, many efforts, such as compositing conductive polymers (CP) with carbon materials, have been made for reducing or eliminating the accumulation of carbon materials. Conducting polymers, such as polyaniline (PAN), polypyrrole (PPy) and polyindole (PIn) are widely used in electrode modified materials. PPy, derived from pyrrole monomers through polymerization, has captured attention from many fields because of its potential applications in electrochromic devices, electrode modified materials and biosensors [5].

In this work, polypyrrole/carbon nanotubes/reduced graphene oxide (CNTs/PPy/RGO) ternary composites were firstly synthesized through a facile method and proved to be a novel composite material with outstanding electrochemical performance, such as high capacity, long cycling life and good stability. The composites were fabricated through a two-step process. Firstly, the surface of pristine multiwalled CNTs was modified with PPy through in situ polymerization, leading to a CNTs/PPy core/shell composites by π-π interaction. The PPy coated on the surface of CNTs could prevent the aggregation of CNTs by electrostatic repulsive interaction between them and lead to a high dispersibility. Secondly, the CNTs/PPy effectively insert into the lamellar structure of graphene. The as-prepared composites could highly disperse in water and show high capacity and excellent cycling stability at high current rates.

EXPERIMENT SECTION

Synthesis of CNTs/PPy

The CNTs/PPy composites were obtained through in situ polymerization, in which 2.5 mL H_2O_2 was used to initiate the CNTs/pyrrole/$FeCl_2$/H_2O (0.1 g/0.1 g/0.01 g/97.5 mL) mixture and the reaction lasted for 6 h. Then, the products were washed by deionized water several times to remove unused reactants. After that, centrifugation was used to collect the products. Finally, the producing composites were dried in vacuum at 80 °C for 12 h.

Synthesis of CNTs/PPy

Graphene oxide (GO) was synthesized from graphite powder by a modified Hummer's method [6], dispersed in distilled water with a concentration of 1.0 mg/1.0 mL by ultrasonication for the following experiments. And the GO were mixed with the prepared CNTs/PPy composites by strong stirring for 6 h at room temperature. As a result, the CNTs/PPy/GO composites were produced, which was then reduced by hydrazine hydrate at 90 °C for 6 h. After that, the CNTs/PPy/RGO composite were obtained, which were washed several times with water.

Apparatus

The morphology of the nanostructures was observed by transmission electron microscopy (TEM) images (JEM 2100 high-resolution TEM). X-ray photoelectron spectroscopy (XPS) measurements were performed on a PHI 5000 VersaProbe. All Raman analysis was carried out using a Jobin Yvon HR800. All electrochemical measurements were carried out on a CHI660d (Shanghai CH Instrument Company, China).

RESULTS AND DISCUSSION

Because of strong π-π stacking and Van der Waals interaction, the carbon based materials like CNTs and graphene tend to form irreversible agglomerates or even stack. Therefore, they usually need pre-treatment or modification with surfactants to prevent the accumulation. As showed in FIGURE 1A, the pristine CNTs have a typically smooth outer surface with diameter of 40−60 nm and are several micrometres in length. By contrast, the surface of nanotubes is rougher after the decoration of PPy, and the diameter is increased due to the introduction of PPy shell (FIGURE 1B). Furthermore, from the morphology of CNTs/PPy/RGO (FIGURE 1C) we can find that the tubular structure of CNTs/PPy are completely coated on RGO sheets, which indicates the modification of RGO by CNTs/PPy. And the dispersity of prepared materials is demonstrated by FIGURE 1D, in which they were placed for 72 h after ultrasonication to reveal the dispersibility directly. CNTs/PPy and CNTs/PPy/RGO are dispersing well, while CNTs and RGO aggregate and precipitate in aqueous solvent. The excellent dispersibility of the CNTs/PPy and CNTs/PPy/RGO in our study can be attributed to the electrostatic repulsions between the PPy.

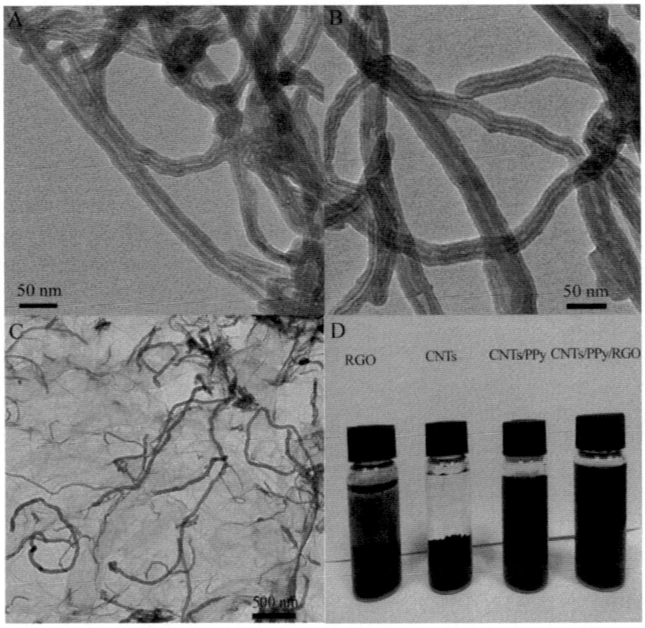

FIGURE 1. TEM images of CNTs (A), CNTs/PPy (B) and CNTs/PPy/RGO (C), Photographs of RGO, CNTs, CNTs/PPy, CNTs/PPy/RGO dispersed in water after ultrasonication and stood for 72 h (D).

The structures of the prepared materials were studied by Raman spectra as shown in FIGURE 2A. Raman spectra of the carbon-based materials (CNTs and RGO) have very distinctive peaks for the D and G bands, which are located at about 1350 and 1584 cm^{-1}, respectively. The D/G band intensity ratio can demonstrate the ratio of sp^3/sp^2 carbons. It is shown that, as the introduction of PPy, a significant decrease in the D/G band intensity ratio can be observed, which imply the successful fabrication of CNTs/PPy and CNTs/PPy/RGO composites [7]. To further investigate the surface chemical ingredients of the prepared composites, XPS was carried out. As shown in FIGURE 2B, The peaks of N 1s and O 1s in the CNTs/PPy and CNTs/PPy/RGO are obviously observed compared with the pure CNTs and RGO, which further confirms that PPy with functional groups has successfully been coated on CNTs and RGO.

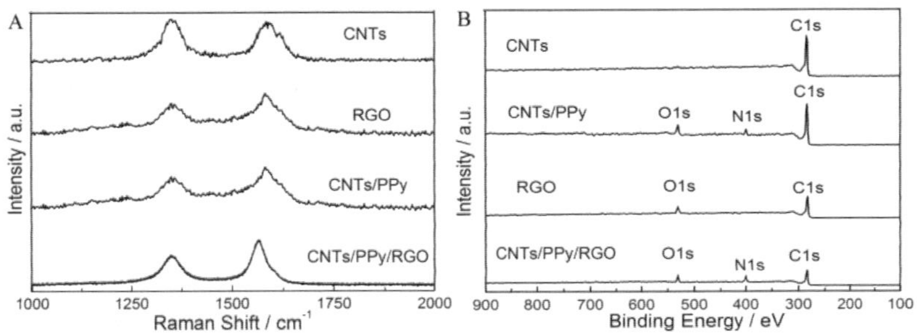

FIGURE 2. (A) Raman spectra of CNTs, RGO, CNTs/PPy and CNTs/PPy/RGO. (B) XPS spectra of CNTs, CNTs/PPy, RGO and CNTs/PPy/RGO composites.

To research the electrochemical performance of the prepared materials, cyclic voltammetry (CV) and galvanostatic charge–discharge measurements were carried out. FIGURE 3A shows the CV curves obtained from CNTs, RGO, CNTs/PPy and CNTs/PPy/RGO electrodes at the same scan rate (500 mV s^{-1}). The area of the CV curve from the CNTs/PPy/RGO electrode is much larger than that from other electrode, which indicates that the electrochemical performance of the carbon materials has been greatly improved by the modification of PPy [8]. In addition, we can see that the curve shape of CNTs/PPy/RGO is normally close to an ideal rectangular shape and the current responds quasi-linearly with increasing potential scan rate, exhibiting remarkable capacitive performance (FIGURE 3B). FIGURE 3C presents the galvanostatic charge–discharge curves of the prepared CNTs/PPy/RGO supercapacitor at different current densities in the range of 1-15 A/g, which are symmetric and linear, demonstrating the good capacitive behavior. The calculations for specific capacitance (C_m) are from the following equation:

$$C_m = I\,\Delta t / m\,\Delta E$$

where I is the discharge current, Δt is the discharge time, ΔE is the potential window during the discharge process, and m is the effective electrode mass [9]. In the case of CNTs/PPy/RGO, a high specific capacitance of 267.5 F/g is achieved at a current density of 1 A/g, which is much higher than those of the CNTs/PPy (176.2 F/g), RGO (109.2 F/g) and CNTs (78.1 F/g) under the same measure conditions. Even as the current density increases at 20 A/g, the CNTs/PPy/RGO composite still retains 84% of the specific capacitance at 1 A/g (FIGURE 3D). These results are probably attributed to the high dispersibility of RGO and CNTs, which are beneficial for the increasing of effective surface area and conductivity, accessible for fast ion transportation [10].

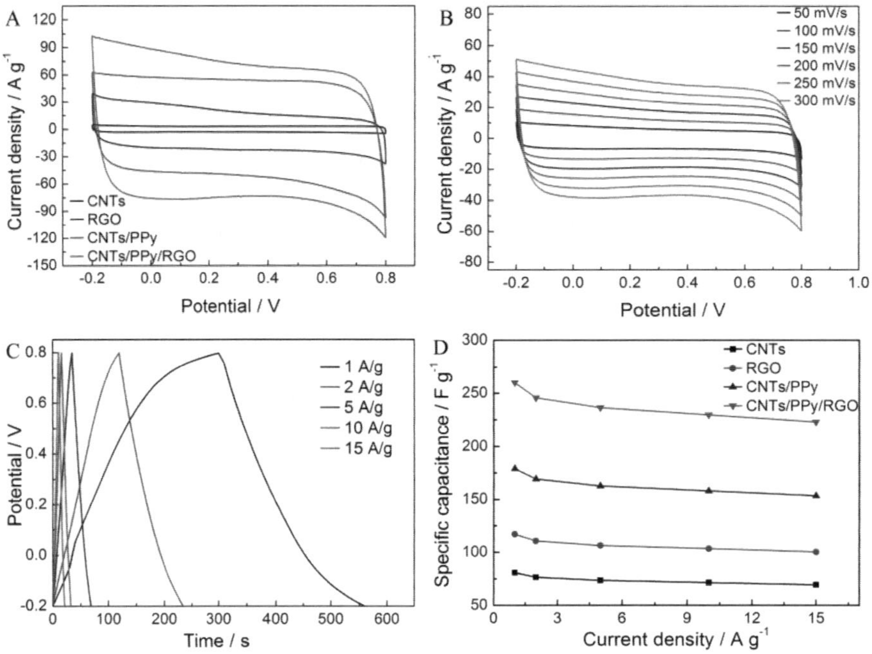

FIGURE 3. (A) CV curves obtained from CNTs, RGO, CNTs/PPy and CNTs/PPy/RGO electrodes at the same scan rate (50 mV s-1). (B) CV curves of CNTs/PPy/RGO composites at different scan rate. (C) Galvanostatic charge/discharge curves of CNTs/PPy/RGO composites at different scan rate. (D) Galvanostatic charge/discharge curves of four different materials at different scan rate.

To research the cycling stability of CNTs/PPy/RGO, the cycling performance within -0.2–0.8 V at a current density of 15 A/g was recorded by galvanostatic charge/discharge for 5000 cycles. The capacitive retention of the composite was 82.2% after 5000 cycles (FIGURE 4), indicating a good cycling ability of the composite material.

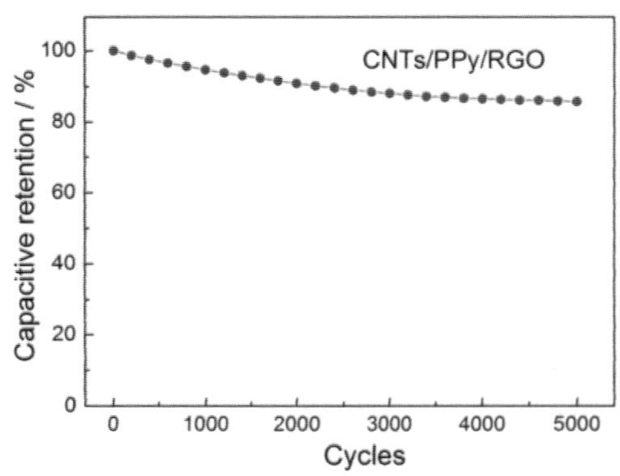

FIGURE 4. (A) Cycling performance of CNTs/PPy/RGO composites.

ACKNOWLEDGMENTS

This work was supported by the National Natural Science Foundation of China (Nos. 51272100 and 51273073).

REFERENCES

1. G. Z. Sun, X. Zhang and R. Z. Lin, et al, Angew. Chem., Int. Ed. 54, 4651–4656 (2015).
2. C. F. Sun, H. L. Zhu and M. H. Okada, et al, Nano Lett. 15, 703–708 (2015).
3. Y. Song, T. Y. Liu and X. X. Xu, et al, Adv. Funct. Mater. 25, 4626–4632 (2015).
4. X. H. Liu, M. H. Yu and T. Zhai, et al, Nano Lett. 13, 2628–2633 (2013).
5. C. X. Guo, G. Yilmaz and S. C. Chen, et al, Nano Energy 12, 76–87 (2015).
6. X. H. Lu, M. H. Yu and G. M. Wang, et al, Energy Environ. Sci. 7, 2160–2181 (2014).
7. W. J. Liu, N. S. Liu and Y. L. Shi, et al, J. Mater. Chem. A 3, 13461–13467 (2015).
8. W. W. Liu, X. Li and M. H. Zhu, et al, J. Power Sources 282, 179–186 (2015).
9. M. H. Yu, Y. Zeng and Y. Han, et al, Adv. Funct. Mater. 25, 3534–3540 (2015).
10. N. Y. Cheng, Q. Liu and J. Q. Tian, et al, Chem. Commun. 51, 1616–1619 (2015).

Crystallization of Polymers in Processing Conditions: An Overview

Jean-Marc Haudin and Séverine A.E. Boyer

MINES ParisTech, PSL – Research University, Centre for Material Forming (CEMEF), UMR CNRS 7635, 1 Rue Claude Daunesse, 06904 Sophia Antipolis, France

Structure development is a key issue in polymer processing, with a view to master the final properties of the products (mechanical, optical, etc.). Among the phenomena involved, crystallization plays a major role. It generally occurs in complex, inhomogeneous and coupled mechanical (flow, pressure), thermal (cooling rate, temperature gradient) and geometrical (surface of processing tools) conditions.

A first route to understand structure development is to design model experiments to isolate the specific influence of a given parameter. The emphasis will be laid here on:
(i) the influence of shear flow through rheo-optical measurements.
(ii) the influence of high cooling rates. High cooling rates are obtained with a modified hot stage named CRISTASPEED (up to 2000 °C/min) [1] and comparison has been made with DSC experiments using conventional as well as flash DSC.
(iii) the influence of high pressures. An original cell named CRISTAPRESS (up to 200 MPa) [2] has been especially designed for high-pressure experiments.

Numerical simulation is also a useful tool to understand and predict the coupled phenomena involved in crystallization. Based on Avrami's ideas and equations, a general differential formulation of overall crystallization kinetics has been proposed [3]. It is able to treat both isothermal and non-isothermal cases, and has been extended to take into account flow as well as surface (transcrystallinity) effects.

.

[1] S.A.E. Boyer et J.M. Haudin, Polym. Test., 29, 445-452 (2010)
[2] S.A.E. Boyer, F.E.J. Fournier, Ch.-A Gandin et J.M. Haudin, Rev. Sci. Instrum., 85, 013906 (2014)
[3].J.M. Haudin et J.L. Chenot, Intern. Polym. Proc., 19, 267-274 (2004)

Super-toughened Polylactide Prepared by Blending with Biobased Copolyesters

Xiaoran Hu [a] and Liqun Zhang [a,*]

[a] *State Key Laboratory of Organic-Inorganic Composites, Beijing University of Chemical Technology, Beijing 100029, P. R. China*

Abstract. Blends of poly(lactic acid) (PLA) and a series of renewable bioelastomers (PLBSIs) were fabricated through melt blending to improve the toughness of PLA. Benefiting from reasonable macromolecular design, the introduction of lactic acid in PLBSIs not only increases the compatibility with PLA but also helps PLBSIs turn from crystalline plastic to amorphous elastomer. The increasing compatibility was evidenced by the decreasing particle size of PLBSIs in PLA and the better toughening effect of elastomer than that of plastic was characterized by the thermal tensile test. Thus, the super toughened PLA blends were obtained and exhibited maximum elongation at break of 324 % and the impact strength of 35.7 KJ/m2, 50 and 15 times higher than neat PLA, respectively. Excellent performances in 3D-printed tensile tests imply the toughened PLA was ideal 3D-printing ink. The super toughened PLA will eliminate the brittleness of PLA for wide application and hold great potential for both engineering and biomedical applications.

Keywords: toughening, compatibility, polylactide
PACS: 83.80.Tc Polymer blends

INTRODUCTION

Bio-based polymers have attracted much attention due to their potential to reduce not only the utilization of petrochemicals, but also the emission of environmental issues[1-2]. Polylactide (PLA), a biodegradable polyester completely derived from renewable resources on commercial scale, has been proven viable in replacing petroleum-based plastics in some applications. However, the inherently brittleness of PLA greatly confines its wide applications. Therefore, extensive efforts were developed to toughen PLA.

Blending PLA with elastomers is one of the promising approaches to improve the toughness of PLA because it is more cost-effective and facile compared to chemical modification. However, most of the elastomers do not constitute long-term solutions as they modify the compostable feature of the final PLA material thus suppressing one of the main advantages of PLLA. A recent trend is to adopt degradable and renewable elastomers to toughen PLA. Unfortunately, most elastomers show poor compatibility with PLA and resulted in unsatisfactory toughening effects. According to many studies, the compatibility between PLA and other components significantly contributed to the excellent toughness and should be taken into consideration in toughening PLA as a dominate factor.

Recently, we have prepared novel copolyesters from polymerizing biobased monomers including lactic acid, sebacic acid, itaconic acid and butanediol, all of which are derived from renewable resources on commercial scale[3]. It is noteworthy that the introduction of lactic acid into these copolyesters provides them similar lactate structure to PLA, implying good compatibility with PLA. In the present work, we focused on highly toughening PLA by compounding PLA with our synthetic bioelastomers.

EXPERIMENT

Preparation of PLBSI:

A series of poly(butanediol/lactate/sebacate/itaconate) bioelastomers (PLBSI) with different content of lactic acid were prepared according our previous work[3]. The chemical structure are presented in Figure 1. In brief, calculated LA, BDO, SeA and IA were heated at 130℃ for 1 h and then at 180℃ for another 2 h in a 100-ml three-neck flask equipped with a mechanical stirrer and under a dry nitrogen atmosphere. Then the polymerization was performed at 220℃ for 6 to 8 hours under reduced pressure (<300 Pa) with TBT (0.1wt% relative to all chemicals) as the catalyst until the Weisenberg effect was observed. The polymerization was carried out in the presence of tri-(4-hydroxy-TEMPO) phosphate (0.0621 g, 0.1wt% relative to all chemicals) as antioxidant and inhibitor. The obtained PLBSIs was dissolved in chloroform and precipitated in excess methanol to remove any unreacted monomers and remaining oligomers. Finally, the PLBSIs were collected by filtration, extensively washed with methanol, and dried in vacuum at 60 °C. The unsaturated copolyester containing40 mol% of lactate relative to all chemical amounts was denoted as PLBSI-40.

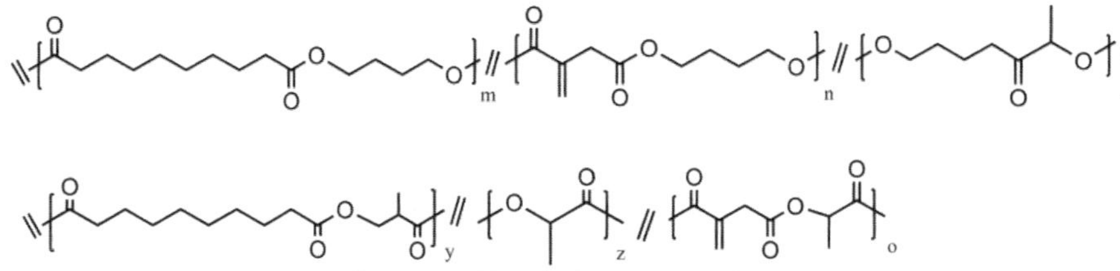

FIGURE 1 Chemical structure of PLBSI

Preparation of PLBSI/PLA Blends: PLA and PLBSIs were dried in a vacuum oven at 60 ℃ for 24 h prior to use. According to our previous study[3], blends were prepared by melt blending PLBSIs with PLA at 15wt% for 10 min in a Haake Remix at 170℃and a rotational speed of 80 rpm. All the samples were finally hot-pressed under 10 MPa at 190℃ for 5 min to form 1-mm thick sheets.

RESULTS and DISCUSSIONS

Morphology of PLA and PLBSI/PLA blends

The compatibility of the components in the blends determines the phase behavior and interfacial interaction of the blends, which has great influence on the mechanical properties of the materials. The good compatibility between the two components leads to uniformly dispersion of dispersed phase in the matrix with relatively small particle size and narrow particle size distribution. SEM was used to observe the size of the PLBSI particles in PLBSI/PLA blends. In Figure 2, where the PLBSI particles are observed as the dark phase in TEM micrographs, a clear phase-separated morphology was found in all blends. The uniform micron dispersion as spheres of the PLBSI particles in the PLA matrix reveals good compatibility of the PLBSI/PLA blends. Statistics from 100 randomly selected particles from several SEM micrographs show that the number-average particle diameter (Dn) decreases from 1000 nm in PLBSI-0/PLA blends to 450 nm in PLBSI-40/PLA blends. The observed variation tendency of particle size implied improved compatibility between lactate-rich PLBSIs and PLA matrix.

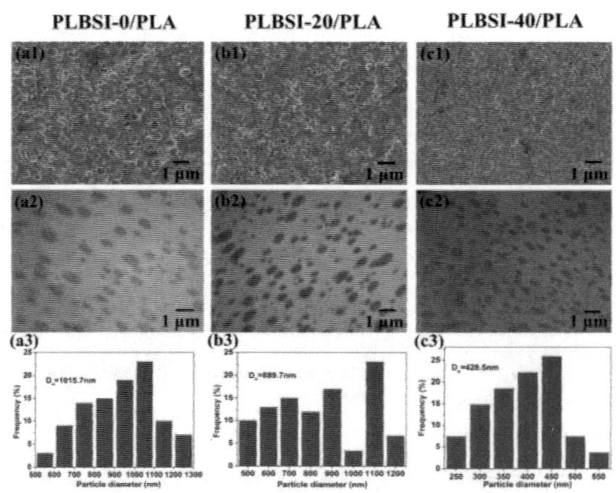

FIGURE 2 SEM cryofractured surface micrographs (a1, b1 and c1), TEM micrographs (a2,b2 and c2) of PLBSI/PLA blends and particle diameter distribution (a3,b3 and c3) of PLBSI/PLA blends

Tensile and impact Properties of PLA and PLBSI/PLA blends

The toughness of PLBSI/PLA blends was measured by both tensile testing and impact testing. As shown in Figure 3 (a), the addition of PLBSIs significantly changes the tensile behavior of PLA. The neat PLA is a brittle polymer with poor ductility evidenced by the elongation at break only around 6%, and deformed in a typical brittle behavior without obvious yielding under tensile stress. Compared to neat PLA, all the blends underwent distinct yielding followed by obvious cold drawing during the tensile test, indicating that the brittle fracture of PLA changed to a ductile fracture with the incorporation of the PLBSI. The maximum elongation at break is obtained by PLBSI-40/PLA blend, which is 50 times higher than neat PLA, indicating excellent toughness of our toughened PLA.

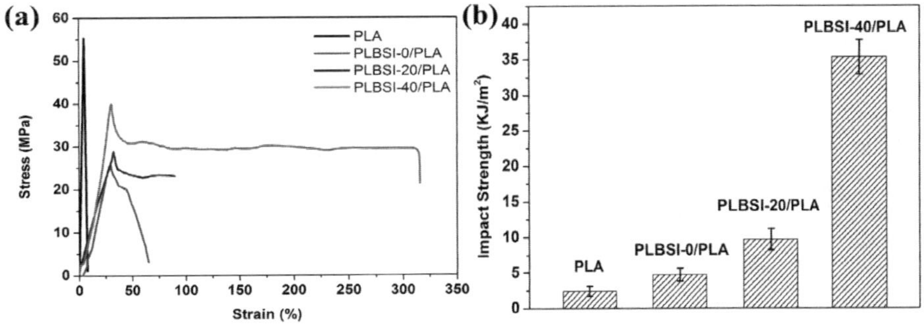

FIGURE 3 (a) Stress-Strain curve and (b) impact strength of PLA blending with different PLBSIs and (c) Stress-Strain curve and (d) impact strength of PLA blending with PLBSI-40 of different weight ratio

Measurement of the notched impact strength is the observation of fracture energy absorption during high loading, which represents a more accurate and useful method. Notched impact strength tests results are presented in Figure 3 (b). An obvious correlation of the lactate content in the PLBSIs with the impact strength can be found. In the case of PLBSI-0, very limited reinforcement in notched impact strength was observed. However, increasing lactate content in PLBSIs resulted in a significant enhancement of the impact strength. The PLBSI-40/PLA blend seems an optimal composition, inasmuch as this blend reaches the maximum impact strength of 35.3 kJ/m^2, which is 15 times higher than that of neat PLA. According to the mechanical properties of PLBSI/PLA blends, it is obvious that both the increasing compatibility of PLBSIs by the introduction of lactic acid into PLBSIs succeed to achieve superior toughness.

CONCLUSION

In this study, super toughened PLA was obtained by blending PLA with PLBSIs, which are completely synthesized from natural renewable resource. The uniformly dispersion of PLBSIs in the PLA matrix with relatively small particle size and narrow particle size distribution indicate good compatibility between the two components. Importantly, super toughened PLBSI-40/PLA blends exhibiting maximum elongation at break of 324 % compared to 3.8% for neat PLA and the impact strength of 35.7 KJ/m^2 compared to 2.4 KJ/m^2 for neat PLA. Our research indicates that these PLBSI/PLA blends hold great potential for both engineering and biomedical applications.

ACKNOWLEDGMENTS

This work was supported by the National Natural Science Foundation of China (Grant No. 50933001 and 51221002).

REFERENCES

1. Gandini, A. Green Chem., 2011, 13, 1061-1083.
2. Gandini, A. Macromolecules, 2008, 41, 9491-9504.
3. Hu X, Kang H, Li Y, Polymer Chemistry, 2015, 6, 8112

Thermal Insulating Nanocomposite Foams with Bimodal Cellular Structure

Pengjian Gong, Shuo Zhai, Minh-Phuong Tran, Piyapong Buahom, Guilong Wang, and Chul B. Park*

Microcellular Plastics Manufacturing Laboratory, Department of Mechanical and Industrial Engineering, University of Toronto, 5 King's College Road, Toronto, Ontario, Canada M5S 3G8

pgong@mie.utoronto.ca
park@mie.utoronto.ca

Abstract. We developed a bimodal polystyrene (PS)/multi-walled carbon nanotube (MWCNT) nanocomposite foam with a very low thermal conductivity of 30 mW/m-K without using any insulation gas. The MWCNTs were critical to the heat transfer because they decreased radiation significantly through the foams despite a marginal increase in bulk solid conduction. The superior heat-blocking characteristics of the bimodal cellular morphology and the MWCNTs as follows: (1) The primary large cells acted to increase the expansion ratio, thereby decreased the solid conduction; (2) The secondary small cells were sufficiently small to induce the Knudsen effect on gas conduction; and (3) In the polymer matrix, the MWCNTs' strong infrared (IR) absorption activity intensively blocked the IR transmission. It clearly shows that we could take advantage of a bimodal cellular structure and use MWCNTs to further decrease the thermal conductivity by tailoring the foam structure.

Keywords: Thermal insulation, MWCNTs, bimodal nanocomposite foam, supercritical CO_2 foaming.
PACS: 01.10.Fv

INTRODUCTION

There is a strong global demand for innovative materials with superthermal insulation properties to help save energy. Numerous countries have invested heavily in developing renewable energy to replace conventional energy, but these large investments have not proven as profitable as expected. For example, there was very little contribution, less than 0.2%, of the wind, tidal, and solar energy to the total produced energy in Canada during 2010 - 2014 [1]. Therefore, it would be more prudent to focus on reducing the energy waste through enhanced insulation. In fact, industrial and residential buildings typically consume more than 60% of the total energy produced [1], and it will be desirable to decrease the waste energy from the heating and cooling system with improved insulation. In this context, superthermal insulation materials are urgently required to reduce energy loss and to address the concerns of scarce energy resources.

Efforts have been made to maintain this thermal conductivity (e.g., below 30 mW/m-K) using various well-known insulation materials that do not use any insulation gas. These initiatives have been undertaken to protect the environment and to reduce costs. To achieve a lower thermal conductivity, complicated foam morphology must be designed to take advantage of both the large expansion ratio and the small cell size. In thermal insulation, Zhang et al. produced bimodal PS foams with a low thermal conductivity of 31.5 mW/m-K [2]. This unique aspect, combined with the sophisticated foaming procedure and the intricate foaming mechanism in the $scCO_2$ foaming method used to produce bimodal foams, means that a specific heat-transfer is urgently required for them.

On the other hand, blocking radiation is one prerequisite for polymeric foams to achieve a low thermal conductivity without using an insulation gas. Therefore, carbonaceous materials are often added to act as a black body. Multi-walled carbon nanotube (MWCNT) is one kind of carbonaceous material that can be brought to an excited electronic state after absorbing radiation in an electromagnetic field. In the region of infrared (IR) radiation, MWCNTs have been used in the application of thermal insulation [3]. In addition, MWCNTs that acted as nucleating agents initiated heterogeneous bubble nucleation during the $scCO_2$ foaming process, and this increased cell density and reduced cell size [4]. To achieve bimodal foams with a low thermal conductivity, an amorphous polymer with a broad foaming window is preferential. Consequently, PS/MWCNT nanocomposites have been selected for the matrix because their foam morphology can be easily tailored by manipulating both the foaming pressure and temperature [5]. In our study, we prepared bimodal PS/MWCNT foams by $scCO_2$ foaming. The large/small cell size ratio of ~ 20, small secondary cells of ~ 5 μm and MWCNTs led to a low thermal conductivity.

EXPERIMENTAL

General purpose polystyrene, GPPS 3100, was purchased at Styrolution (melt flow rate = 10 g/10 min at 200°C/5 kg, density = 1.04 g/cm^3) and was used as received. MWCNTs (average outer diameter = 10 nm, average length = 1.5 μm, surface area = 250 ~ 300 m^2/g, carbon purity 90%, density = 1.44 g/cm^3) were supplied by NanocylTM, Belgium, and were produced by Catalytic Carbon Vapor Deposition (CCVD). Carbon dioxide (Linde Gas, purity over 99%) and pentane (Caledon Laboratories Ltd., water content less than 0.02%) were used as physical blowing agents.

TABLE i). Sample ID at each foaming condition and MWCNT concentration.

Foaming Condition	Foam morphology	MWCNT wt% in PS/MWCNT nanocomposite					
		0	0.1	0.25	0.5	1.0	2.0
13.8MPa, 0 ml Pentane	Unimodal	U1-0	U1-0.1	U1-0.25	U1-0.5	U1-1.0	U1-2.0
13.8MPa, 10 ml Pentane	Unimodal	U2-0	U2-0.1	U2-0.25	U2-0.5	U2-1.0	U2-2.0
13.8MPa, 30 ml Pentane	Unimodal	U3-0	U3-0.1	U3-0.25	U3-0.5	U3-1.0	U3-2.0
20.7MPa, 30 ml Pentane	Unimodal	U4-0	U4-0.1	U4-0.25	U4-0.5	U4-1.0	U4-2.0
20.7MPa, two-step depressurization	Bimodal	B1-0	B1-0.1	B1-0.25	B1-0.5	B1-1.0	B1-2.0
27.6MPa, two-step depressurization	Bimodal	B2-0	B2-0.1	B2-0.25	B2-0.5	B2-1.0	B2-2.0
13.8MPa, 60 ml Pentane	Bimodal	B3-0	B3-0.1	B3-0.25	B3-0.5	B3-1.0	B3-2.0
20.7MPa, 60 ml Pentane	Bimodal	B4-0	B4-0.1	B4-0.25	B4-0.5	B4-1.0	B4-2.0

CO_2 and pentane co-blowing agents were used for foaming; U and B series samples indicate unimodal and bimodal foam morphology, respectively; The number after U or B in the sample ID means a different foaming condition; The number after the dash in the sample ID indicates the MWCNT concentration in wt%.

PS/MWCNT 10 wt% masterbatch was supplied by Nanocyl^{TM}, Belgium. The 1.0 and 2.0 wt% nanocomposites were then prepared by mixing the masterbatch with pristine PS using a DSM twin-screw compounder (Berstorff ZE25, screw length 48D, chamber volume 15 ml) at 200°C and 100 rpm for 10 min. The 0.1 wt%, 0.25 wt%, and 0.5 wt% nanocomposites were prepared by mixing the 1.0 wt% nanocomposite with pristine PS at 200°C and at 100 rpm for 5 min. to make sure that the MWCNTs were well dispersed in the PS matrix.

Pristine PS and PS/MWCNT rod samples were placed into a high-pressure autoclave to dissolve both CO_2 and pentane. Before pressurizing CO_2, liquid pentane was poured into the autoclave. The autoclave was then heated to 50°C. When the required pressure (13.8 MPa, 20.7 MPa or 27.6 MPa) and temperature (50°C) were achieved, the system was held for three days to absorb a sufficient amount of the blend blowing agents. **TABLE 1** summarizes the detailed foaming conditions and the nanocomposite component. Due to the strong plasticization effect of the blowing agents, the glass transition temperature (T_g) was decreased to below the gas-saturation temperature of 50°C, and foaming was induced by rapid depressurization alone in 2 s. All of the foamed samples were then placed in 100°C boiling water for 2.5 min. to further expand the cells. Before any characterizations were done, all of the foam samples were vacuumed for one week, and then they were left for over one month to ensure that air permeated into the cells and was the only gas in them.

Scanning Electron Microscopy (SEM) was used to observe the foam morphology. The samples were first frozen in liquid nitrogen to create cryogenic fractural surfaces, and then were coated with gold for 180 s. The SEM images were collected at an acceleration voltage of 25 kV and at a wide distance of 30 mm using the SEM (JEOL JSM-606). The image processing software, Image-J, was used to calculate the number average of the cell diameter, d, and the cell density, with respect to the solid polymer, N_f. The samples' bulk densities before and after foaming were measured using the water-displacement method (the ASTM D792-00).

The transient plane source (TPS) Hot Disk thermal constants analyzer (Therm Test Inc., TPS 2500, Sweden) was used to measure the thermal conductivity of the PS/MWCNT nanocomposites and foams. In this method, the Hot Disk sensor is used both as a heat source to increase the temperature of the sensor and as a dynamic temperature sensor to record the sample temperature's increase as a function of time. The Hot Disk sensor is placed between two pieces of the sample to mimic the situation inside the sample. By using an electrical current to heat the sensor for a certain amount of time, the surrounding sample's temperature increase is a function of the thermal conductivity. The samples' thermal conductivity will then be obtained using the Hot Disk thermal constants analyzer software to solve the function. In this study, two sample pieces around 15 mm in diameter were used to conduct the test. According to the testing samples' thermal characteristics, the power output (0.005 ~ 0.020 W), test duration (5 ~ 20 s) and sensor size (2 or 3 mm in radius) were determined. The measurement was performed 3 times for each sample, and the experimental deviation was less than 5%. Details of the TPS method can be found in Refs [6, 7].

Fourier transform infrared spectroscopy (FTIR) (PerkinElmer spectrum one) was used to measure the spectral transmittance of IR radiation through the foams. We collected the spectral transmittance by averaging 8 scans in a spectral range of 4000 to 400 cm^{-1} (wavelength from 2.5 to 25 μm). The spectral resolution was 4 cm^{-1}. The airborne H_2O and CO_2 background noise was registered before recording the samples' IR transmittance. To eliminate the sample thickness's effect on the IR transmittance, the foams were cut into plate shapes of 15 mm in diameter, with a thickness ranging from 0.2 to 2 mm. At least 6 plate samples of different thicknesses were measured to calculate the radiative thermal conductivity.

RESULTS AND DISCUSSION

The large degree of CO_2 supersaturation in the PS matrix would have enhanced bubble nucleation, and the pentane's strong plasticization effect on the PS matrix would have promoted bubble growth. When less than 30 ml of pentane was added to the foaming chamber to plasticize the PS matrix, all of the samples foamed in this condition had a unimodal foam morphology. **FIGURE 1** summarizes the cell size, the cell density and the expansion ratios of the unimodal foam samples in the U1 to U4 series, and **TABLE 1** shows the corresponding foaming conditions. At a foaming pressure of 13.8 MPa (U1, U2 and U3 series), the increased amount of pentane increased the expansion ratio from 9- to 19-fold. The foamed samples' cell density was ~ 1×10^{11} #/cm^3. Thus, the increased expansion ratio led to an increase in the cell size from 2 μm to 6 μm. When 30 ml of pentane was added to the foaming system in the U3 and U4 series, the foaming pressure increased from 13.8 MPa to 20.7 MPa and this increased the cell density to ~ 3×10^{11} #/cm^3. But it decreased the expansion ratio to ~ 15-fold and also reduced the cell size. With an increased gas content (that is, a higher gas pressure), the cell density increased, and both the cell size and the expansion ratio decreased. It seems that the higher cell density, which had resulted from a larger gas content, had decreased the cell wall thickness. Thus, cell opening was promoted, and eventually, the gas loss was accelerated, and this induced a lower expansion ratio [8].

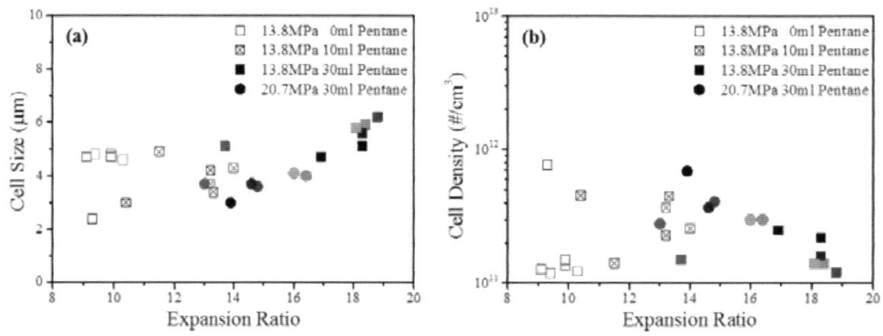

FIGURE 1. Expansion ratio, cell size and cell density of unimodal foams: black – PS foams; red – PS/MWCNT 0.1 wt% foams; green - PS/MWCNT 0.25 wt% foams; blue - PS/MWCNT 0.5 wt% foams; cyan - PS/MWCNT 1.0 wt% foams; pink - PS/MWCNT 2.0 wt% foams

When the amount of pentane added to the foaming chamber was increased from 30 ml to 60 ml, it not only induced stronger plasticization in the PS matrix, but it also changed the bubble nucleation behavior. The result was a bimodal cellular structure with a large expansion ratio. Two-step depressurization was also applied to prepare bimodal foams with a relatively small expansion ratio. **FIGURE 2** shows the SEM micrographs of the fabricated pristine PS and PS/MWCNT bimodal foams. **TABLE 2** summarizes the expansion ratio of the bimodal foam, and shows the thickness of the secondary cell layer, and the cell size, the cell density and the volume ratio of the primary and secondary cells (voids only) for the B1 to B4 series. **TABLE 1** shows the corresponding foaming conditions. In our study, the bimodal foams' primary large cells were 6 ~ 25 times larger in diameter than the secondary small cells. This led to an increased expansion ratio in the PS/MWCNT bimodal foams from 8- to 31-fold. At the same time, the corresponding secondary cells remained similar in size to those of the unimodal foams.

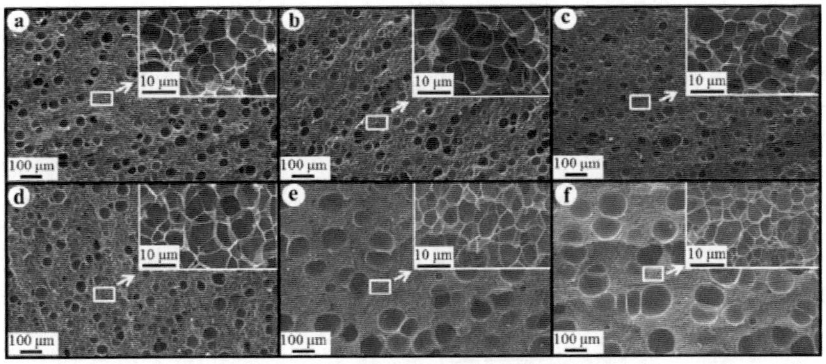

FIGURE 2. SEM micrographs of pristine PS and PS/MWCNT bimodal foams (foaming condition: 13.8 MPa with 60 ml pentane added): (a) PS foam; (b) PS/MWCNT 0.1 wt% foam; (c) PS/MWCNT 0.25 wt% foam; (d) PS/MWCNT 0.5 wt% foam; (e) PS/MWCNT 1.0 wt% foam; and (f) PS/MWCNT 2.0 wt% foam

TABLE ii). Sample ID, cell size, cell density, volume ratio, and expansion ratio of pristine PS and PS/MWCNT bimodal foams

Sample ID	Primary Cell Size (μm)	Cell Density (#/cm³)	Volume Ratio	Secondary Cell Size (μm)	Cell Density (#/cm³)	Volume Ratio	Expansion Ratio	Thickness of secondary cell layer (μm)
B1-0	71.7	3.5×10^6	8.9%	3.3	3.7×10^{11}	91.1%	12.1	93.2
B1-0.1	66.5	6.2×10^6	12.2%	2.8	6.1×10^{11}	87.8%	12.1	71.4
B1-0.25	75.4	5.0×10^6	10.6%	3.4	4.8×10^{11}	89.4%	14.8	87.4
B1-0.5	69.8	7.3×10^6	15.2%	3.2	4.1×10^{11}	84.8%	12.8	64.7
B1-1.0	77.7	4.1×10^6	11.1%	3.2	5.0×10^{11}	88.9%	13.6	88.1
B1-2.0	56.5	2.3×10^6	2.9%	3.1	4.7×10^{11}	97.1%	12.2	133.4
B2-0	63.7	3.7×10^6	9.4%	2.8	4.2×10^{11}	90.6%	9.3	81.7
B2-0.1	62.4	6.9×10^6	14.6%	2.5	6.3×10^{11}	85.4%	10.6	60.2
B2-0.25	65.9	5.4×10^6	10.3%	2.9	5.5×10^{11}	89.7%	10.4	79.4
B2-0.5	61.7	6.6×10^6	15.4%	2.6	4.9×10^{11}	84.6%	9.1	57.9
B2-1.0	60.2	4.8×10^6	11.3%	2.5	5.2×10^{11}	88.7%	8.9	69.5
B2-2.0	52.8	3.0×10^6	5.4%	2.5	5.0×10^{11}	94.6%	8.2	93.1
B3-0	51.5	4.5×10^7	24.7%	5.2	1.3×10^{11}	75.3%	21.8	30.0
B3-0.1	43.3	8.3×10^7	19.3%	7.6	7.2×10^{10}	80.7%	30.5	34.3
B3-0.25	44.8	6.9×10^7	18.5%	6.5	8.3×10^{10}	81.5%	24.8	30.1
B3-0.5	42.0	6.9×10^7	21.1%	6.9	5.8×10^{10}	78.9%	24.1	28.5
B3-1.0	115.4	6.6×10^6	37.7%	5.8	8.6×10^{10}	62.3%	28.2	44.4
B3-2.0	112.4	5.5×10^6	42.5%	5.1	8.1×10^{10}	57.5%	19.6	37.6
B4-0	43.2	4.2×10^7	17.4%	4.6	1.6×10^{11}	82.6%	16.7	33.4
B4-0.1	43.3	1.3×10^8	24.7%	6.9	9.0×10^{10}	75.3%	31.0	24.0
B4-0.25	41.1	6.6×10^7	18.4%	6.5	9.2×10^{10}	81.6%	21.9	35.6
B4-0.5	34.5	1.3×10^8	20.7%	5.9	1.0×10^{11}	79.3%	24.5	23.9
B4-1.0	93.5	7.4×10^6	33.9%	5.3	8.0×10^{10}	66.1%	19.6	40.9
B4-2.0	119.6	3.4×10^6	34.5%	4.6	1.1×10^{11}	65.5%	19.3	49.8

The total thermal conductivities of the unimodal and bimodal PS/MWCNT foams are shown in **FIGURE 3**. Because of their limited expansion ratio, the conduction through the unimodal foams was dominant. And the radiative absorption capacities of the MWCNTs had only a limited effect in reducing the total thermal conductivity. On the other hand, bimodal foams with a large expansion ratio were easily achieved. Therefore, the contribution of the conductivity via conduction to the total thermal conductivity in these foams was decreased. But when the expansion ratio became sizable, the radiative thermal conductivity also became significantly large. So there was an optimal expansion ratio. But this increase in the radiative thermal conductivity (and in the total thermal conductivity) along with the expansion ratio's increase was arrested when the MWCNTs were introduced, as happened in the case of the unimodal foams. In the bimodal foams, the MWCNTs dramatically absorbed radiation. Consequently, the total thermal conductivity was decreased at larger expansion ratios. Correspondingly, the optimal expansion ratio required to minimize the total thermal conductivity will increase with an increased MWCNT content.

By adjusting the expansion ratio and the MWCNT concentration, the lowest thermal conductivity of the unimodal PS/MWCNT foams obtainable in this study was 32.8 mW/m-K. The application of a bimodal cellular structure was effective in further decreasing the total thermal conductivity. This was because the primary large cells present in the bimodal foams induced a larger expansion ratio than they would do in unimodal foams. This expansion reduced the solid conduction while the secondary small cells in the bimodal foams also experienced the Knudsen effect, which reduced gas conduction. Furthermore, the MWCNTs in the bimodal foam with a large expansion ratio absorbed more radiation and contributed to less solid conduction than they did in the unimodal foams. As a result, by tailoring the bimodal cellular structure and the MWCNT concentration, the PS/MWCNT 1.0 wt% bimodal foam achieved a minimum thermal conductivity of 30.2 mW/m-K.

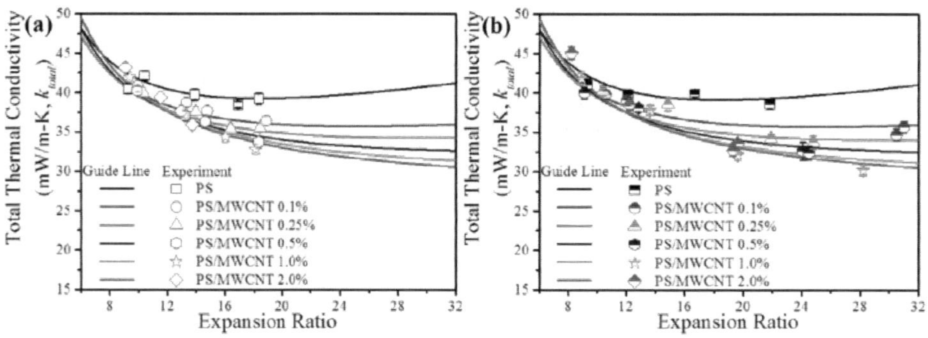

FIGURE 3. Total thermal conductivity of (a) unimodal foams and (b) bimodal foams

CONCLUSION

In this study, we prepared PS/MWCNT nanocomposite foams using the $scCO_2$ foaming system. Both unimodal and bimodal cellular structures were obtained by using co-blowing agents with an amount of plasticizing agent or by using two-step depressurization method in the foaming process. Their thermal insulation performances were evaluated. We optimized the cellular structure and the MWCNT's concentration, and then we achieved a minimum total thermal conductivity in the bimodal PS/MWCNT 1.0 wt% foam system with a 28-fold expansion ratio and 5.8 μm secondary cells. The primary large cells reduced solid conduction while secondary small cells reduced gas conduction, and the MWCNTs reduced radiation. Because of this action, a very low thermal conductivity of 30.2 mW/m-K was successfully obtained without using any insulation gas.

REFERENCES

1. Canadian Energy Overview 2014; *National Energy Board*: 2014
2. C. Zhang, B. Zhu, D. Li and L. J. Lee, *Polymer* **53**, 2435-2442 (2012).
3. P. Gong, P. Buahom, M.-P. Tran, M. Saniei, C. B. Park and P. Pötschke, *Carbon* **93**, 819-829 (2015).
4. A. Rizvi, R. K. M. Chu, J. H. Lee and C. B. Park, *ACS Appl. Mat. Interfaces* **6**, 21131-21140 (2014).
5. X. Xu and C. B. Park, *J. Appl. Polym. Sci.* **109**, 3329-3336 (2008).
6. R. Coquard and D. Baillis, *J. Heat Transfer* **128**, 538-549 (2005).
7. T. Log and S. E. Gustafsson, *Fire Mater.* **19**, 43-49 (1995).
8. C. B. Park, A. H. Behravesh and R. D. Venter, *Polym. Eng. Sci.* **38**, 1812-1823 (1998).

A comparison between several commercial polymer hollow fiber membranes for gas separation

Xiao Yuan Chen[ab], Serge Kaliaguine[a] and Denis Rodrigue[a,*],

[a] *Department of Chemical Engineering, Université Laval, Quebec City, QC, G1V 0A6, Canada*

[b] *Centre National en Électrochimie et en Technologies Environnementales, Collège de Shawinigan, Shawinigan, QC, G9N 6V8, Canada*

Abstract. Polyethersulfone (PES), polyetherimide (Ultem® 1000), and polyimides (Matrimid® 5218 and P84) are common commercial polymers used to produce hollow fibre membranes for different gas separation applications. In this work, asymmetric hollow fiber membranes were prepared using these polymers by a phase inversion technique. The effects of spinning parameters (composition of the dope and bore solution, bore flow rate, air gap distance, temperature of the spinneret and coagulation bath, as well as take-up speed) on the membrane structure and gas permeation properties were investigated. The membrane separation performances were characterized by measuring their gas permeation properties (permeability and selectivity) for different gases (H_2, CO_2, O_2, N_2, and CH_4) and by their cross-sectional morphology using scanning electron microscopy (SEM). The relationship between the hollow fiber gas separation performance and the intrinsic gas properties of dense flat membranes made of the same materials were also studied. A comparison between the average apparent skin layer thickness calculated from O_2 permeability/permeance, and the results based on SEM images was made and good agreement was obtained between both results.

Keywords: polymer, membrane, gas separation, permeance, selectivity.
PACS: 81

INTRODUCTION

Gas separation membranes are generally made from organic and inorganic materials. Inorganic membranes include metals, ceramics, metal organic frameworks (MOF) or carbon molecular sieves (CMS), but the majority of the membranes today are organic made from polymers. This trend is associated to the fact that polymer membranes have good mechanical properties, good processability and economical operation [1]. When the membranes are used for industrial gas separation, large surface area is highly needed for high process capacity. Hollow fiber membranes are known to have higher effective surface area per unit volume of membrane module (10000 m^2/m^3) than spiral wound (200-1000 m^2/m^3) and envelope type (30-500 m^2/m^3) modules [2]. Also, hollow fibers provide mechanical support (module) and are easier to handle in fabrication, as well as in gas separation operation. Therefore, hollow fiber modules are more commonly used for industrial applications. Polyethersulfone (PES), polyetherimide (Ultem® 1000), and polyimides (Matrimid® 5218 and P84) are common commercial polymers used to produce flat and hollow fiber membranes for different applications, especially gas separation. They are compared in this work.

BACKGROUND AND THEORY

Permeance (Q) is defined through the steady state permeation flux N via the pressure difference (Δp) across the membrane of thickness (l) as [3]:

$$Q = \frac{P}{l} = \frac{N}{\Delta p} \tag{1}$$

In this work, the pure gas transport properties were measured using the variable pressure (constant volume) method [4]. So the permeance (10^{-6} cm^3(STP) cm^{-2} s^{-1} cmHg) is determined by:

$$Q = 22412 \times \frac{V}{ART} \times \frac{1}{\Delta p} \times \frac{dp}{dt} \tag{2}$$

where A is the membrane area (cm^2); p is the pressure (psi); V is the downstream volume (cm^3); R is the universal gas constant (6236.56 cm^3 cmHg/mol K); T is the absolute temperature (K); and dp/dt is the permeation rate (psi/s).

The values are commonly reported in units of GPU (gas permeation units) defined as: 1 GPU = 1 Barrer/1 micron (1 Barrer = 10^{-10} cm^3(STP) cm^{-2} s^{-1} cmHg). The ideal selectivity of hollow fiber membrane (α_H) is defined as the ratio of gas permeability or permeance where Q_A, P_A and Q_B, P_B are the permeability coefficients and permeance of gases A and B, respectively as:

$$\alpha_H = Q_A / Q_B = (P/l)_A / (P/l)_B \tag{3}$$

EXPERIMENTAL

The asymmetric structure of hollow fiber membranes was produced by dry-wet spinning using a complete system of manufacturing hollow fiber membranes from EMI (European Membrane Institute) including two gear pumps, a spinning nozzle, a working tank, a control box, a coagulation bath, a steel rinsing bath (stainless isolated), a heating system, a rolling device and a filtration unit. Two metering pumps transfer precise quantities of the polymer solution dope and the bore fluid to the spinneret. Then, the polymer solution and non-solvent fluid go from the spinneret and solvent evaporation in the air-gap region. Finally, the take-up unit collects the fibers from the coagulation bath. Fig. 1 illustrates the complete production line of solution spinning via phase inversion.

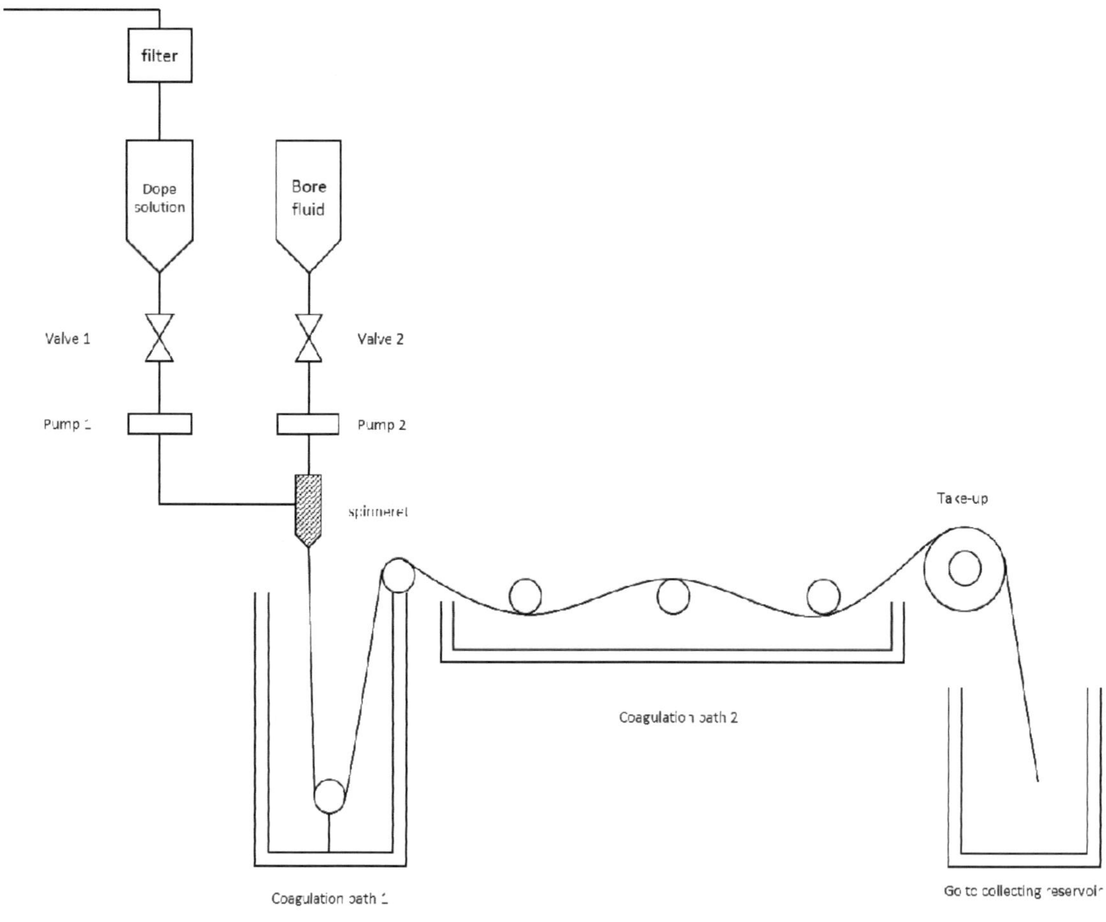

FIGURE 1. Schematic diagram of the spinning apparatus for hollow fiber membrane production.

Materials

Ultem® 1000 polymer, Matrimid® 5218 and P84 powder was purchased from GE Plastics (Pittsfield, MA), Huntsman Advanced Materials Americas Inc. and HP Polymer Inc., respectively. Polyethersulfone (Ultrason

E6020P) was graciously provided by BASF. All the polymers were dried overnight at 120°C before use. Their glass transition temperature and chemical structure are listed in Table 1.

TABLE 1. Glass transition temperature and chemical structure of the polymers used.

Materials	Glass transition temperature (°C)	Chemical structure
Ultem® 1000 (polyimide)	217	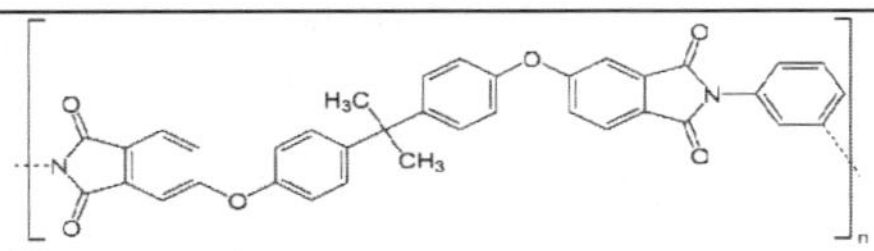
Matrimid® 5218 (polyimide)	319	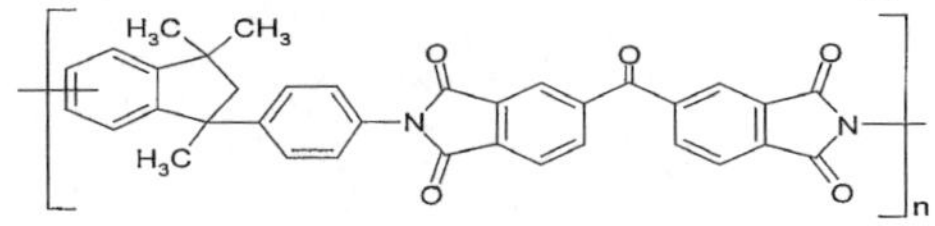
Ultrason E6020P (PES)	225	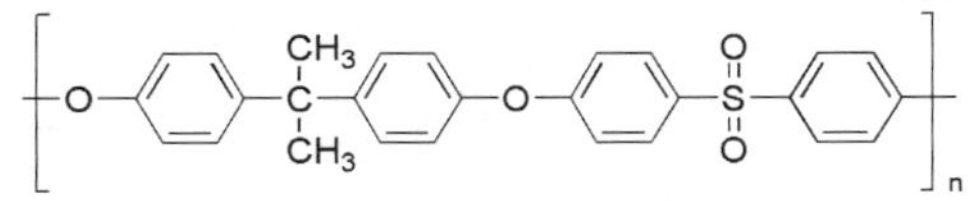
P84 (polyimide)	315	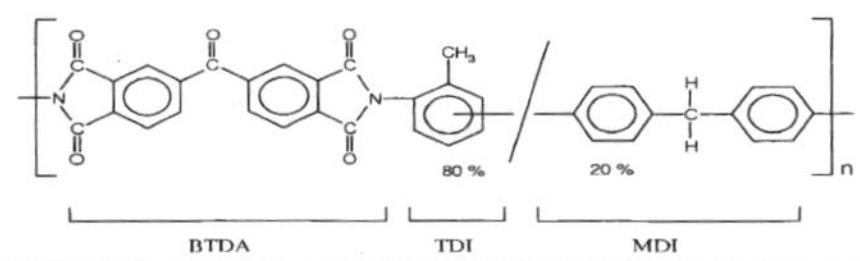

Characterization

TGA: The weight loss curves (TGA-DTG) were recorded using a TA Instruments model Q5000IR from 50 to 850°C at a heating rate of 10°C/min under air.
DMA: The glass transition temperature and tensile modulus of the hollow fiber membranes were determined using a dynamic mechanical thermal analyzer (RSA-3, TA Instruments, USA) at a frequency of 1 Hz. The temperature was increased from 30 to 350°C at a rate of 10°C/min.
SEM: scanning electron micrographs (SEM) were recorded to determine the skin layer thickness and characterize the morphology of the hollow fiber membranes asymmetric structure using a JEOL JSM-840A operated at 15-20 kV.
Gas permeation measurements by the high throughput gas separation equipment (HTGS 1000) was provided by POROMETER which is a Belgian-German membrane and filter characterization equipment manufacturer.

RESULTS AND DISCUSSION

Morphology of the Hollow Fibers

The fiber morphology was examined by SEM and typical images are reported in Fig. 2. These images show the PES hollow fibers produced under the following conditions: dope composition of PES = 28% wt. in NMP 72% wt.; bore fluid composition of 20% wt. water in 80% wt. NMP; dope flow rate of 3.0-4.0 ml/min; bore flow rate of 3.0-4.0 ml/min; air-gap distance of 3 cm, a spinneret temperature of 23°C and a take-up speed of 10 m/min. Similar structures were obtained for Ultem, Matrimid and P84, but not shown here due to page limitation.

FIGURE 2. SEM images of PES hollow fibers asymmetric structure. (A) Outer surface of the dense skin; (B) overall cross-section of a fiber; (C) membrane thickness; and (D) skin layer thickness.

Gas Transport Properties

From the sample produced, Table 2 presents the gas transport properties measured at 30°C and 2 bar. Comparison of the gas transport results from these four different polymers shows that the highest permeance for all the gases was obtained from PES. However, selectivities for H_2/N_2, H_2/CH_4, CO_2/CH_4 and O_2/N_2 are smaller than the other three polymers with the exception of CO_2/N_2. The lowest gas permeance for all gases was obtained from P84. However, the selectivities were not higher than for the others polymer membranes. Therefore, P84 needs improvement of its properties by blending with materials having higher permeance. But P84 has a very important advantage which is good spinnability. Matrimid was the best material for gas separation, especially for CO_2/CH_4 and CO_2/N_2. Ultem has lower cost compared to Matrimid, P84, and other commercial polyimides. Ultem also has higher mechanical properties and performs in continuous use up to 170°C and even up to its glass transition temperature (217°C). The selectivity of H_2/N_2 and H_2/CH_4 can reach 117 and 91.7 which are quite high. So the final material selection must be made depending on the industrial application and which gas separation needs to be performed.

TABLE 2. Pure gas permeation data for the asymmetric hollow fiber membranes measured at 30°C and 2 bar.

Membranes	Gases	Ultem	Matrimid	Polyethersulfone	P84
Permeance (GPU)	H_2	38.5	56.4	58.7	4.44
	CO_2	7.53	17.7	27.9	0.65
	O_2	3.34	3.88	8.38	0.41
	CH_4	0.42	0.73	3.05	0.06
	N_2	0.33	0.52	1.59	0.06
Selectivity	H_2/N_2	117	109	37.1	73.0
	H_2/CH_4	91.7	77.3	19.2	78.0
	CO_2/N_2	22.8	34.0	17.6	10.7
	CO_2/CH_4	17.9	24.3	9.15	11.4
	O_2/N_2	10.4	7.46	5.29	7.20

CONCLUSION

In this work, four different commercial polymers (Ultem, Matrimid, Ultrason and P84) were used to prepare asymmetric hollow fiber membranes. From the samples produced, their permeance were obtained for different gases (H_2, CO_2, O_2, N_2 and CH_4). The results showed that no single polymer can be universally used for any gas separation. So a careful selection must be made based on the type of gas separation to perform. In our case, based on selectivity values, Ultem performed best for H_2/N_2, H_2/CH_4 and O_2/N_2, while Matrimid performed better for CO_2/N_2 and CO_2/CH_4. Nevertheless, these conclusions only apply for the samples produced and under the conditions tested (30°C and 2 bar). Different results might be obtained for other conditions and this is currently being investigated.

ACKNOWLEDGMENTS

The authors would like to thank the Natural Science and Engineering Research Council of Canada (NSERC) for financial support through a strategic grant. X.Y. Chen also thanks the Mitacs Accelerate program for a postdoctoral scholarship.

REFERENCES

1. X. Y. Chen, H. Vinh-Thang, A. A. Ramirez, D. Rodrigue and S. Kaliaguine, *RSC Adv.* **5**, 24399-24448 (2015).
2. R. W. Baker, in *Advanced membrane technology and applications*, ed. N. N. Li, A. G. Fane, W. S. Winston Ho and T. Matsuura, John Wiley & Sons, Inc., New Jersey, 2008, pp. 557-580.
3. K. Ghosal and B. D. Freeman, *Polym. Adv. Technol.* **5**, 673-697 (1994).
4. A. E. Barnabeo, W. S. Creasy and L. M. Robeson, *Polym. Sci. Polym. Chem. Ed.* **13**, 1979-1986 (1975).

Study on different high-density polyethylenes during stress and photo-oxidative aging

Xiancheng Ren[1*], Dafu Luo[1]

[1]College of Polymer Science and Engineering, Sichuan University, Chengdu 610065, Sichuan, China

Abstract: An exiperimental study has been conducted to determine the effect of molecular architecture on the stress and photo-oxidative aging behavior of high-density polyethylene(HDPE). These were carried out using the most representative grades of HDPE: 8920, 6100, 9070, 7600, TR480, 2480NT, TR144 and 4801. The comonomer content and the average molecular were measured by succesive self-nucleation and annealing technique and high-temperature gel permeation chromatography. Besides, the mechanical behavior was tested to acquire the toughness of different materials. The samples were exposed to 5 MPa stress and ultraviolet irradiation in an aging apparatus, and observed at time intervals. The changes in stucture and properties were charaterized by metallurgical microscope, gel content measurements, differential scanning calorimetry(DSC), attenuated total reflection–Fourier transform infrared spectroscopy. It was found that with increasing time, the melting point and oxygen induction times decreased, whereas the crytallinity increased. It was also observed polar groups in the polyethylene were generated, and gel content reached about 10% during degradation. The results indicated that the resistance to cracking of different HDPES followed the order: 8920<6100<9070<7600<TR480<2480NT<TR144<4801. This study also may be useful for the rational selection of HDPE for the sheath material of bridge cable.

Keywords: aging; degradation; properties and characterization; high-density polyethylene

PACS: 81.05.Lg

INTRODUCTION

High-density polyethylene(HDPE) has been widely employed as pipes, fibers, fishing nets, containers and woven bags, due to its excellent and comprehensive properties, such as high rigidity, chemical stability, excellent dielectric properties and corrosion resistance[1]. Over the past decades, so many researchers have studied about photo-oxidative aging on HDPE and resolved changes in chemical strcture and mechanical properties, and other aspects of the process[2-6].

However, only a few researchers have studied the strss and photo-oxidative aging behavior of HDPE. In this work, the stucture changes of different HDPEs during stress and photo-oxidative aging and the relationship between materials and cracking time were studied, through the experiments of the surface morphology observation and structure analysisi. This study may be useful for the rational selection of HDPE for the sheath material of bridge cable.

EXPERIMENTAL

Materials and Sample Preparation

The eight white pellet materials were supplied by Dow Chemical. Antioxidant Irganox 1010 and auxiliary antioxidant 168 were obtained from Ciba Specialty Chemicals(Switzerland). The parameters of above eight materials were listed in Table 1.

Sample preparation

Eight HDPE samples were mixed with 2.5 wt‰ antioxidant irganox 1010 and 2.5 wt‰ auxiliary irganox 168, then extruded in TSSI-25 twin-screw extruder at 190℃. Then the extruding samples were processed by compression molding into sheets with a thickness of 2.1mm. The sheets were allowed to cool down to ambient temperature.

Thin sample layers were cut into standard dumbbell-shaped specimens(length 120 mm, width 25 mm).

Table 1. Material data of eight HDPE samples used

Sample	Mn ($g\ mol^{-1}$)	Mw ($g\ mol^{-1}$)	Mw/ Mn	Density($g\ cm^{-3}$)
8920(1#)	18440	63158	3.4446	0.954
6100(2#)	14863	223352	15.0274	0.954
9070(3#)	19753	178272	9.0251	0.945
7600(4#)	12950	285358	22.0354	0.947
TR480(5#)	24635	262015	10.6359	0.944
2480NT(6#)	23792	261394	10.9866	0.951
TR144(7#)	23191	176061	7.5918	0.946
4801(8#)	25988	226574	8.7184	0.945

Aging Condition

Stress and photo-oxidative aging experiments of HDPE specimens were performed in a self-made aging apparatus (Chengdu yibang analysis instrument Co., Ltd) at 50 °C ± 1 °C with air circulation and relative humidity 30% ± 1%. The samples were applied 5MPa stress in the apparatus, which was equipped with a 1000W Ga-In source lamp with a spectrum range of 300~405 nm, and with a maximum intensity at 365 nm. The irradiation intensity was 3.5 Wm^{-2} tested by a UV irradiance meter model UV-A Tester (Photoelectric Instrument Factory of Beijing Normal University, Beijing, China). The samples were taken out to observe at interval time.

Characterization

Successive Self-Nucleation and Annealing Technique

The SSA analysisi comprised of successive heating and refrigerated cooling pograms operating on a TA instrument DSC Q200 under a rate of 50mL min^{-1} N_2 flow.

Metallurgical Microscopy Observations

Surface morphologies of stress and photo-oxidative aged samples were observed using a UM200i metallurgical microscope (Chongqing UOP Photoelectric Technology Corp. Ltd, China)

Fourier Transform Infrared Spectroscopy

Infrared spectra of aged samples were recorded by using a Nicolet 6700 instrument equipped with a diamond ATR accessory. Carbonyl index (CI) value was defined as the ratio of the carbonyl stretching peak area range 1670 to 1800 cm^{-1} and CH_3 stretching peak area at 2917 cm^{-1}.

$$CI = \frac{A_{C=O}}{A_{2917}} \tag{1}$$

where $A_{C=O}$, A_{2917} are respectively carbonyl group absorption peak areas and internal reference peak areas.

Differential Scanning Calorimetry

A TA instrument DSC Q200 was used for the thermal analysis of different aging time specimens. Approximately 5mg powder samples scraped from the surface of specimens were heated from 40 °C to 210 °C at rate of 10 °C min^{-1}. The heat of fusion (ΔH) of experimental melting peak area was derived to calculate crystallinity (Xc):

$$X_C = \frac{\Delta H}{\Delta H_{100}} \times 100\% \tag{2}$$

Here ΔH_{100} is defined as the heat of fusion of 100% crystalline HDPE (293 J/g[7-8])

In addition, oxidation induction time (OIT) was measured on the same instrument. 5mg samples were heated to 210 °C at the same rate, with 50 ml min^{-1} N_2 flow. After thermal equilibration for 5 min, the purge gas was switched to O_2. OIT was the time interval between O_2 insertion and oxidation onset point measured within ± 0.1 min.

RESULTS AND DISCUSSION

Surface Morphology Observation

0h

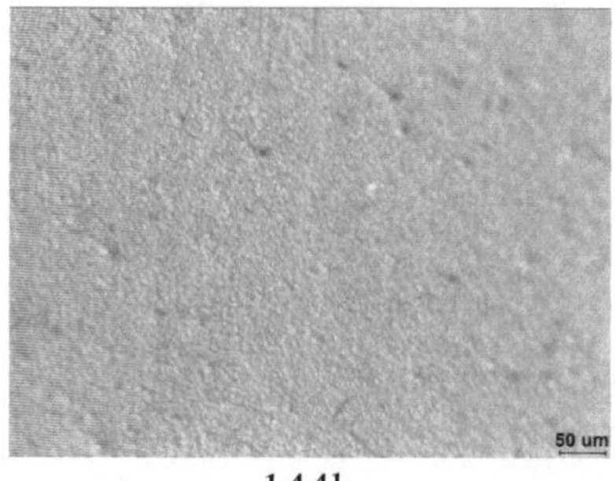

144h

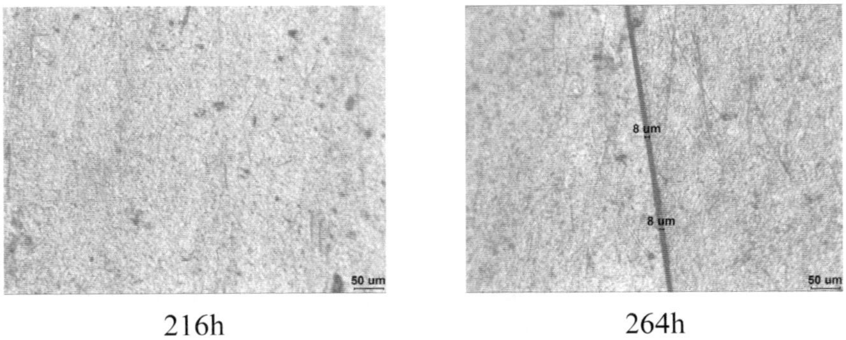

Figure 1. The morphologies of 7600 surfaces during stress and photo-oxidative aging.

The surfaces of eight HDPE samples were observed with metallurgical mircoscopy.7600 was taken as example, and the micrographs of 7600 were shown in Fig. 1. From Fig. 1 , we found that in 7600 samples microcracks appeared after 264h of aging, and the direction of crack propagation was vertical to the stress dirction.

Chemical Changes on the Surface of HDPE Samples after Aging

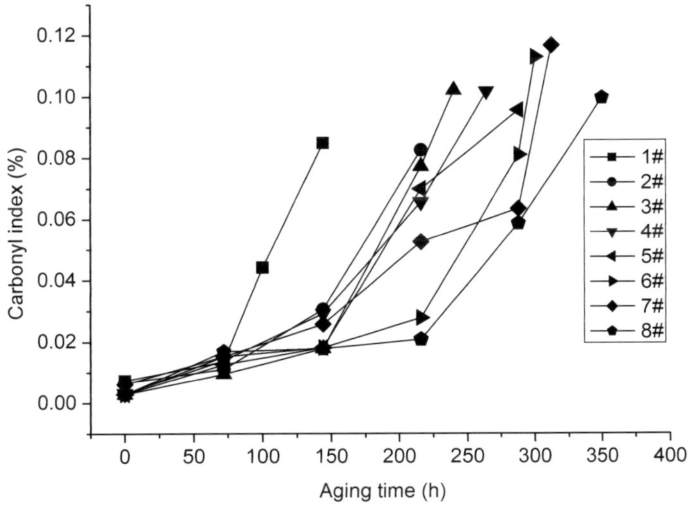

Figure 2. Changes in carbonyl index for eight samples during stress and photo-oxidative aging.

IR spectroscopy was widely used to confirm the presence of various oxidation products which were formed as a result of HDPE samples degradation reactions[9-11]. In our study, ATR-FTIR spectroscopy was used to characterize the chemical changes on the surface of samples after degradation. It was clear that carbonyl groups were formed after aging from the absorption band at 1715 cm-1 which was the C=O stretching vibration.

The degree of oxidation for HDPE samples during aging was represented by the carbonyl index. Figure 2 showed the results for eight HDPE materials after different aging times. It could be observed that CI increased with aging time, which indicated that the materials performed gradual oxidation reaction to produce more free radicals and chemical changes.

Differential Scanning Calorimetry

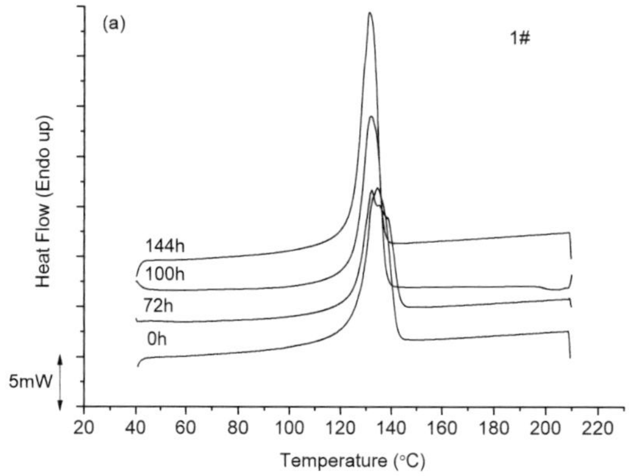

Figure 3. DSC crystallization cuve of 8920 during stress and photo-oxidative aging

In this work, the TA instrument DSC Q200 was used for the thermal analysisi of specimens processed at different aging times. 8920 was taken as example, and the DSC thermograms under different aging times was shown in figure 3.

CONCLUSION

Eight most commercially available HDPE grades during stress and photo-oxidative aging were systematically studied. It was found that with increasing time, the melting point and oxygen induction times decreased, whereas the crytallinity increased. It was also observed polar groups in the polyethylene were generated, and gel content reached about 10% during degradation. The results indicated that the resistance to cracking of different HDPES followed the order: 8920<6100<9070<7600<TR480<2480NT<TR144<4801. This study also may be useful for the rational selection of HDPE for the sheath material of bridge cable.

ACKNOWLEDGMENTS

The author would like to thank the National Natural Science Foundation of China (grant 51273130) for financial support of this work.

REFERENCE

1. S.S. Wu, J. Zhang, and X. Xu, *Polym. Int.* 52, 1527-1530 (2003).
2. X.M. Shi, J.D. Wang, S. Stapf, C. Mattea, W. Li, and Y.R. Yang, *Polym. Eng. Sci.* 51, 2171-2177 (2011).
3. F. Carrasco, P. Pages, S. Pascual, and X. Colom, *Polym. J.* 37, 1457-1464 (2001).
4. S. A. Jabrain and A. L. Elizabeth, *J. Appl. Ploym. Sci.* 53, 411-423 (1994).
5. R. Yang, Y. Liu, J. Yu, and D. Q. Zhang, *Polym. Eng. Sci.* 48, 2270-2276 (2008).
6. J. Pabiot and J. Verdu, *Polym. Eng. Sci.* 21, 32-38 (1981).
7. Wunderlich, B. and Dole, M, *J. Polym. Sci.* 24, 201-213(1957).
8. Wunderlich, B. and Czornyj, G, *Macromol.* 10, 906-913(1977).
9. G.R. Rideal and J.C. Padget, *Polym. Degrad. Stab.*57, 1-15 (1976).
10. F. Gugumus, *Polym. Degrad. Stab.* 67, 35-47 (2000).
11. T. Zanasi, E. Fabbri, and F. Pilati, *Polym. Test.* 28, 96-102 (2009).

Due to the development of the synthesis technique, novel NPs are now being fabricated, for instance, Janus particles, which are characterized by two regions of different surface chemical composition. It is important to study their self-assembly in the polymer matrix. The visco-elastic properties are important aspects of PNCs, since they are used mostly under conditions of dynamic deformation (this is of special importance for elastomer-based nanocomposites such as automobile tires). We adopt molecular-dynamics simulations, both on the coarse-grained and the chemistry-specific levels, to study the influence of the morphology on the mechanical properties of polymer nanocomposites (PNCs) filled with uniform spherical nanoparticles (which means without chemical modification) and patchy spherical nanoparticles (with discrete, attractive interaction sitesat prescribed locations on the particle surface). Through the coarse-grained model, the non-linear decrease of the elastic modulus (G') and the maximum of the viscous modulus (G") around the shear strain of 10% is clearly reproduced. By turning to the polybutadiene model, we examine the effect of the shear amplitude and the interaction strength among uniform NPs on the aggregation kinetics. Interestingly, the change of the G' as a function of the aggregation time exhibited a maximum value at intermediate time attributed to the formation of a polymer-bridged filler network in the case of strong interaction between NPs. By imposing a dynamic periodic shear, we probe the change of the G' as a function of the strain amplitude while varying the interaction strength between uniform NPs and its weight fraction. A continuous filler network is developed at a moderate shear amplitude, which is critically related to the interaction strength between NPs and the weight fraction of the fillers. In addition, we study the self-assembly of the patchy NPs, which form the typical chain-like and sheet-like structures. For the first time, the effect of these self-assembled structures on the visco-elastic and stress-strain behavior of PNCs is compared. In general, in the coarse-grained model we focus on the size effect of the rough NPs on the Payne effect, while some other parameters such as the dynamic shear flow, the interaction strength between NPs, the weight fraction and the chemically heterogeneous surface of the NPs are explored for the chemistry-specific model.

Numerical Simulations for Injection Molding of Postconsumer Waste

Frederik Desplentere [a,*], Sofie Deceur[a]

[a] *Research group ProPoLiS, KU Leuven Technology Campus Ostend, Zeedijk 101 8400 Ostend Belgium*

Abstract. In current polymer postconsumer waste streams, more and more incompatible blends are found which are nowadays non valorized. Already some companies are applying compatibilizers to increase the value of these kind of waste streams; sometimes upcycling is used to describe this process. To be able to handle these type of materials within numerical simulations tools for the injection molding process, accurate rheological information is needed. Within this investigation, a dual bore rheometer and a reverse engineering technique on an injection molding machine are used to determine industrial relevant rheological data. On the injection molding machine, the characterization is performed using a tensile bar mold equipped with pressure and temperature sensors. Finally, the compatibilized material is used to produce an industrial part through injection molding. During the processing, pressure and temperatures are registered within the mold. Good correlation is found between experimental data and numerical simulation results using the new rheological data.

Keywords: Numerical simulations for Injection Molding, Compatibilizer, Postconsumer waste
PACS: 83.50.Uv, 83.60.Fg, 83.85.Cg

INTRODUCTION

One of the biggest challenges of the industry is to produce in a sustainable way. As a consequence, the reduction, re-use, and upgrade of waste is becoming increasingly important. In current polymer postconsumer waste streams, polymers are very often contaminated by amounts of other polymers. In these cases, their recyclates often demonstrate poor mechanical properties, since most polymers are not compatible. Because of the unsatisfactory quality of mixed recyclates, they are recycled into less valuable products (downgrading) or not recycled at all. In this respect, the development of a processing road for postconsumer recycled material has become an important challenge. A route to improve processability and to optimize the polymer quality is offered by the so-called "compatibilization".

The results presented in this paper are generated within a IWT (Flemish funding agency) supported running project (IWT VIS 135100). Within this project, the aim is to find a good compatibilizer for major waste streams. One of those streams is the immiscible blend PE/PP. The contribution of the KU Leuven Technology Campus Ostend is to set up the appropriate workflow to be able to perform numerical simulations that cope well with compatibilized blends.

The main process of interest in the project is injection molding. To prepare the correct modelling for the injection molding process, first of all, accurate rheological information is needed. Within this investigation, a dual bore rheometer and a reverse engineering technique on an injection molding machine are used to determine industrial relevant rheological data. On the injection molding machine, the rheological characterization is performed using a tensile bar mold equipped with pressure and temperature sensors. Using this data should allow to correctly predict the pressure field for different injection molding conditions and geometries.

With this input, a comparison between the experimental data and numerical simulation results will be performed during the injection molding process for a small box.

RHEOLOGICAL CHARACTERIZATION

Introduction

In order to perform numerical simulations, there is a need for shear viscosity data. Rheological characterization of an immiscible blend can be performed with the standard dual bore capillary rheometer, only if the blend is well compatibilized. In this case, the blend behaves like a homogeneous material. However, if the compatibilization is not of a good quality, different characterization methods should be applied. [1] Within this contribution a reverse engineering technique on an injection molding machine is applied. In that way, the rheological characterization is performed under real processing conditions and offers industrial relevant data. To validate this alternative method, the shear viscosity of a virgin material is determined on the one hand with the capillary rheometer and on the other hand with the alternative method. A Ceast dual bore capillary rheometer is used in combination with capillary dies with a diameter of 1 mm and lengths of 5 and 20 mm.

Reverse engineering method

Injection molding experiments are performed on a tensile bar mold equipped with pressure and temperature sensors. Figure 1 (a) shows the locations of these sensors. Pressure as function of time is measured. (Figure 2(a)) As the shear viscosity of the polymer melt influences the pressure in the mold, the goal of the method is to adjust the maximum pressure measured in the experiment with the maximum pressure in the simulations by changing the rheological properties. A common PP grade material has been used (Ineos 402 CB12).. Despite the presence of this material in the Autodesk Simulation Moldflow Insight 2016 material database, there is still a need to characterize the rheological behavior. Figure 2(b) shows too much deviation between capillary rheometer results and cross WLF data from the Moldflow database. The numerical simulation model used in this investigation is shown in figure 1(b) [2]. A midplane mesh (5840 elements) is built up, each element contains 20 laminae. Cool-fill-pack analyses are performed for different flow rates. A full 3D simulation able to take into account transient mold behavior (temperatures), is not performed, as the influence of the thermal properties on the mold filling behavior is limited because of the large thickness and the plain geometry of the tensile bar.

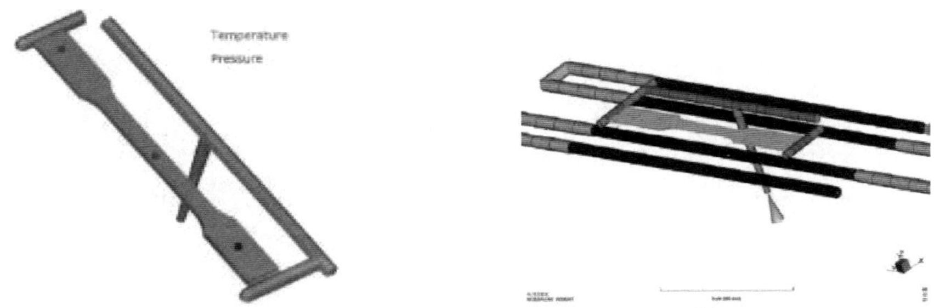

(a) (b)

FIGURE 1. (a) Locations of pressure and temperature sensors of the tensile bar (b) Example of numerical simulation model, for clarity, the cooling line symmetrically positioned to the parting plane are not visualized

A first order approach to describe the shear viscosity is applied. Rheological properties are changed by adapting coefficients A, B and C.

$$\eta = A \cdot \dot{\gamma}^{B} \cdot e^{C \cdot T} \tag{1}$$

Experiments are performed at two flow rates : 20 and 60 cc/s, respectively corresponding with shear rates in the parallel zone of the test sample of 800 and 2500 s^{-1}. The cooling water temperature was set to 25°C and a reference melt temperature of 230°C was used. The comparison between the experimental data and the numerical simulation result for the two injection flow rates is shown in Figure 2 (a).

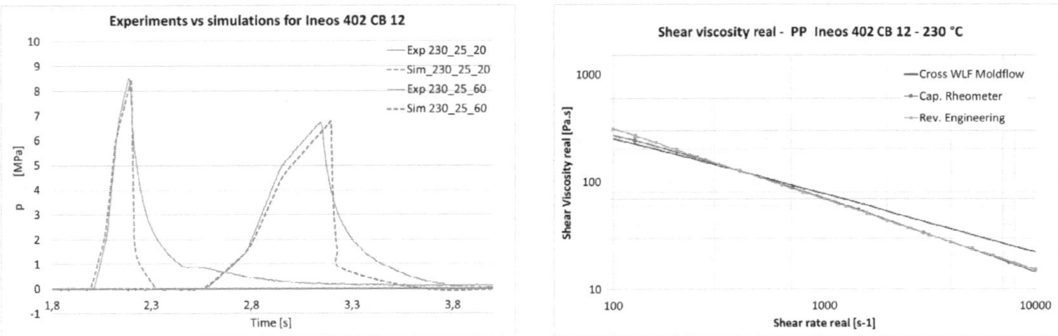

(a) (b)

FIGURE 2. (a) Pressure as function of time : experiment vs. simulations. (b) Shear viscosity data for virgin PP Ineos 402 CB 12 at 230°C

The shear viscosity results, both from capillary rheometer and reverse engineering method are shown in Figure 2 (b). In the shear rate ranges valid for injection molding, one can consider a perfect correlation between the two methods. For this reason, it is allowed to apply the reverse engineering method to characterize the rheological behavior. This method will also be applied on industrial blends.

Rheological data for industrial compatibilized blend

Similar tests as discussed above are performed for an industrial blend of PE/PP with compatibilizer. The results are shown in figure 3.

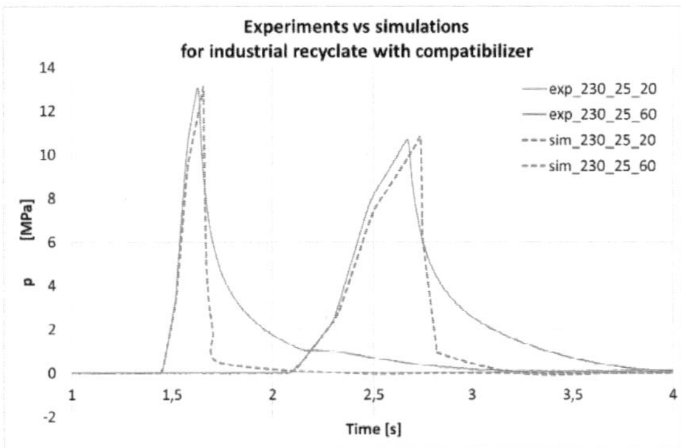

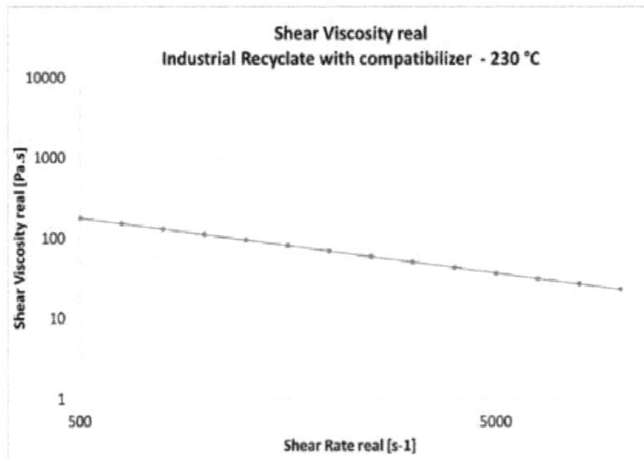

FIGURE 3. (a) Pressure as function of time : experiment vs. simulations. (b) Shear viscosity data for industrial blend with compatibilizer at 230°C

INJECTION MOULDING OF AN INDUSTRIAL PART

This data input should allow to correctly predict the pressure field for different injection molding conditions and geometries. To check this, injection molding experiments on one hand, and simulations on the other hand are performed on an industrial part and the results of both are compared. Figure 4 shows a picture of the small industrial part geometry. The dimensions of this box are 100x50x50 mm and the wall thickness is 1.5 mm. The mold is equipped with 3 temperature sensors and one pressure sensor. The pressure is registered close to the injection point of the product. Cavity temperatures are also registered which allows to judge about the quality of the process simulation.

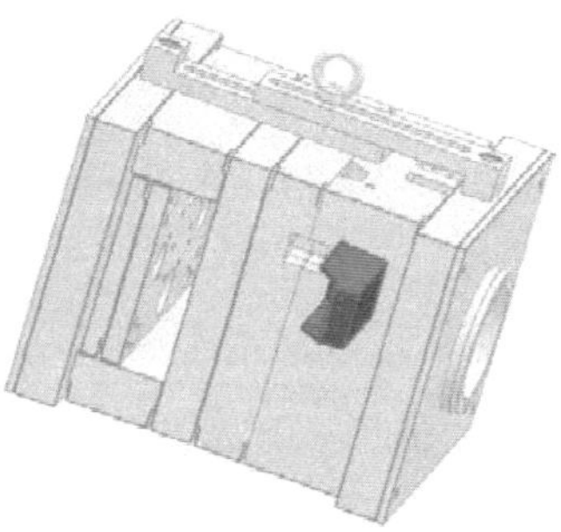

(a)

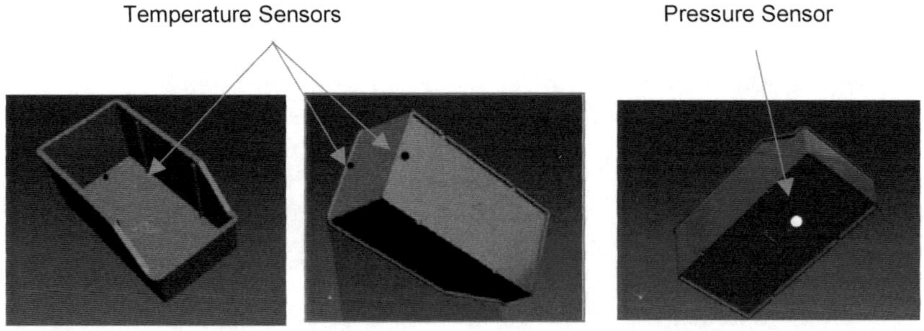

(b)

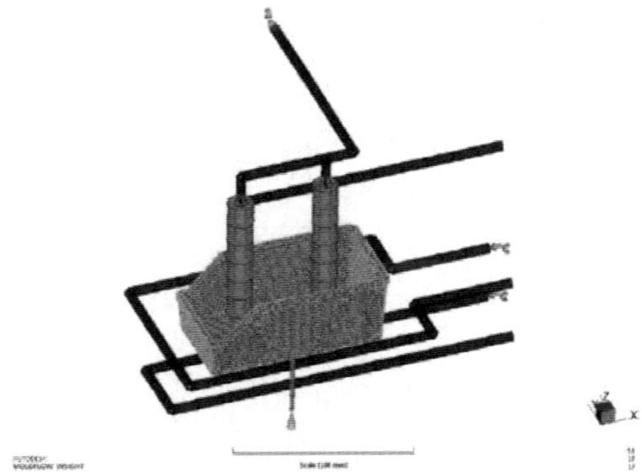

(c)

FIGURE 4. (a) Numerical simulations model of the small box. (b) Locations of the temperature and pressure sensors (c) numerical model built within Autodesk Moldflow Insight 2016

Experiments are performed with cooling water of 25 °C and barrel temperature of 230°C. Flow rates of 20, 40 and 60 cc/s are tested while the cycle time was kept constant at 15 s.

A dual domain mesh with 12 layers has been used together with a 3D mold representation allowing to predict the transient mold temperature behavior. The mesh for the product description contains 20979 elements while the mold representation contains 445985 elements. The cooling geometry contains bubblers to cool the core of the industrial box. The model is also shown in Figure 4 (c).

To optimize the simulations results, the processing parameters has been modelled as accurate as possible [3]. Within the numerical simulation, for example, the real screw speed has been modelled. Also the correct thermal data valid for the mold has been implemented, namely the thermal conductivity λ and the specific heat c_p. The heat transfer coefficient between the melt and the mold and melt temperature have been altered during the different simulation runs to aim at a better temperature prediction compared to the measured cavity temperatures. In this way, a melt temperature of 240°C, and the default HTC values used in the simulations deliver acceptable data. The correlation is shown in Figure 5(a) for a melt temperature of 230°C and in Figure 5(b) for a melt temperature of 240°C.

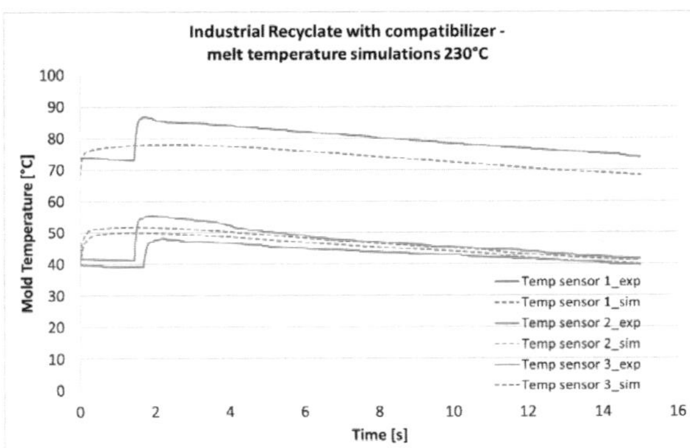

(a)

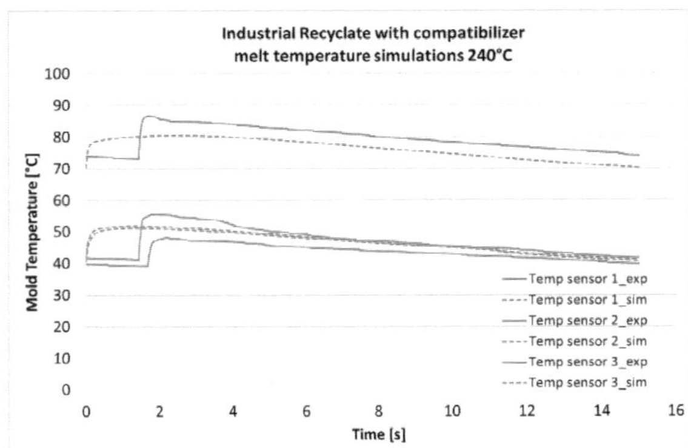

(b)

FIGURE 5. (a) Mold temperature as function of time with melt temperature within simulations at 230°C (a) and 240°C (b)

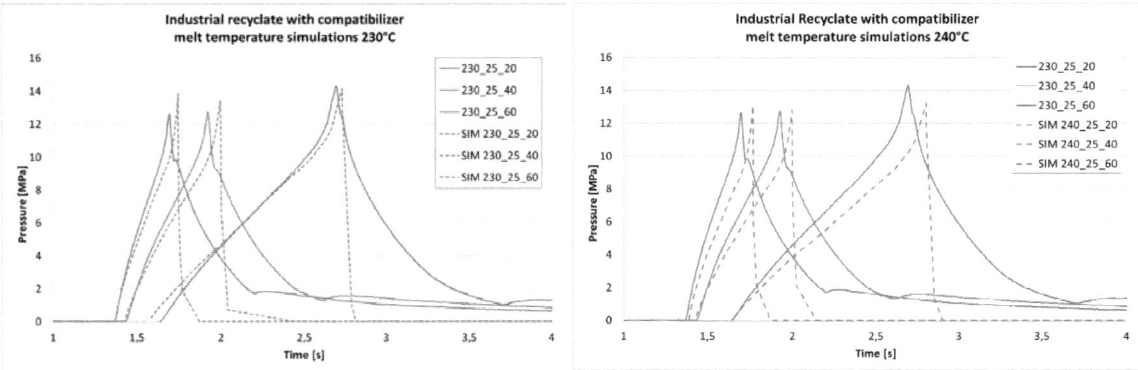

(a) (b)

FIGURE 6. (a) Pressure as function of time with melt temperature within simulations at 230°C (a) and 240°C (b)

Figure 6 (a) and (b) give a comparison for the pressure prediction with the experimental characterization. There is a better correlation between the simulations and the experiments if the melt temperature is put at 240°C. This increase compared to the machine setting of 230°C is mainly due to shear heating combined with a rather short cycle time.

CONCLUSIONS

Within this investigation a reverse engineering technique is validated to characterize the rheological behavior of an industrial compatibilized blend. With this data, an accurate prediction of temperatures and pressures during the injection molding process is achieved by adapting the appropriate parameters within the numerical simulation model.

ACKNOWLEDGMENTS

The authors gratefully acknowledge the funding by IWT for the Plasticom Project (IWT 135100).

REFERENCES

1. F. Desplentere and S. Deceur, *Comparison of measurement techniques for shear viscosity of non-compatibilized immiscible blends*, Polymer Processing Society Europe 2015 Location: Graz date 21-25 September 2016: Hanser Publishers
2. P. K. Kennedy, *Practical and scientific aspects of injection molding simulation,* Eindhoven, 2008
3. L.M. Galantucci, *Evaluation of filling conditions of injection moulding by integrating numerical simulations and experimental tests*, Journal of Materials Processing Technology 141 (2003) 266-275

Polymer nanocomposites (PNCs) filled with various kinds of nanoparticles (NPs) have been a hot topic in the academy and industry, due to the promising mechanical, electrical, thermal properties, gas permeability and so on [1-3]. This is realized by combining the functionality of NPs with the good processability of polymers. In most cases, it is commonly accepted that a homogeneous dispersion of NPs is needed to achieve improved physical properties such as reduced permeability and enhanced mechanical strength. Conversely, to obtain excellent thermal and electrical conductivity, precise assembly and ordering of NPs to form a continuous network should be developed. Therefore, to effectively tailor the properties of PNCs inspires us to establish some strategies to finely manipulate the spatial arrangement and localization of NPs in polymer matrices [4-6].

One effective approach is to uniformly graft the NPs with short polymer chains [3,7,8]. For instance, Akcora et al [9] have observed that the grafted spherical NPs can robustly self-assemble into a variety of anisotropic superstructures in the corresponding homopolymer matrix, attributed to the competition between the energy gain when the NPs approach each other and the entropy of distorting the grafted polymers. Kumar et al [8] also pointed out that using the polymer grafted NPs is effective to control the spatial dispersion of the NPs in the polymer matrices, which is ignited by the development of the synthesis techniques to controllably functionalize NPs with polymer chains. For example, Chen-Yang et al [10] have successfully prepared precisely tailored polystyrene (PS)-grafted NPs, including those with varied shapes and sizes of cores but identical PS shells, and NPs with fixed shapes but varied PS shells, which are considered ideal model NPs to probe the dynamics and reinforced mechanism of PNCs. It as well commonly accepted that the morphological behavior of grafted NPs in polymer matrices is similar to the block copolymers or amphiphiles in selective solvents [11]. Meanwhile, Green PF [12] reviewed the morphological phase diagrams defining the miscible, incompatible or partially miscible regions, by systematically considering the effects of the nanoparticle curvature, the chain grafting density and length, the length of the polymer matrix and the thermodynamic (Flory-Huggins) interaction parameter between the host and grafted chains. The effects of the grafted chain flexibility [13], the polymer matrix bidispersity with the equal number of short and long chains [14] on the morphology of the grafted NPs filled PNCs are also studied. Meanwhile, through the microscopic polymer reference interaction site model integral equation theory, Jayaraman et al [15] have calculated the potential of mean force (PMF) between grafted NPs in the homopolymer matrix. Pryamtisyn et al [16] have predicted that the polymer grafted NPs can readily form the self-assembled structures when the nanoparticle size is comparable to the radius of the gyration of the polymer brushes. By extending to much higher loading of NPs, Ginzburg VV [17] adopted the combined self-consistent field theory (SCFT)-density functional theory(DFT) approach to study the dependence of the morphology on the length and density of grafted chains, finding a new morphology of bundles of wires compared to the lower loading, and this method can be also applicable for some other cases that the matrix or the ligand is a kind of polymer blend or block copolymer.

However, although lots of research work have been carried out to study the effects of the structural parameters on the spatial distribution and organization of the end-grafted NPs in the polymer matrices, little efforts have been devoted to establish the relation between the morphology and the resulting mechanical properties. Through the rheology analysis of polymer-grafted NPs filled polymer nanocomposites, Moll et al [3] found that the formation of a transient, long-lived, percolating polymer-NPs network with the NPs serving as the network junctions contributes to the maximum mechanical reinforcement. And experimentally they also studied the effect of the shear flow on the morphological evolution of the assembles of grafted spherical NPs [18]. In particular, Riggleman et al [19] have simulated the effect of the curvature of the NPs and the grafting density on the mechanical properties of PNCs, finding that the elastic constants and yield properties are enhanced nearly uniformly, while the strain hardening modulus is weakly affected.

In our work, by varying the grafted chain length, we have observed various self-assembled structures formed by grafted NPs in the homo-polymer matrix. By comparing the stress-strain behavior of each morphology, we find that the sheet structure possesses the best mechanical reinforcing efficiency, as well validated by the occurrence of the stress overshoot at large strain. Meanwhile, we also simulate the effect of the grafted chain flexibility on the self-assembled morphology and the resulting mechanical properties. Compared to the flexible grafted polymer chain, we observe that the stress-strain behavior of the semi-flexible grafted polymer chains becomes much better at long grafted chain length, attributed to the physical inter-locking and wetting between matrix polymer chains and grafted polymer chains, which has the same trend in the presence of the interfacial chemical bonding between grafted and matrix polymer chains. In general, it is anticipated that this work could provide some guidance on tailoring the mechanical performance of grafted NPs filled polymer nanocomposites, by taking advantage of varying the grafted chain length and flexibility.

References

[1] Jancar, J. et al. Current issues in research on structure-property relationships in polymer nanocomposites. *Polymer* 51 (2010), 3321-3343.

[2] Ganesan, V. & Jayaraman, A. Theory and simulation studies of effective interactions, phase behavior and morphology in polymer nanocomposites. *Soft Matter* 10 (2014), 13-38.

[3] Moll, J. F. et al. Mechanical Reinforcement in Polymer Melts Filled with Polymer Grafted Nanoparticles. *Macromolecules* 44 (2011), 7473-7477.

[4] Jayaraman, A. Polymer Grafted Nanoparticles: Effect of Chemical and Physical Heterogeneity in Polymer Grafts on Particle Assembly and Dispersion. *Journal of Polymer Science Part B-Polymer Physics* 51 (2013), 524-534.

[5] Balazs, A. C., Emrick, T. & Russell, T. P. Nanoparticle polymer composites: Where two small worlds meet. *Science* 314 (2006), 1107-1110.

[6] Yan, L. T. & Xie, X. M. Computational modeling and simulation of nanoparticle self-assembly in polymeric systems: Structures, properties and external field effects. *Progress in Polymer Science* 38 (2013), 369-405.

[7] Kalb, J., Dukes, D., Kumar, S. K., Hoy, R. S. & Grest, G. S. End grafted polymer nanoparticles in a polymeric matrix: Effect of coverage and curvature. *Soft Matter* 7 (2011), 1418-1425.

[8] Kumar, S. K., Jouault, N., Benicewicz, B. & Neely, T. Nanocomposites with Polymer Grafted Nanoparticles. *Macromolecules* 46 (2013), 3199-3214.

[9] Akcora, P. et al. Anisotropic self-assembly of spherical polymer-grafted nanoparticles. *Nature Materials* 8 (2009), 354-U121.

[10] Ruan, Y. B. et al. Polymer-Grafted Nanoparticles with Precisely Controlled Structures. *Acs Macro Letters* 4 (2015), 1067-1071.

[11] Maillard, D., Kumar, S. K., Rungta, A., Benicewicz, B. C. & Prud'homme, R. E. Polymer-Grafted-Nanoparticle Surfactants. *Nano Letters* 11 (2011), 4569-4573.

[12] Green, P. F. The structure of chain end-grafted nanoparticle/homopolymer nanocomposites. *Soft Matter* 7 (2011), 7914-7926.

[13] Palli, B. & Padmanabhan, V. Chain flexibility for tuning effective interactions in blends of polymers and polymer-grafted nanoparticles. *Soft Matter* 10 (2014), 6777-6782.

[14] Martin, T. B. & Jayaraman, A. Effect of Matrix Bidispersity on the Morphology of Polymer-Grafted Nanoparticle-Filled Polymer Nanocomposites. *Journal of Polymer Science Part B-Polymer Physics* 52 (2014), 1661-1668.

[15] Jayaraman, A. & Schweizer, K. S. Effective Interactions and Self-Assembly of Hybrid Polymer Grafted Nanoparticles in a Homopolymer Matrix. *Macromolecules* 42 (2009), 8423-8434.

[16] Pryamtisyn, V., Ganesan, V., Panagiotopoulos, A. Z., Liu, H. J. & Kumar, S. K. Modeling the anisotropic self-assembly of spherical polymer-grafted nanoparticles. *Journal of Chemical Physics* 131 (2009), DOI: 10.1063/1.3267729.

[17] Ginzburg, V. V. Polymer-Grafted Nanoparticles in Polymer Melts: Modeling Using the Combined SCFT-DFT Approach. *Macromolecules* 46 (2013), 9798-9805.

[18] Moll, J. et al. Dispersing Grafted Nanoparticle Assemblies into Polymer Melts through Flow Fields. *Acs Macro Letters* 2 (2013), 1051-1055.

[19] Chao, H. K. & Riggleman, R. A. Effect of particle size and grafting density on the mechanical properties of polymer nanocomposites. *Polymer* 54 (2013), 5222-5229.

Fatigue Design Methodology for Short Fibre Reinforced Polymers Considering Fibre Orientation and Weld Lines

Andreas Primetzhofer [a,*], Andreas Mösenbacher[a] and Gerald Pinter[b,c]

[a] *Department of Product Engineering, Montanuniversitaet Leoben*
Franz-Josef-Strasse 18, 8700 Leoben, Austria
[b] *Department of Polymer Engineering and Science, Montanuniversitaet Leoben*
Otto Gloeckel-Strasse 2, 8700 Leoben, Austria
[c] *Polymer Competence Center Leoben GmbH; Roseggerstrasse 12, 8700 Leoben, Austria*

Abstract. Due to high specific strength and the cost efficient manufacturing process enabled by injection moulding, reinforced polymeric materials are increasingly popular in structural applications in automotive and other industrial applications. As a result of the injection moulding as manufacturing process, reinforced polymers show a strong anisotropic material behaviour. To consider the lifetime of a part in an early state of a development process, it is indispensable to deal with technological effects on the material ranging from the manufacturing process to the real load history. A closed simulation chain is established to estimate the life time of a part based on local mechanical properties, depending on the fibre orientation from an injection moulding process. Other stress based influences like the mean stress or the stress peaks at notches are also covered. The behaviour of polymer based materials shows a strong dependency on the fibre orientation. Therefore it is also necessary to understand whether and how the anisotropy influences the mechanical behaviour of the material. This work focuses on how the fibre orientation and the weld-line formation affects the mechanical properties of a short glass fibre reinforced polyamide.

Therefore quasi-static and cyclical tests were performed on specimen with different fibre orientation as well as weld lines. The test results show a significant change of the mechanical properties depending on the fibre orientation. Weld lines lead to a significant strength decrease. While CT-scans show a good correlation between moulding simulation and the averaged fibre orientation, the core, the surface layer and weld lines cannot not sufficiently be predicted. An improvement in moulding simulation technique is encouraged to gather local fatigue results based on numerical predicted fibre orientation.

Keywords: fatigue, weld lines, fibre orientation
PACS: 81.40.Np

INTRODUCTION

Due to injection moulding as manufacturing process of short glass fibre reinforced (sgfr) materials an anisotropic material behaviour is obtained. This behaviour is mainly affected by the flow of the mould through the tie, which defines the fibre orientation. Beside the orientation, the formation of weld lines is driven by the flow and the geometry of the tool. Weld lines occur whenever the mould is split by a mould core or more than one gate is used. Weld lines can also be caused by a partial gap in flow velocity due to varying wall thickness. Based on their formation, two types of weld lines have to be distinguished (FIGURE 1) [1, 2]. On the one hand the two melt flows meet laterally and keep on flowing as a combined weld line (Type 1). On the other hand the two melt flows head-on frontally and stay at the meeting point (Type 2).

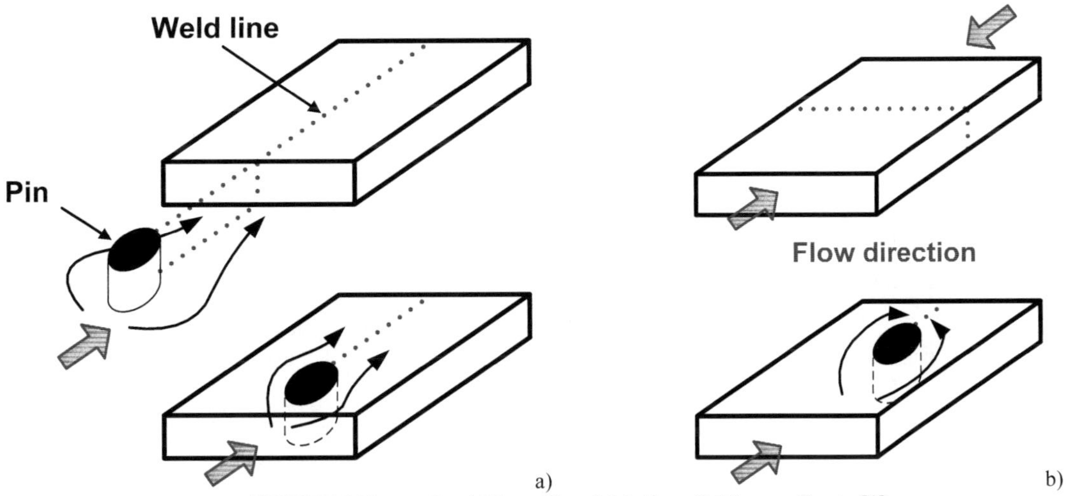

FIGURE 1 Types of weld lines: Type I (a), Type II (b) according to [1]

While the fibre orientation can be more or less well predicted by a filling simulation, the prediction of weld lines is limited to certain information on the surface like the meeting angle. Since the formation of a weld line influences the material behaviour along the tracing path over the whole thickness it is indispensable to get details about the anisotropy through the whole volume of the weld line for a life time prediction based on a local S/N approach. Beside a change in material behaviour the formation of weld lines can lead to a notch on the surface and therefore a stress concentration. Further results on the morphology of weld lines are covered in literature [2–5].

EXPERIMENTAL

To characterise the material, quasi-static as well as cyclic tests were performed on an injection moulded cross shaped specimen. Due to the geometry of the specimen it is possible to run tests longitudinal as well as transversal to the main fibre orientation on the same specimen without a manufacturing influence. By inserting a pin, a weld line Type I can be realized. All tests were performed on a servo hydraulic test rig at room temperature and a relative humidity of 50%. The investigated material, manly used for automotive applications, is a partial aromatic polyamide containing 50 m% of glass fibres.

Results

Quasi-static as well as cyclic tests show a significant reduction in material behaviour perpendicular to the fibre orientation and also in case of a weld line. The ultimate strength as well as the fatigue limit at $N = 10^6$ decline to about 40 % compared to properties to the longitudinal orientation. Differently than expected, a present weld line doesn't lead to a further significant reduction in the material behaviour (FIGURE 2). The slightly increased slope of *S/N* – curves for transversal oriented specimen and specimen with weld line is caused by a change in failure mechanisms. While rupture in case of parallel oriented fibres is probably driven by fibre debonding and fibre failure; transversal, the formation and merging of micro cracks in the matrix leads to final failure. Since the test results for transversal tested specimen and specimen with weld line show quite similar results, also the failure mechanisms should be comparable. Further microscopic investigations (light microscopy and SEM) on the fracture surface microstructure will provide additional information to the test results and the failure mechanisms.

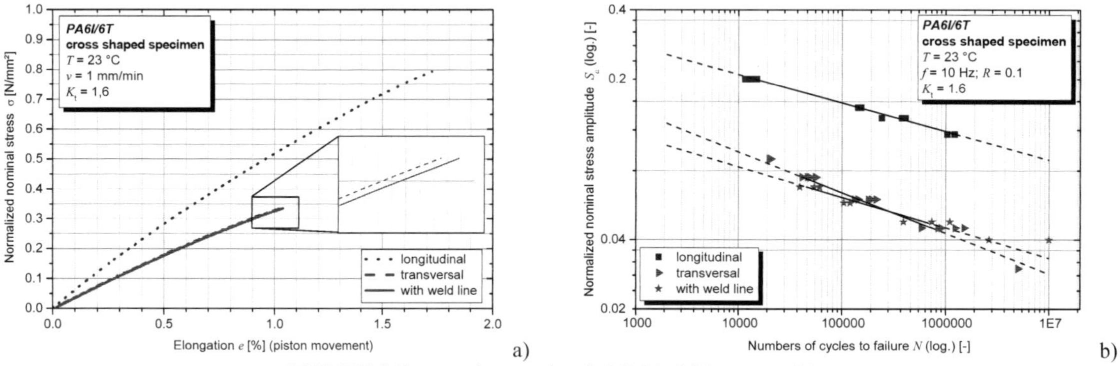

FIGURE 2 Test results: quasistatic [6] (a), *S/N* – curves (b)

SIMULATION

Based on the concept of local *S/N* – curves, first developed for metals, a simulation chain for a life time calculation of sgfr polymers has been adapted and implemented in a commercial post processing tool (FEMFAT®) [7–9]. Originating from a given geometry, the load case and a basic material characterisation with *S/N* – curves, the life time of a part can be estimated by the linear damage accumulation according to Miner [10]. Since the material behaviour of reinforced materials is strongly influenced by the manufacturing process (e.g. injection moulding), the resulting anisotropy has to be taken into account. Therefore the fibre orientation as well as the fibre distribution is computed in a filling simulation based on commonly used algorithms like the Folgar-Tucker or RSC model [11]. Including the fibre orientation in each element of the FE-mesh, the local anisotropic material behaviour for the FE-analysis is computed by an external solver (e.g. DIGIMAT®) by generalized homogenization methods (RVE method) [12, 13]. Additionally, the endurable number of cycles at certain stress level for a life time calculation is given according to Guster [14] and Gaier [7].

Filling simulation vs. CT-scans

Fibre orientation is one of the most significant influences on material strength and therefore the filling simulation is utmost important. To gain a better understanding and to check the accuracy of the simulated fibre orientation, CT-scans were performed at three positions on an injection moulded plate (100 x 100 x 2 mm). FIGURE 3 (b) and (c) show the orientation given by the trace values a_{xx}, a_{yy} and a_{zz}. Samples are located along the median plane perpendicular to the flow direction as shown in FIGURE 3 (a) represented as red squares. The evaluated fibre orientation indicates a strong dependency on the layer build up from the position along the width of the plate. While there is no pronounced core layer in the outer regions of the plate (FIGURE 3 (c)), a layered structure can be found around the symmetry axis parallel to the flow direction (FIGURE 3 (b)). This fact leads to a varying material strength along the width of the plate. Due to the higher averaged fibre orientation, specimens taken out form the outer parts of the plate display a considerable improved material behaviour.

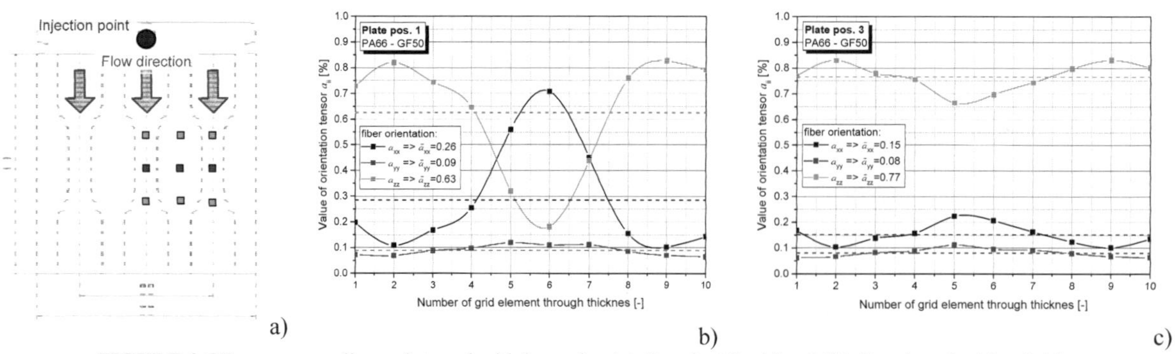

FIGURE 3 CT-scans: sampling points and withdraw plan (a), Result at Position 1 (b), Result at Position 2 (c)

In addition crack surfaces, form quasi static as well as cyclic tests performed on specimen taken from plates were analysed optically. Fracture surfaces also show a strong dependency on the layer structure form the distance of the symmetry axis, similar to the results form CT-scans. Specimen, taken out at the middle position (FIGURE 4), exhibit a core layer while in the left and right position no layered structure can be obtained. In contrast to the measured fibre orientation and the optical analysis of the fracture surfaces, the predicted orientation does not show a spatial distribution (FIGURE 4).

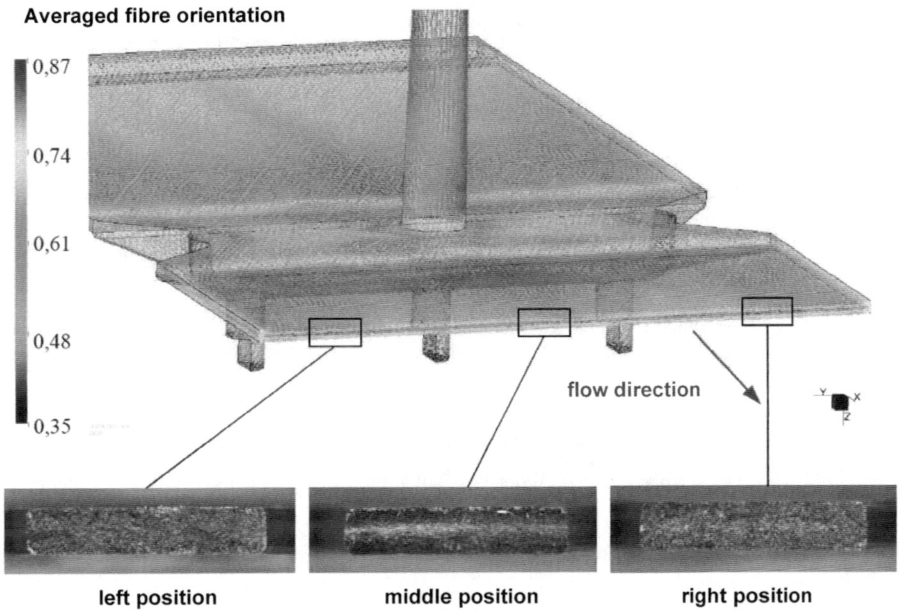

FIGURE 4 Comparison of layered structure from a simulation result and corresponding crack surfaces at three positions

Weld lines

Besides the prediction of fibre orientation the formation of weld lines is essential in an injection moulding simulation. Therefore numerical tools are provided in all commercially used software packages (in this paper Moldflow®). The formation and the position of weld lines can be principally predicted by these tools. Compared to measurements the length of the computed weld line is much shorter. After a certain distance behind an obstacle no formation is detected in a simulation [15]. In addition the deviation of the calculated weld line angle increases in this limited region [16]. Another weak point in the prediction is the limited information of spatial geometry within the weld line volume, which may even cause crack-like formations. For a reliable forecast in a life time calculation an adequate image of the weld line through the whole part is essential.

CONCLUSION

The fibre orientation as well as the formation of weld lines has a great influence on the mechanical behaviour of sgfr polymers. Quasi-static as well as cyclic tests show no significant differences between weld lines and specimen tested perpendicular to the main fibre orientation.

In general the averaged fibre orientation, predicted in a filling simulation, is comparable to the real orientation obtained in CT-scans, although the distribution of the core, the boundary layer and weld lines are not correctly computed over the whole part. In case of weld lines an improvement is necessary to cover weld lines not only on the surface and in a narrow region behind obstacles.

Based on the described simulation flow an accurate life time calculation is possible in principle. Since this assessment is significantly influenced by the fibre orientation, improved algorithm should be provided for injection moulding simulation. Especially the correct prediction of weld lines is mandatory because they constitute potential weak points in most parts.

ACKNOWLEDGEMENT

The research work of this paper was performed at the Department of Product Engineering at the Montanuniversiaet of Leoben in collaboration with the Polymer Competence Center Leoben GmbH (PCCL, Austria) within the framework of the COMET-program of the Austrian Ministry of Traffic, Innovation and Technology with contribution by BMW-Group, BOSCH-AG, Engineering Center Steyr (MAGNA Powertrain ECS), EMS-Grivory, VW-Group, Schaeffler Technologies AG & Co KG and Qualitech AG. The PCCL is founded by the Austrian Government and the State Governments of Styria, Lower and Upper Austria.

REFERENCES

1. T. Nguyen-Chung: Strömungsanalyse der Bindenahtformation beim Spritzgießen von thermoplastischen Kunststoffen, Technische Universität Chemnitz, 2001.

2. T. Nguyen-Chung: Flow analysis of the weld line formation during injection mold filling of thermoplastics; *Rheologica Acta* 43 (2004) 3, pp. 240–245.

3. J. K. Kim; S. H. Park; H. T. O and H. K. Jeon: The effect of weld-lines on the morphology and mechanical properties of amorphous polyamide/poly(ethylene-ran-propylene) blend with various amounts of an in situ compatibilizer; *Polymer* 42 (2001) 5, pp. 2209–2221.

4. C. Lu; S. Guo; L. Wen and J. Wang: Weld line morphology and strength of polystyrene/polyamide-6/poly(styrene-co-maleic anhydride) blends; *European Polymer Journal* 40 (2004) 11, pp. 2565–2572.

5. K. Yamada; K. Tomari; U. S. Ishiaku and H. Hamada: Evaluation of mechanical properties of adjacent flow weldline; *Polymer Engineering & Science* 45 (2005) 8, pp. 1180–1186.

6. C. Guster: Ansätze zur Lebensdauerberechnung von kurzglasfaserverstärkten Polymeren, PhD thesis, Montanuniversität Leoben, 2009.

7. C. Gaier; H. Dannbauer; A. Werkhausen and R. Wahlmüller: Fatigue life prediction of short fiber reinforced plastic components; in *Proceedings of the sixth Annual Automotive Composites Conferences & Exhibition* (Society of Plastics Engeneers Ed.), 2008.

8. C. Gaier and H. Dannbauer: Ein effizientes kritisches Schnittebenenverfahren für duktile, semi-duktile und spröde Werkstoffe; in *1. Leobener Betriebsfestigkeitstage* (Eichlseder, W. Ed.), Lehrstuhl für Allgemeinen Maschinenbau Montanuniversität, Leoben, 2006.

9. C. Guster; G. Pinter; A. Mösenbacher and W. Eichlseder: Evaluation of a Simulation Process for Fatigue Life Calculation of Short Fibre Reinforced Plastic Components; *Procedia Engineering* 10 (2011) 0, pp. 2104–2109.

10. M. A. Miner: Cumulative Damage in Fatigue; *Journal of Applied Mechanics-Transactions of the Asme* 12 (1945) 3, pp. A159-A164.

11. J. Wang and X. Jin: Comparison of Recent Fiber Orientation Models in Autodesk Moldflow Insight Simulations with Measured Fiber Orientation Data; in *Austrian-Slovenian Polymer Meeting,* 2010.

12. L. Brassart; I. Doghri and L. Delannay: Homogenization of elasto-plastic composites coupled with a nonlinear finite element analysis of the equivalent inclusion problem; *International Journal of Solids and Structures* 47 (2010) 5, pp. 716–729.

13. George Z. Voyiadjis; Amir Shojaei and Guoqiang Li: A generalized coupled viscoplastic–viscodamage–viscohealing theory for glassy polymers; *International Journal of Plasticity* 28 (2012) 1, pp. 21–45.

14. C. Guster: Fiber orientation and fatigue behavior of a short glass-fiber reinforced partial aromatic polyamide; *2. Fatigue Symposium Leoben* (2008).

15. T. Nguyen-Chung; C. Plichta and G. Mennig: Flow disturbance in polymer melt behind an obstacle; *Rheologica Acta* 37 (1998) 3, pp. 299–305.

16. J. G. Kovács and B. Sikló: Experimental validation of simulated weld line formation in injection moulded parts; *Polymer Testing* 29 (2010) 7, pp. 910–914.

Special Materials and Technologies for Fused Filament Fabrication

Martin Spörk [a], Joamin Gonzalez-Gutierrez [a], Christian Kukla [b] and Stephan Schuschnigg [a], Clemens Holzer [a,*]

[a] *Montanuniversitaet Leoben, Department of Polymer Engineering and Science, Chair of Polymer Processing, Otto Gloeckel-Strasse 2, 8700 Leoben, Austria*

[b] *Montanuniversitaet Leoben, Industrial Liaison Department, Peter Tunner Strasse 27, 8700 Leoben, Austria*

Abstract. In additive manufacturing Fused Filament Fabrication (FFF) is outstanding since common thermoplastics can be used. Thus, this method can meet the industry's demands of producing complex designs in limited series with familiar materials. To fully exploit the technique's potential, new printing materials have to be developed, as currently only a few types of materials (mostly PLA and ABS) are commercially available for FFF. Therefore, the Chair of Polymer Processing is systematically developing novel materials for their use in FFF.

For example, specially filled polyolefins with low shrinkage coefficients and good dimensional stabilities are investigated. These materials provide several advantages compared to existing filament types, such as higher impact strength and better temperature stability. Moreover, wood fibre reinforced materials for FFF are under investigation. The biggest challenge with these materials is an accurate production of perfectly round filaments, as the additional fibres prevent elongation and therefore a proper calibration. Another field of research are highly filled materials, wherein the polymer acts as the binder system. After printing, the parts are debound in a solvent and sintered in a furnace similar to powder injection moulded parts. Currently, compounds filled with highest possible amounts of metal and ceramic powders are being developed. So far, compounds with a maximum of 91 wt.% (55 vol.%) of stain-less steel are still processable.

For the systematic development of new materials for FFF, mechanical, flow, adhesion and shrinkage properties of materials have been characterised, compared and linked to printability.

Keywords: additive manufacturing, adhesion, fused filament fabrication, highly-filled polymers, mechanical properties, polyolefin, polylactic acid, shrinkage, viscosity, wood polymer composites

PACS: 62.20.de, 62.20.fq, 66.20.Ej 81.05.Bx, 81.05.Lg, 81.05.Qk, 81.05.Rm, 81.20.Ev, 81.40.Jj, 81.70.Bt, 83.50.Uv, 83.60.Fg, 83.80.Hj, 83.80.Tc, 83.85.Jn

INTRODUCTION

Fused filament fabrication (FFF), an extrusion based additive manufacturing technology that is also referred to as Fused Deposition Modelling (FDM), has the highest potential for polymer manufacturing among the different available rapid prototyping techniques. The FFF technique was developed by S. Crump in 1989 (Patent: US5121329A). Only recently, however, FFF was able to compete with conventional polymer processing techniques, such as injection moulding, due to greatly increased research activities [2]. In the state-of-the-art FFF process, a solid thermoplastic filament is pushed into a hot die by two counter-rotating wheels. The filament is extruded through a moving nozzle head onto a heated surface resulting in a layer-by-layer fabrication of the desired part [1]. Figure 1 shows a schematic representation of the FFF process.

The range of applications of FFF is extensive, ranging for example from medical [3] to automotive applications [1]. However, to fully exploit the technique's potential, it has to be further improved. A case in point is the development and optimisation of building materials as nowadays only a few types of materials (based mostly on PLA and ABS) are commercially available for FFF. In order to meet the industries' increasing demands, novel materials for FFF are the focus of current research activities at the Chair of Polymer Processing in Leoben.

Polyolefins represent an interesting group of materials to be used in FFF. They provide several advantages compared to existing filament types, such as higher impact strength, better temperature stability, good modifiability, food safety and comparability with industrial products [4]. Furthermore, polyolefins show no problems like hydrolysis, moisture absorption, and long-time stability. Nevertheless, the main problem encountered in polyolefins is their high shrinkage coefficient, which negatively affects the dimensional stability of the printed parts [4]. Consequently, low shrinkage compounds based on polyolefins are being investigated.

Wood plastic composites (WPC) are considered environmental friendly materials, which are mainly used as a substitution for expensive tropical woods for outdoor decking and flooring [5]. However, FFF should open up new fields of application for WPC. The biggest challenge with these materials is the accurate production of perfectly round filaments. Filaments with constant dimensions and shape are really important to ensure good quality of printed parts. Moreover, adding fibres to polymers dramatically affects the properties required for having printable filaments; the filaments become more brittle and the viscosity increases.

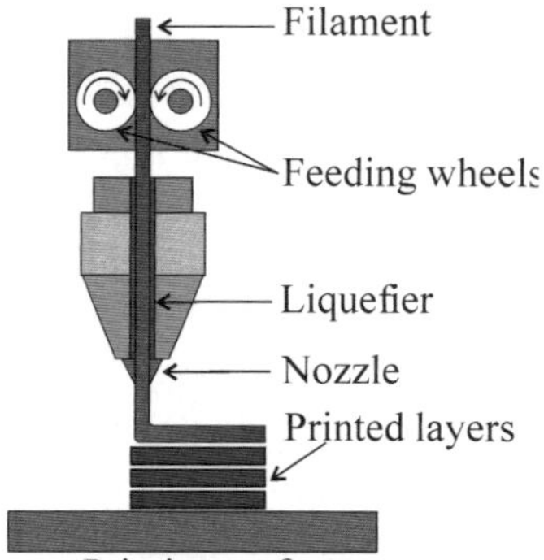

FIGURE 1. Cross cut view of fused filament fabrication process.

It is believed that FFF could be used to replace in some applications injection moulded parts, when the number of parts needed is small or when the geometrical requirements are very demanding. This includes also Powder Injection Moulding (PIM). In PIM a highly-filled polymeric system with metal or ceramic powders is shaped by injection moulding, the polymeric system is then removed, and finally the part is sintered to obtain a solid part of metal or ceramic. The final part is smaller than the original moulded part due to closing of the pores between particles during debinding and sintering [6]. The main problem when dealing with highly filled systems in FFF is to maintain the good flow properties without affecting the filament's mechanical strength and flexibility.

As it can be expected, the range of properties for polyolefins, WPC and highly-filled polymers is very large, but there must be minimum requirements that all of these materials have to fulfil in order to be printable and to lead to printed parts with good quality. The material properties that need to be taken into account include: (i) filament mechanical properties, (ii) flow properties of molten filament, (iii) adhesion between printed material and printing surface as well as on the printed parts itself, and (iv) shrinkage upon solidification. In this paper several materials usable in FFF have been characterised in terms of these four groups of properties. The main contribution of this paper is to provide the required level of properties needed for selecting new materials that can be processed via FFF.

MATERIALS

Five materials have been selected for this investigation. Their short name and description is given in Table 1.

TABLE 1. List of materials investigated and their description.

Short name	Type of material	Short description
PLA_00	Unfilled biopolymer	Black polylactic acid commercially available as filaments for FFF.
PP_00	Unfilled polyolefin	Neat polypropylene commercially available used for cast film.
PP_GS	Filled polyolefin	Compound of PP_00 and compatibilised PP with coated hollow glass spheres prepared in laboratory at 30 vol% (17.9 wt%).
PLA_WF	Wood polymer composite	Compound of PLA and wood fibres commercially available as filaments for FFF at 25 wt%.
HFP_SS	Highly-filled polymer	Blend of two compatibilised polymers filled with spherical 316L stainless steel powder prepared in laboratory at 55 vol% (91 wt%)

MATERIAL CHARACTERISATION METHODS AND RESULTS

Filament Mechanical Properties

The filament is pulled by the driving wheels and then pushed by those wheels into a liquefier and a nozzle (Figure 1). Therefore, sufficient mechanical strength of the filament is required to retain its shape after being forced through the drive wheels. At the same time, the filament must be strong enough to avoid shearing due to the friction from the wheels. Finally, the filament must not buckle between the drive wheels and the liquefier, where it also acts as a plunger. The force transferred from the drive wheels to the filament should be efficiently transferred into the centre of the liquefier in the direction of the melt flow, with minimal loss due to filament buckling and compression [7]. In addition to these requirements, the filament should also be flexible enough so that it can be spooled.

The mechanical properties of the filaments were tested with a material tensile testing machine with pneumatic grips (Zwick Z001 2.5 kN load cell, Zwick GmbH & Co. KG, Germany). Measurements were performed at 23 °C and 50 % r.h. Gauge length was set at 50 mm. Between 0 and 0.25% extension, an extension rate of 1 $mm \cdot min^{-1}$ was used, afterwards the rate was increased to 10 $mm \cdot min^{-1}$ until the sample broke. Five repetitions were performed.

Tensile testing results for the filaments investigated are shown in Figure 2a. Two important parameters have been selected to describe mechanical response of the filaments: the secant modulus between 10 and 40% of yield stress and the elongation at break. Filaments should have sufficient modulus or sufficient elongation at break. For example, the easiest material to print is PLA_00 which has the highest modulus (~26 MPa) and a medium elongation at break of ~40 %. The addition of wood fibres to PLA (PLA_WF) leads to a decrease in modulus to ~11 MPa and an elongation at break of ~5 %. Printing trials with this material showed that filaments could break at the feeding wheels if not properly adjusted, but they are still printable. The two filaments based on PP have very distinct tensile behaviour: The filled material with glass spheres (PP_GS) has a modulus of ~10 MPa and a very short elongation at break (~5 %), while the unfilled PP_00 has a very large elongation at break of more than 1500 % and a modulus of ~8.5 MPa. PP_00 is a printable material while PP_GS still needs to be tested but due to its brittle nature problems during printing are expected. Finally, the highly-filled polymer (HFP_SS) has the lowest modulus (~8 MPa), but a medium elongation at break (~40 %) and it is still printable. These results indicate that for a filament to be printable a minimum modulus of ~10 MPa may be necessary, but filaments with lower modulus (~8 MPa) can be still printable if medium range elongation at break is present (>40 %). Further investigation with other materials is still under way to clarify these results.

Flow Properties

The flow properties are the second set of properties of interest. After the material is liquefied in the print nozzle it should be able to be squeezed out of the nozzle by the pressure coming from the filament itself. The main flow property of interest is the viscosity (Figure 2b). If the material has very high viscosity, the mechanical properties of the filament need to be higher too and the force exerted on the filament will need to be increased for the material to come out of the nozzle. However the viscosity should not be extremely low otherwise the liquid polymer cannot be deposited in a controlled manner after leaving the nozzle. Simulations showed that at usual printing speeds the shear rate at the die walls (Ø=0.5 mm) are about 800 s^{-1}.

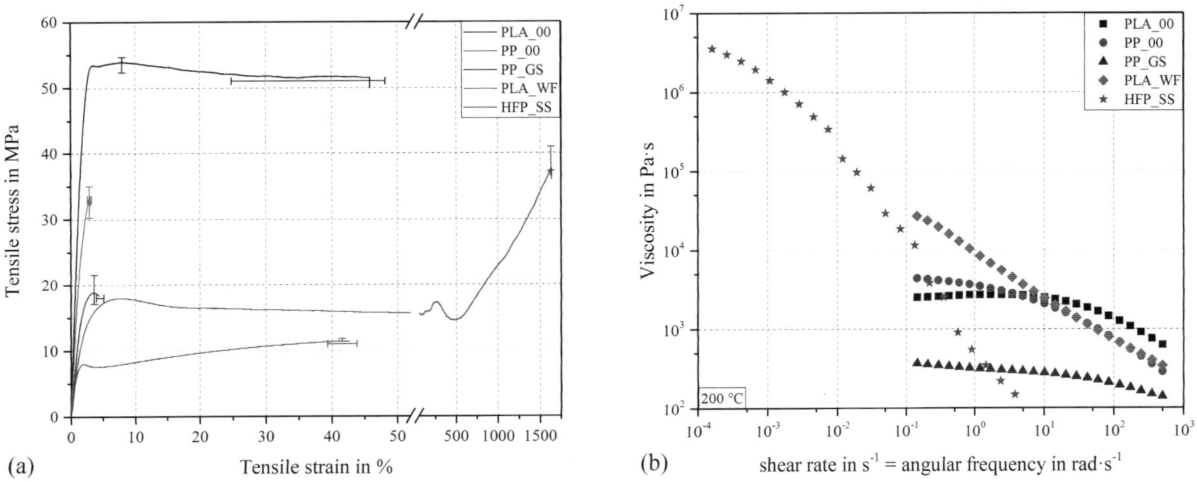

FIGURE 2. (a) Tensile stress as a function of tensile strain of filaments and (b) viscosity versus angular frequency or shear rate.

Viscosity measurements were performed in a rotational rheometer (MCR-500, Anton Paar GmbH, Graz, Austria). The selected measuring geometry was parallel plate system with a 25 mm diameter. Samples were shaped into discs 30 mm in diameter and 1 mm in thickness. Measurements were performed at 200 °C and were made in oscillatory shear controlled mode at frequencies between 10^{-2} and 500 $rad \cdot s^{-1}$ for all materials except for HFP_SS, which was measured in constant rotational mode at shear rates from 10^{-4} to 10 s^{-1}. Three repetitions were made per material. Assuming that the Cox-Merz rule applies all average viscosity curves are shown in Figure 2b.

The viscosity curve for HFP_SS shows very high values at the lowest shear rates, but there is a very strong shear thinning effect as the shear rate increases, which facilitates the flow and thus this material is still printable. PLA_WF has also a strong shear thinning effect, which is expected since it also has a significant filler content and printing is also possible. The two unfilled polymers (PP_00 and PLA_00) show viscosity values independent of angular frequency at lowest angular frequencies, followed by a shear thinning effect that contributes to the printability of these materials. Finally, PP_GS has the overall lowest viscosity from all the tested materials, which is unexpected considering that the load of glass spheres is as high as 30 vol%. The reduction on viscosity can be attributed to the compatibiliser used that also acts as a lubricant. The viscosity of all materials investigated is sufficiently low (< 1000 Pa·s) at higher shear rates (> 100 s^{-1}); therefore it should not be the limiting factor in determining the printability of these materials.

Adhesion

After the liquid material is extruded out of the nozzle, it should be able to adhere to the printing surface. If the material does not adhere to the printing surface the first layer of the part cannot be made. Adhesion for the various polymers is very different; the optimal settings for one polymer might be bad for another. Consequently, it is important to have the right combination of printable material, printing surface and cleaning agent.

The adhesion between the printed material and the printing surface can be investigated by performing contact angle measurements. In our project three commonly used surfaces were investigated: glass mirror, glass covered with Kapton (polyimide) film and masking Bluetape. The glass mirror and the Kapton film were cleaned with a cleaner consisting of water, surfactant, butane, methoxy-propanol, ethanol and propane. The Bluetape was used as supplied. The measurements were done with two polymers (PLA_00 and HFP_SS), which were pressed to plates with a vacuum press. The samples were analysed with a contact angle measuring instrument (DSA100, Krüss GmbH, Hamburg, Germany). Water, diiodomethane and ethylene glycol were the test-liquids selected, since their polar and disperse fractions are well known. Ten repetitions were performed per combination. The resulting contact angle at each surface was measured and the interfacial tensions of these material combinations were calculated. Out of the equation for the interfacial tension [8], it can be described that the adhesion of two solids is increasing with decreasing interfacial tension values.

Figure 3a summarises the adhesion results for the two investigated polymers (PLA_00 and HFP_SS) on three different printing surfaces. For PLA_00 the best adhesion results could be achieved with the Kapton surface, whereas the interfacial tension between PLA_00 and the Bluetape is significantly higher than on the mirror or on Kapton, resulting in worse adhesion. On the contrary, HFP_SS indicates very good adhesion on the Bluetape, but is not adhering at all to the mirror, because of their very different polar fractions. Extended printing tests of these materials on the different surfaces confirm the results displayed in Figure 3a.

These results highlight the importance of selecting the appropriate printing surface for a given material and how important surface tension measurements are for the selection of the appropriate printing surface.

Shrinkage

Finally, after the material is extruded and has adhered to the printing surface, it should not shrink too much during the solidification process. Excessive shrinkage leads to warpage of the printed part and loss of adhesion between the printing bed and/or the printed layers. Therefore, materials with low shrinkage should be used for FFF.

The shrinkage behaviour can be studied by measuring the change of specific volume as a function of temperature and was measured with a SWO pvT100 (SWO Polymertechnik GmbH, Krefeld, Germany). Temperatures between 40 and 200 °C, cooling rate of 6 K·min^{-1} and a constant pressure of 200 bar were used. Three repetitions were made.

In Figure 3b the relative specific volume, normalised by the specific volume at 40 °C is depicted as a function of temperature for the investigated materials at 200 bar. The standard filament PLA_00 exhibits only small problems due to shrinkage. Hence, the optimal value for the maximum shrinkage in materials used for FFF should be similar or lower than the one for PLA. According to this plot, PLA_00 shrunk around 10 %, when cooled down from 200 to 40 °C. Both highly filled materials show much lower shrinkage values than PLA, as the high filler load prevents the polymer to change its volume. PP_00 exhibits the maximum shrinkage (15.2 %) of all the selected materials, and the printing trials revealed that the printed part had warpage due to the excessive shrinkage of the material. The high shrinkage value for PP_00 can be reduced by 25 %, down to 11.4 % by filling it with 30 vol% of compatibilised glass spheres, resulting in a compound with shrinkage values close to that of neat PLA_00. Printing trials with PP_GS are still pending, only then we can confirm if the achieved shrinkage is low enough to prevent warpage.

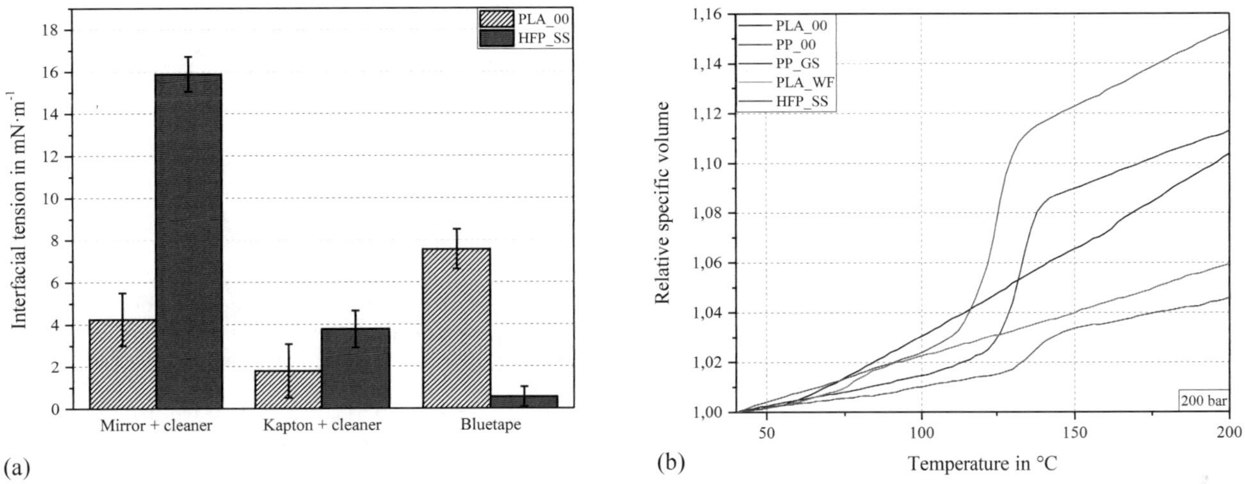

FIGURE 3. (a) Interfacial tension between the investigated surfaces and compounds and (b) the relative specific volume, normalised by the specific volume at 40 °C, versus the temperature for the investigated materials at 200 bar.

CONCLUSIONS

New materials for FFF have been developed and characterised. These materials include polyolefins, wood polymer composites and highly-filled polymeric compounds. Important properties to consider include the mechanical, flow, adhesion and shrinkage properties. All of them have to be considered in determining if a material will be printable and lead to printed parts with acceptable quality.

ACKNOWLEDGMENTS

The authors acknowledge the financial support of the European Commission with grant agreement 636881 (REProMag Project), the Austrian Research Promotion Agency (FFG) with grant agreement 848624 (NextGen3D project) and grant agreement 849297(Addmanu project). Authors also acknowledge the technical support of Florian Arbeiter during the mechanical characterisation of filaments, Phillip Huber during the contact angle measurements and Martin Hanzlík for drawing the schematics.

REFERENCES

[1] A. Gebhardt, 3D-Drucken, Grundlagen und Anwendungen des Additive Manufacturing (AM), Munich: Hanser, 2014.

[2] O. S. Carneiro, A. F. Silva and R. Gomes, "Fused deposition modeling with polypropylene," *Materials & Design,* vol. 83, p. 768–776, 2015.

[3] I. Gibson, D. W. Rosen and B. Stucker, Additive Manufacturing Technologies, Rapid Prototyping to Direct Digital Manufacturing, New York: Springer, 2010.

[4] J. L. White and D. D. Choi, Polyolefins, Processing, Structure Development, and Properties, Munich: Hanser, 2005.

[5] A. Gooneie, S. Schuschnigg, I. Duretek and C. Holzer, "Numerical simulations of the flow of wood polypropylene composites with wall slipping in a profile die: The significance of material data," *AIP Conf. Proc. (AIP Conference Proceedings),* vol. 1664, no. 1, p. 050014, 2015.

[6] J. Gonzalez-Gutierrez, G. B. Stringari and I. Emri, "Powder injection moulding of metal and ceramic parts," in *Some Critical Issues for Injection Moulding*, Rijeka, InTech, 2012, pp. 65-88.

[7] K. Elkins, H. Janak, R. W. Nordby, R. W. Gray, J. H. Bøhn and D. G. Baird, "Soft Elastomers for Fused Deposition Modeling," in *Proceedings of 8th Solid Freeform Fabrication Symposium*, Austin, TX, 1997.

[8] G. R. Berger, C. Steffel and W. Friesenbichler, "A study on the role of wetting parameters on friction in injection moulding," *International Journal of Materials and Product Technology, Vol. 52,* pp. 193-211, 2016.

Selective Laser Irradiation to Immiscible Polymer Blend for Dispersion Control

Takushi Saito [a,*], Tatsuya Kawaguchi[a] and Isao Satoh[a]

[a] *Department of Mechanical Engineering, Tokyo Institute of Technology, Ookayama, Meguro-ku, Tokyo, JAPAN*

Abstract. In this study, effect of the infrared laser irradiation on the blend morphology was investigated using immiscible polymer blend. Single-screw extruder with transparent window for laser beam transmission and fiber laser system having the wavelength of 1063 nm were used to investigate the impact of the laser irradiation on the blend dispersion. The test blend consisted of polystyrene (major phase) and polyethylene (minor phase), and small amount of carbon black was added to polyethylene to effectively increase the radiation absorption coefficient. Extruded sample was quenched by water bath and the dispersed polyethylene phase was taken out from the sample with xylene washing. Droplet shape and number were investigated through the SEM observation to quantitatively discuss the dispersion situation. Averaged size of the dispersed phase was decreased if the laser intensity was lower than 30 W/cm^2, but it was increased above 30 W/cm^2.

Keywords: Laser irradiation, Immiscible polymer blend, Dispersion control

INTRODUCTION

Polymer blend is a mixing process of different polymer material at molten state, and it is often used to improve the physical properties of polymer materials. When the mixing pair is immiscible, dispersion situation affects the physical properties of the blend material[1)]. If the viscosity ratio between the dispersed phase and matrix phase is large, dispersion progress of the minor phase will be delayed especially in the early stage of the blend mixing.

Saito et al.[2)] investigated the realization of inhomogeneous temperature field for a blend system in molten state by utilizing the difference of radiation absorption coefficient and the irradiation of infrared radiation. As the result, dispersion situation was controlled by the viscosity ratio change between the dispersed phase and matrix phase.

In this study, selective laser irradiation method was adopted to the extrusion process to investigate the effect of several process parameters on the dispersion situation. Temperature distribution around a dispersed droplet was investigated by numerical simulation to estimate the impact of the radiation heating on the system.

EXPERIMENTAL APPARATUS AND PROCEDURES

Materials

The test blend material consisted of polyethylene (dispersed phase) and polystyrene (matrix phase). Carbon black powder was mixed with polyethylene in advance so as to increase the radiation absorption coefficient of the dispersed phase. Physical properties of the used materials were summarized in Table 1. Weight ratio of the test blend was polystyrene 98% and polyethylene 2%.

Table 1 Physical properties used in this study

	PE + CB	PS
Density [kg/m^3]	922	1,050
Specific heat [J/(kg K)]	2,300	1,300
Thermal conductivity [W/(mK)]	0.36	0.19
Absorption coefficient @ 1063 nm [1/m]	102,000	46.4

Shear Viscosity Measurement

Temperature dependent viscosity change of the polymer materials used in this study were measured by shear type rheometer, PHYSICA MCR300 by Anton Paar. The measured shear viscosity is shown in Figure 1. The shear rate in the experiment was determined by the rotation speed and the groove depth of the screw.

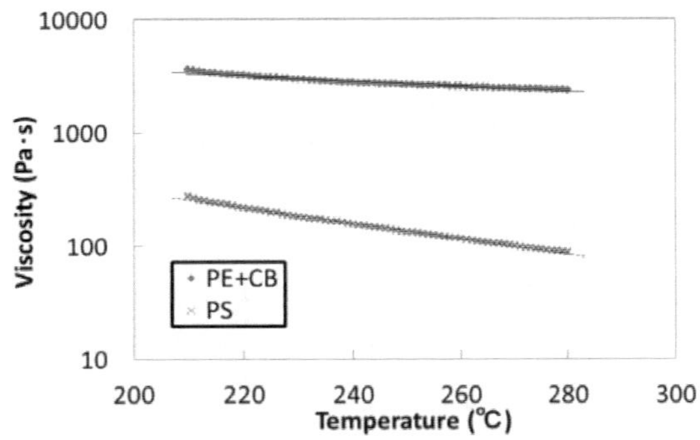

Figure 1 Temperature dependent viscosity change of the used material

Extrusion Process and Evaluation

Experimental conditions applied in this study are summarized in Table 2. Effect of the laser irradiation and melt temperature were investigated by comparing the change of the minor phase (PE) diameter.

Extruded sample was quenched by water bath to fix the dispersion situation. Matrix phase of PS was taken away by using xylene washing to obtain the dispersed phase of PE. The equilibrium diameter of PE droplets was obtained from the image taken by scanning electron microscope (SEM). Evaluated droplet number was more than 200 for each condition. Simple heat transfer simulation was also performed to estimate the temperature distribution around the dispersed droplet.

Table 2 Experimental conditions

Melt temperature [°C]	Intensity of laser irradiation [W/cm^2]	Screw speed [rpm]
210, 220, 230, 240	0, 10.2, 25.5, 50.9	60

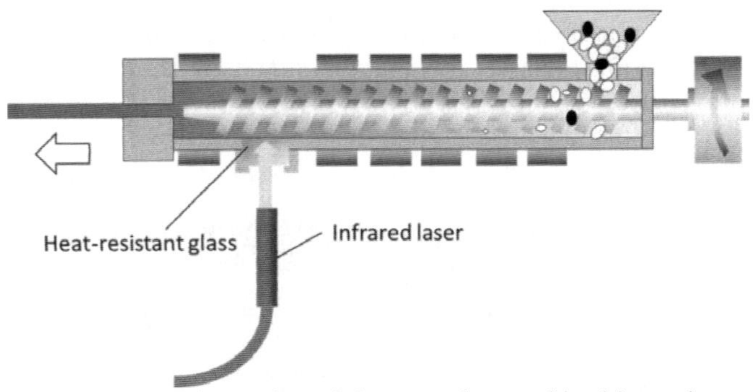

Figure 2 Schematics of the extruder used in this study

RESULTS AND DISCUSSION

Figure 3 is the SEM image showing the dispersed PE droplets. Variation of the equilibrium diameter at different conditions is shown in Figure 4. At the radiation intensity less than 30 W/cm^2, droplet diameter tended to be decreased with the radiation intensity increase. However, the diameter was contrary increased with the radiation intensity if its value exceeded 30 W/cm^2. From these results, effect of the selective radiation heating on the dispersion control was experimentally confirmed even in the extrusion process.

Careful observation of Figure 4 showed that the droplet size tended to become large with the melt temperature increase even at the constant radiation intensity. This result meant that the increase of the melt temperature disturbed the minor phase dispersion, and this was opposite to the effect of the selective radiation heating being discussed in this study. This was caused by the fact that the viscosity ratio of the used blend system (minor phase/matrix phase) increased with the temperature (Figure 1). Important point was that the excess radiation heating might bring the temperature increase of the whole blend system and it would result in the droplet diameter increase.

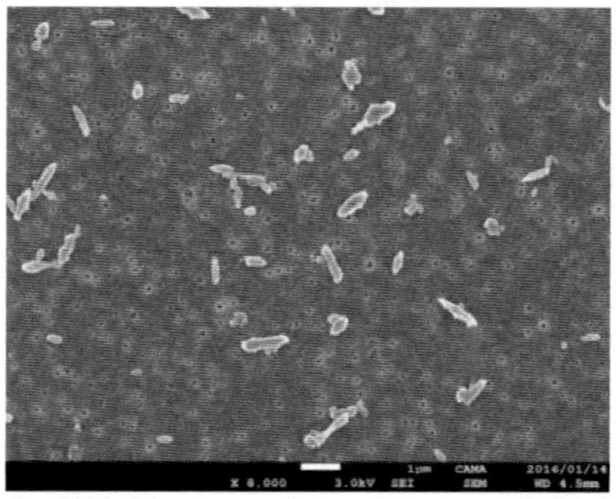

Figure 3 SEM image of the dispersed PE droplets

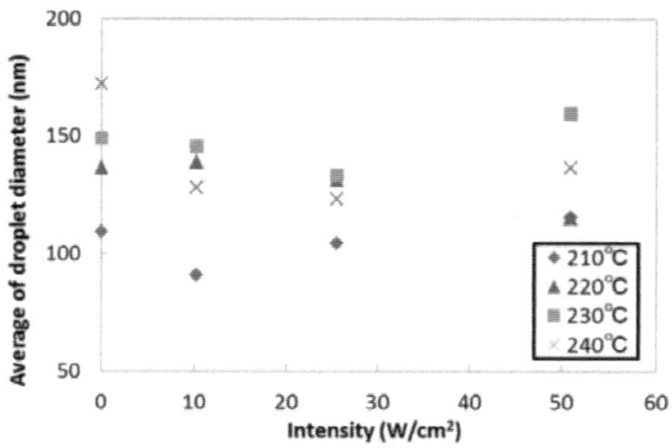

Figure 4 Variation of the dispersed droplet diamter

Figure 5 shows the numerical simulation result of a typical temperature distribution under the selective radiation heating. Radiation energy was coming from the bottom side of the domain, and the selective heating of several tens degree Celsius was observed in the lower half of the droplet. It was also shown that the temperature of the surrounding matrix phase was increased by the thermal conduction effect. Design of the effective viscosity ratio change of the blend system considering the temperature distribution will be important for the next step of the discussing technique.

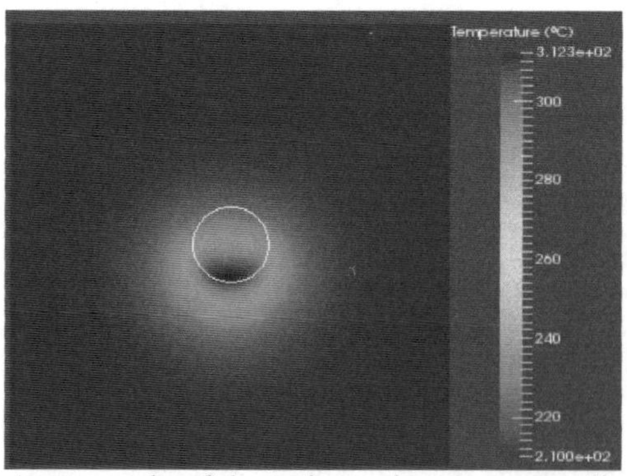

Figure 5 An example of the estimated temperature distribution;
Droplet diameter 100 μm, Initial melt temperature 210°C, Irradiation Intensity 50.9 W/cm^2, Heating time 0.106 s

CONCLUSIONS

In this study, selective radiation heating was adopted to the extrusion process, and its effect on the dispersion situation of minor phase was investigated. The obtained results were summarized as below.

1. Selective radiation heating successfully changed the blend dispersion extruded from the single screw extruder equipping transparent window for the laser introduction.
2. Equilibrium diameter of the dispersed phase was controlled with the radiation intensity. There was a threshold intensity for the dispersion control, and it was 30 W/cm^2 regarding the used PE/PS blend system.

REFERENCES

1. L. A. Utracki, *Polymer Engineering & Science*, **35**, 2-17 (1995)
2. T. Saito, T. Kawaguchi and I. Satoh, *Journal of Applied Polymer Science*, **129**, 3606-3612 (2013)

Hydrophobic Nanocoating of Cellulose by Solventless Mechanical Milling

Yong Huang [a,*], Min Wu [a], Mengmeng Zhao [a, b], and Shigenori Kuga [a]

[a] *Technical Institute of Physics and Chemistry, Chinese Academy of Sciences, 29 Zhongguancun East Road, Haidian District, Beijing 100190, People's Republic of China.*

[b] *University of Chinese Academy of Sciences, Beijing 100039, People's Republic of China.*

Abstract. Solventless mechanical treatment of native cellulose in polytetrafluoroethylene (PTFE) vessel was found to cause friction transfer of PTFE to comminuted cellulose particles, resulting in their nanocoating. The PTFE-coated particles were highly hydrophobic, completely repelled from water, with water contact angle of 110~121°. SEM-EDS and XPS analysis showed nearly continuous coverage of cellulose particles by PTFE layer of approx. 10 nm thick. This finding may open a new class of solid-phase processing for nanocoating of particles.

Keywords: Cellulose • Hydrophobic nanocoating • Polytetrafluoroethylene (PTFE) •Solventless mechanical milling

PACS: Replace this text with PACS numbers; choose from this list:
http://www.aip.org/publishing/pacs/pacs-2010-regular-edition

INTRODUCTION

Cellulose is the commonest organic polymer with inexhaustible natural supply.[1] However, its hydrophilic nature makes it sensitive to water, limiting its durability and utility.[2] Hydrophobic modification of cellulose is therefore of great interest. As modifying functionality, fluorinated groups provide the most inert and lowest surface-energy materials.[3] Esterification of cellulose using fluorinated agents is a common process to prepare hydrophobic cellulose.[4-7] The process, however, suffer from requirement of expensive chemicals or special equipment.

Recently, reactive ball milling was found to give hydrophobic, surface-esterified nanofibers potentially useful as composite elements.[8] Dry milling of graphite with cellulose causes exfoliation-dispersion of graphene-like platelets in cellulose matrix.[9] Endo et al. showed usefulness of cellulose milling with synthetic polymers to prepare novel composites.[10-13] These phenomena exemplify the effectiveness of mechanical impacts in processing polymer solids by generating instantaneous high temperature and pressure.[14]

Here we introduce a simple and facile preparation of hydrophobic cellulose particles, in which the comminution and PTFE coating of cellulose take place simultaneously by solventless mechanical milling.

EXPERIMENTAL

A bleached kraft pulp made from Radiata pine (Kinleith Mill, New Zealand) was used as starting cellulose. The degree of polymerization was determined as 1100 from intrinsic viscosity of aqueous LiOH/urea solution. The cellulose sample was vacuum-dried at 80°C for 4h before use.

Fifty grams of cellulose was loaded to a 500-ml PTFE pot. Milling was done by a planetary mill QM-3SP4 (Nanjing Nanda Instruments) at 540 rpm for 4 h-28 h. For comparison, cellulose was milled in an agate pot with zirconia balls.

RESULTS AND DISCUSSION

Solid cellulose is a hydrophilic material due to hydroxyl groups exposed on microfibril surfaces.[15] Therefore, the starting cellulose and cellulose powder milled in agate pot were easily dispersible in water and sedimented soon. In contrast, the cellulose powder milled in the PTFE pot became gradually hydrophobic, as seen by repulsion from water after shaking (Figure 1). The amount of water-repelled fraction increased by elongated milling, and total repulsion was attained after approx. 22-h milling. This behavior suggests that certain amount of PTFE was transferred from the pot wall and adhered to surface of cellulose particles.

The amount of PTFE adhered to cellulose could be determined by weight loss of the pot. Figure 2 shows the time course of weight gain of the milled powder and water contact angle of compacted pellet from it. The weight gain was fast in early stage of milling, slowed down after about 4 h, and leveled off after about 28 h. The rise of contact angle was steep at first, and nearly leveled off after 10 h at 115-120°, close to 117° of neat PTFE. These features show that transfer of PTFE takes place most effectively when the bare cellulose surface hits the wall by ball impact. Because ball milling causes simultaneous comminution of cellulose, the increased surface would catch more and more PTFE. After about 28 h of milling this process seems to stop due to complete coverage of comminuted cellulose.

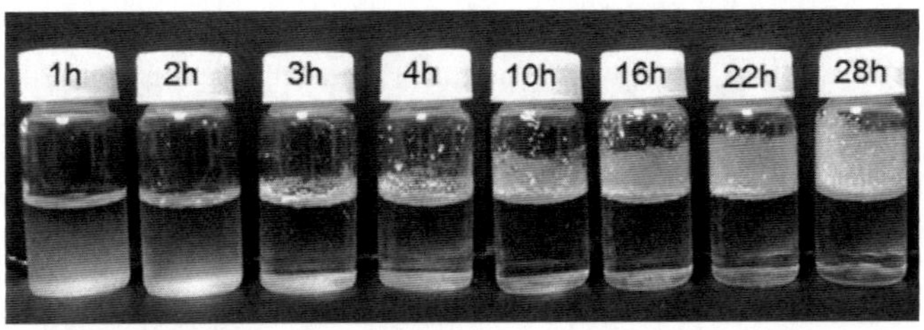

FIGURE 1. Time course of PTFE-milled cellulose showing progressive repellence from water due to PTFE coating.

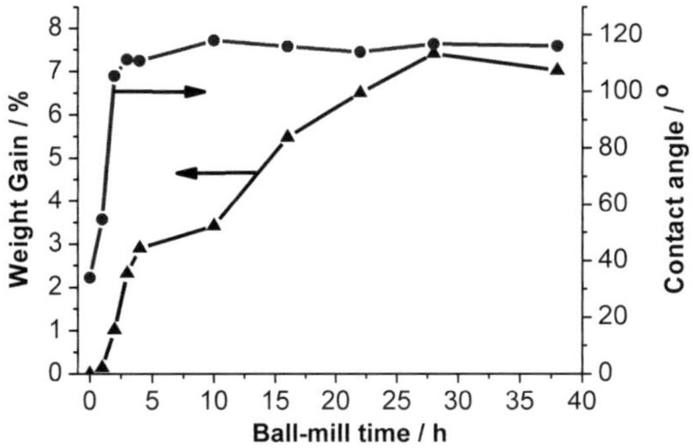

FIGURE 2. Time course of weight gain (add-on percentage) and water contact angle of compacted pellet from PTFE-milled cellulose powder.

Morphology of PTFE-milled Cellulose

The morphology and structure of PTFE-milled particles were examined by SEM. The PTFE-milled cellulose became smooth-surfaced platelets after 10-h milling; thereafter their lateral size did not decrease remarkably, and typical particles after 28-h milling was 10-20 μm wide.

AFM observation of PTFE-milled particles gave information on particle shape from height measurements. Together with the SEM data, average lateral size and thickness were determined. The lateral size decreased quickly from 33 μm±4 μm (width of pulp fiber) to 11 μm±8 μm after 4-h milling, followed by gradual decrease to 5 μm±4 μm after 28-h milling. The thickness measurement by AFM showed a similar pattern, decreasing from 1.4 μm to ~400 nm after 28 h milling. Interestingly, the aspect ratio from average width and height remained approximately constant at 7~9.

Mode of PTFE-coated cellulose

The mode of coverage by PTFE after 28-h milling was examined by SEM-EDS (Figure 3) and XPS. In EDS (Figure 3a), the line scan pattern for fluorine shows continuous coverage of PTFE layers on the surface and small peaks at both edges of the particle, suggesting a situation illustrated in Figure 3b. From XPS, the relative PTFE content of the whole material was 60.9 wt% (Avantage 5.52 software). This value is much higher than that from weight gain, 7.4 wt%, due to limitation of detection depth of XPS (about 10nm), indicating localization of PTFE on the surface of cellulose.

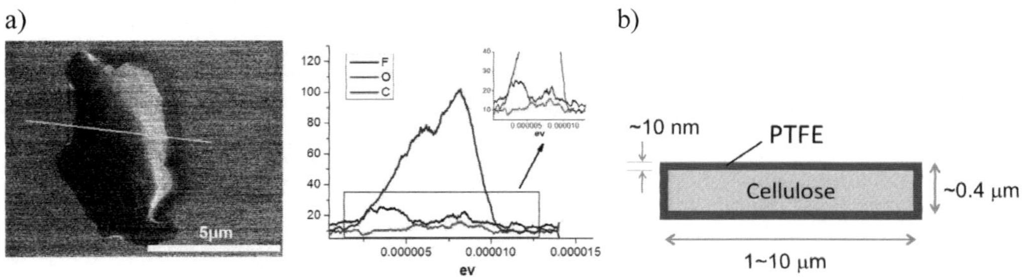

FIGURE 3. a) line-scan of PTFE-milled cellulose (28 h). b) Simplified model of PTFE-milled cellulose particle with continuous coverage by PTFE.

Thickness of the PTFE coating layer

The PTFE layer thickness can be assessed from PTFE/cellulose weight ratio. By assuming aspect ratio of 7~9, one can ignore the contribution of edges. The average thickness of PTFE layer, L, is:

$$\frac{weight\ of\ PTFE}{weight\ of\ cellulose} = 2L * \rho p \,/\, T * \rho c \qquad (1)$$

Where ρc (1.6 g/cm3) and ρp (2.1 g/cm3) are densities of cellulose and PTFE, respectively. Equating the weight ratio in Eq. (1) to the weight gain data in Figure 2 and using thickness of platelets, T, one obtains time course of thickness of PTFE layer, L, as in Figure 4. It is remarkable that the thickness of PTFE layer is nearly constant at ~10 nm after 10 h of milling.

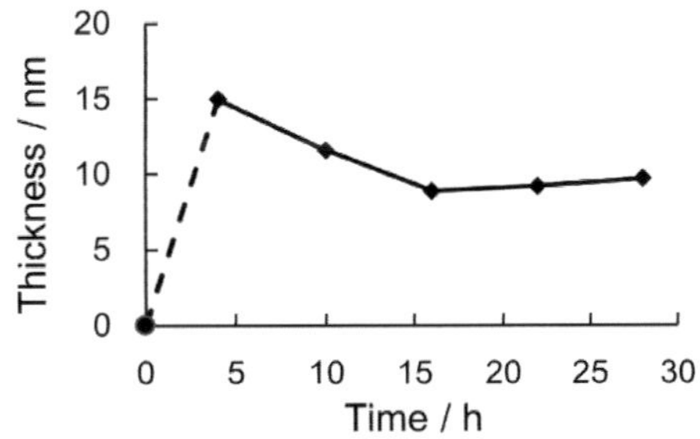

FIGURE 4. Thickness of PTFE layer estimated from platelets' thickness and weight gain by Eq. (1).

Mechanism of the PTFE nanocoating of cellulose

These results as a whole indicate that PTFE is transferred from the wall by mechanical milling, and then adhered to cellulose surface to form an ultrathin coating layer. This phenomenon can be understood as the friction transfer process of polymer, which involves the transfer of a polymer to another surface when the two materials are rubbed together.[16] PTFE is the most commonly studied polymer for its smooth molecular profiles.[17] The friction transfer (deposition) of PTFE has been known for macroscopic and controlled setups depending on the friction conditions (load, sliding speed, temperature).

Native cellulose of higher plants is formed as 3~4 nm wide crystalline microfibrils, which have hydrophilic surfaces exposing (110) or $1\bar{1}0$ crystallographic planes. Since ball milling causes partial disruption of crystalline order, it can cause exposure of (200) hydrophobic plane formed by glucopyranoside rings. Actually, water contact angle of the pellet from decrystallized cellulose by milling in agate pot was 69.2°, higher than that of original cellulose, 34.1°. This behaviour also may contribute to the firm adhesion of PTFE onto cellulose.

Potential application of the PTFE-coated cellulose

The cellulose powder milled in PTFE pot is very smooth and slippery. Together with its hydrophobicity, several properties arise. One is anti-ultraviolet capability: We dispersed the PTFE-milled cellulose into Vaseline. The resultant white emulsion showed obvious ultraviolet absorption in UVA (320-400nm) and UVB (280-320 nm), suggesting usefulness as sunscreen.

Another property is hydrophobic modification of painted surfaces. The PTFE-milled cellulose could be readily blended with a xylene-based commercial aluminum anticorrosive paint. The painted surface on slide glass appeared similar to that of the original paint, but its water contact angle increased with loading level, reaching a maximum of 139 ° at 6% loading, significantly higher than original 97°. Durability of water repellency was tested by change in contact angle of the 2% loading sample after continued immersion in water. The decrease was moderate, keeping the angle at about 122 ° after a 5 day-immersion. Thus the water sensitivity of cellulose was sufficiently suppressed.

CONCLUSION

The present finding of PTFE nanocoating of cellulose powder may be useful to prepare "bulked up" PTFE powder maintaining the unique surface properties of PTFE. This method is simple and advantageous in avoiding use of solvent and no-loss utilization of PTFE. The PTFE-coated cellulose powder is potentially useful as hydrophobic material, lubricant, painting, cosmetics, composite element, etc. New types of ball mill are continually developed,[18] which may be applied to scaling up of the friction transfer by ball milling for industrial practice.

ACKNOWLEDGMENTS

This study was supported by the National Natural Science Foundation of China (No.51373191, 51172247, 51472253) and the Chinese Academy of Sciences Visiting Professorships.

REFERENCES

1. A. C. Osullivan, *Cellulose* **4**, 173-207 (1997).
2. P. Samyn, G. Schoukens, D. Stanssens, L. Vonck and H. den Abbeele, *Cellulose* **20**, 2625-2646 (2013).
3. T. Nishino, M. Meguro, K. Nakamae, M. Matsushita and Y. Ueda, *Langmuir* **15**, 4321-4323 (1999).
4. A. G. Cunha, C. S. R. Freire, A. J. D. Silvestre, C. P. Neto and A. Gandini, *Journal of Colloid and Interface Science* **301**, 333-336 (2006).
5. A. G. Cunha, C. S. R. Freire, A. J. D. Silvestre, C. P. Neto, A. Gandini, E. Orblin and P. Fardim, *Biomacromolecules* **8**, 1347-1352 (2007).
6. A. G. Cunha, C. S. R. Freire, A. J. D. Silvestre, C. P. Neto, A. Gandini, E. Orblin and P. Fardim, *Langmuir* **23**, 10801-10806 (2007).
7. A. G. Cunha, C. S. R. Freire, A. J. D. Silvestre, C. P. Neto, A. Gandini, E. Orblin and P. Fardim, *Journal of Colloid and Interface Science* **316**, 360-366 (2007).
8. P. Huang, M. Wu, S. Kuga, D. Wang, D. Wu and Y. Huang, *ChemSusChem* **5**, 2319-2322 (2012).
9. P. Sun, S. Kuga, M. Wu and Y. Huang, *Cellulose* **21**, 2469-2478 (2014).
10. A. Nagatani, T. Endo, T. Hirotsu and M. Furukawa, *Journal of Applied Polymer Science* **95**, 144-148 (2005).
11. W. Qiu, T. Endo and T. Hirotsu, *Journal of Applied Polymer Science* **94**, 1326-1335 (2004).
12. W. Qiu, T. Endo and T. Hirotsu, *European Polymer Journal* **41**, 1979-1984 (2005).
13. W. Qiu and T. Hirotsu, *Macromolecular Chemistry and Physics* **206**, 2470-2482 (2005).
14. R. Janot and D. Guerard, *Progress in Materials Science* **50**, 1-92 (2005).
15. Y. Nishiyama, *Journal of Wood Science* **55**, 241-249 (2009).

16. N. K. Myshkin, M. I. Petrokovets and A. V. Kovalev, *Tribology International* **38**, 910-921 (2005).

17. C. M. Pooley and D. Tabor, *Proceedings of the Royal Society A Mathematical Physical & Engineering Sciences* **329**, 251-274 (1972).

18. F. J. Gotor, M. Achimovicova, C. Real and P. Balaz, *Powder Technology* **233**, 1-7 (2013).

A Water-soluble Branched Polymer for Enhanced Oil Recovery

Nanjun Lai [a,*], Xin Guo [a]

[a] *School of Chemistry and Chemical Engineering, Southwest Petroleum University Chengdu Sichuan 610500*

Abstract. to address the problem regarding poor shear resistance of commonly used polymers for oil displacement, a modified branched functional monomer (named BA)was synthesized from dodecylamine, methyl acrylate, ethanediamine, maleic anhydride and utilized to react with acrylamide (AM), acrylate (AA) to synthesize a water-soluble branched copolymer oil displacement agent by free-radical polymerization using redox system as initiator. The optimum synthetic conditions were determined as follows: initiator 0.3 %, reaction temperature 40 ℃, monomer concentration 24 %, of which n (AA): n (AM): n (BA) = 3: 7: 0.3 %. The structure of HPBA was characterized by means of FTIR, ^{1}H NMR. Compared with HPAM, the anti-shearing properties of HPBA were investigated, which demonstrated that the introduced branched structure and hydrophobic chains could endow the polymer with great viscosity retention at lower generation. Moreover, on the basic of the indoor displacement experiment, HPBA exhibited excellent mobility control capability that was 66.73 of resistance factor, 11.93 of residual resistance factor.

Keywords: polymerizable branched monomer; branched copolymer; anti-shearing; mobility control capability.
PACS: 61.25.he, 82.35.Jk, 82.35.Lr, 82.35.Gh.

INTRODUCTION

Polymer flooding is an important technology to extend oilfield life and increase oilfield production[1–5]. Due to shearing action of the stirrer, pipeline valve, injection pump, shot hole and porous media during preparation and injection of polymer solution, however, the molecular chains of the traditional linear polymer will be cut off, so the viscosity of the polymer solution will be significantly reduced[6,7], thus seriously reducing its effects of enhancing oil recovery[8–10]. Among dendritic macromolecules, the structure of polyamidoamine (PAMAM) is similar to that of a three-dimensional (3D) sphere and branches, which be of highly shear resistance [11]. Moreover, its terminal groups, amidogen, which could be easily modified, has been widely used in several practices, like biological medicine, [12, 13] surfactants, [14] catalysis, [15] nanomaterials, [16] and so forth. However, such polymers have a poor molecular weight and the highly branched structures exhibit lower viscosities compared to linear polymers with similar molecular weight, which will reduce its thickening efficiency[17] and limit its application in EOR.

Consequently, reference to PAMAM, we synthesized a low generation polymerizable branched functional monomer (BA) which was of unsaturated double bonds and hydrophobic chains (SCHEME 1(a)). The branched polymer (HPBA, SCHEME 1(b)) was obtained by the copolymerization of acrylamide (AM), acrylic acid (AA), and the surfmer BA, that could improve its thickening efficiency through the increase of the molecular weight, decrease of generation and a certain degree of hydrophobic association. Meanwhile, assure that its anti-shearing properties through hydrophobic association of hydrophobic chains in HPBA whose branched structure fewer than high generation dendritic macromolecules. The optimum synthetic conditions were determined though orthogonal and single factor experiments. The structure of HPBA was characterized by FTIR, ^{1}H NMR. The thickening efficiency, anti-shearing properties and mobility control capability of HPBA were assessed for investigating its feasibility in EOR.

EXPERIMENTAL

Reagents and Materials

Dodecylamine, methyl acrylate (MA), ethanediamine (EDA), maleic anhydride (MAH), methanol (CH_3OH), acrylamide (AM) acrylic acid (AA), absolute ethyl alcohol (C_2H_5OH), ammonium persulfate ($(NH_4)_2S_2O_8$) and sodium hydrogen sulfite ($NaHSO_3$), sodium hydroxide (NaOH) and other chemicals were purchased from Chengdu KeLong Chemical Reagent Co., Ltd (China) were of analytical grade and used without further purification

Synthesis of BA

A certain amount of dodecylamine was firstly dissolved in CH_3OH, then, MA was added into the flask. The mixture was stirred 24 h at 25 ℃ under a nitrogen atmosphere. The product was distilled under reduced pressure at 45 ℃ and a light-yellow liquid (GA0.5) was obtained. Similarly, GA0.5 was dissolved in CH_3OH, EDA was added into the flask. After reacting for 24 h at 25 ℃ under a nitrogen atmosphere, The solvent and unreacted EDA were removed under reduced pressure at 65 ℃. A light-yellow wax solid (G1.0) was obtained. Then, GA1.0 dissolved by N, N-dimethylformamide (DMF) was added slowly into the MAH solution which was dissolved in DMF. The reaction was carried out at 70 ℃ for 6 h, and BA was obtained through precipitation with trichloromethane and vacuum filtration. The synthesis route was shown in SCHEME 1(a).

Synthesis of the Branched Copolymer HPBA

HPBA was synthesized by free radical polymerization in aqueous solution and its structure was shown in SCHEME 1(b). The indicated amounts of raw materials AA, AM, DA were added into a flask with mechanical stirring in deionized water under an inert nitrogen atmosphere. After heating at indicated temperature, then the indicated loading of the initiator [n ($(NH_4)_2S_2O_8$): n ($NaHSO_3$) = 1: 1] was added slowly into the flask. After that the copolymerization was carried out under a nitrogen atmosphere for an indicated time. Finally, the branched copolymer HPBA could be obtained by washing (with absolute ethyl alcohol), smashing and drying the reaction products. The optimal synthetic conditions were determined though orthogonal and single factor experiments, as follows: initiator 0.3 %, reaction temperature 40 ℃, monomer concentration 24 %, of which n (AA): n (AM): n (BA) = 3: 7: 0.3 %.

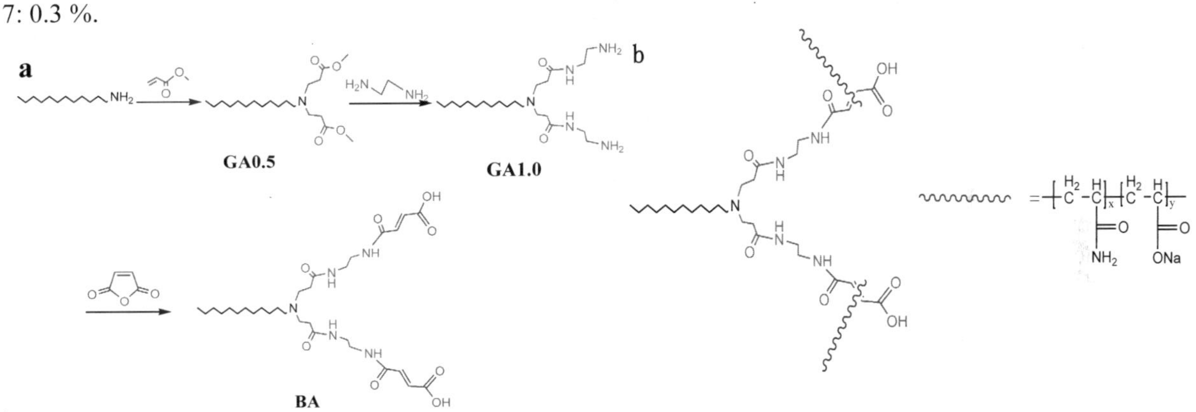

SCHEME 1. (a) The synthesis route of BA, (b) the structure of HPBA.

Measurements and Characterization

The infrared spectrum was recorded on a Perkin Elmer RX-1 Spectrophotometer (Beijing Rayleigh Analytical Instrument) using a KBr pellets. The ^{1}H-NMR was recorded on a Brucker 400 MHz nuclear magnetic resonance spectrometer (Brucker Daltonics, U.S.A.). The apparent viscosities of HPBA solutions before/after shearing (sheared by Mixing Speed Governor (WT-VSA2000B)) were measured using a Brookfield DV-III Ultra Programmable Rheometer (Brookfield Engineering Laboratories, U.S.A.). To investigate the effect of the shear force on the polymer apparent viscosity, we measured the apparent viscosity of the samples with diverse concentrations (500-2500 mg/L) before and after being sheared at 3500 rpm for 20 seconds. Then we changed the shear time varying from 10 to 60 s, and tested the apparent viscosity of the HPBA and HPAM solutions of 1500 mg L^{-1} to obtain the effect of shear time on the polymer solutions. The influence of shear strength was studied by setting the shear strength at 3500, 7000, 11500, 14500, 17000 and 19000 rpm. The mobility control capability of HPBA was assessed though the indoor displacement experiment. The tubes (30 cm in length and 2.5 cm in inner diameter) were packed with 100–120 mesh quartz sand. The brine with a TDS of 8766.28 mg/L was used in the experiments whose ionic composition was shown in TABLE 1. The flooding rates both of the polymer flooding and subsequent water flooding were set as 1.0 mL/min and Its mobility control ability can be achieved through analysing the resistance factor (RF) and residual resistance factor (RRF) which were calculated using the following equations:[18]

$$RF = \frac{K_W / \mu_W}{K_P / \mu_P} \tag{1}$$

$$RFF = \frac{K_{Wb}}{K_{Wa}} \tag{2}$$

where K_W and K_P are the aqueous and polymer phase permeability, respectively ($\times 10^{-3}$ μm^2), μ_W and μ_P are the aqueous and polymer viscosity, respectively (mPa s), and K_{Wb} and K_{Wa} are the aqueous phase permeability before and after polymer flooding, respectively, where $K_{Wb} = K_W$ ($\times 10^{-3}$ μm^2).

TABLE 1 Ionic composition of the brine

Ions	**Na^+, K^+**	**Ca^{2+}**	**Mg^{2+}**	**HCO_3^-**	**SO_4^{2-}**	**Cl^-**
Concentration (cm/L)	3167	99	98	200	100	5099

RESULTS AND DISCUSSION

IR Spectra and ^{1}H-NMR Spectra Analysis

The structures of BA and HPBA were confirmed by IR spectroscopy as shown in Fig. 2a. In the spectrum curve of BA, the peaks at 3423 cm^{-1} was the stretching vibration of N–H, and the peak at 1554 cm^{-1} and 3075 cm^{-1} were due to its in-plane bending vibration. The peak at 2925 cm^{-1} and 2853 cm^{-1} indicated the presence of $-CH_2-$ and $-CH_3-$ groups stretching vibration. The peak at 1710 cm^{-1} suggested that the C=O groups stretching vibration. The peak at 1657 cm^{-1} indicated the presence of C=C. Similarly, it could be seen from HPBA curve that the absorption peak of $-CH_2-$ and $-CH_3-$ groups presented at 2932 cm^{-1} and 2862 cm^{-1}. The broad peak ranging from 3187 cm^{-1} to 3453 cm^{-1} was the overlap of N–H in the skeleton and $-CONH_2$ in the polymeric main chain. The stretching vibration of C=O displayed at 1685 cm^{-1}gave a stronger absorption as it existed in both structures of BA and AM. The peaks at 1406 cm^{-1} corresponded to the structure of –COOH. These results indicated that the target copolymer was successfully synthesized.

The ^{1}H NMR spectra of BA was shown in FIGURE 1(b). The peak at 0.81 and 1.19 ppm were assigned to the -CH_3 and $-CH_2-$ proton in the hydrophobic chains. The peak at 2.97 and 3.60 ppm were due to the $-CH_2-$bonded with tertiary amine group. The signals observed at 7.58 ppm were attributed to the proton peak of $-NH-$ bonded with the carbonyl C=O. At 6.14 ppm, the proton peak of $-CH=CH-$ was monitored. The ^{1}H NMR spectrum of HPBA in FIGURE 1(b) exhibited the following chemical shifts: the peak at 1.10 ppm was due to the $-CH_2-$ in the hydrophobic chains. The proton peaks of $-CH_2-$ and $-CH-$ in the polymeric main chain were 1.56 and 2.15 ppm, respectively. The peak at 7.67 ppm was the proton signal of $-NH-$ associated with the branched skeleton. The peak at 6.88 ppm was due to the $-CH-NH_2$ group existing in the monomer AM. Therefore, the structure of the copolymer HPBA was consistent with the original design by ^{1}H NMR analysis.

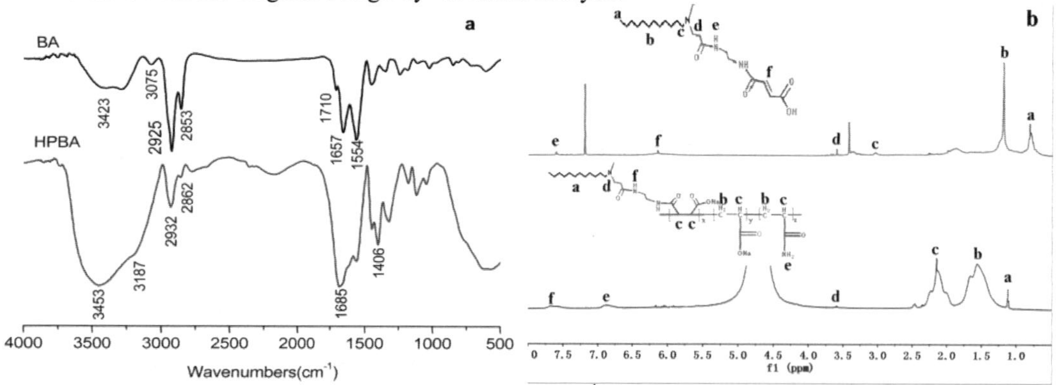

FITURE 2 (a) IR spectra of BA and HPBA, (b) ^{1}H NMR spectra of BA and HPBA.

Anti-shearing Properties

Due to the effects of injection pipelines, equipment and process, completion method (shot hole) and the high-speed shear degradation in the near-wellbore area, the viscosity of polymer solution may significantly reduce. Therefore, the experimental evaluation for the shear resistance of the copolymer HPBA should be studied for its application in EOR. The effect of the concentration of polymer, the mechanical shear rate and shear time on the apparent viscosity of the polymer solutions was investigated.

In the FIGURE 2(a), the apparent viscosity of HPBA before and after sheared by Mixing Speed Governor both displayed a rising trend with increasing concentration. But at low concentration, the unsheared HPBA solutions' viscosity was relatively lower than that of HPAM due to the branched structure. With the increase of concentration, HPAM were overcome gradually, attribute to the hydrophobic association occur in the HPBA solutions. For the sheared polymer solution, with the branched structure which could promote shear resistance, HPDA had better viscosity and viscosity retention than those of HPAM. The maximum retention rate of HPDA was 92.05% while that of HPAM was 61.03%.

As shown in the FIGURE 2(b). With the increasing shear time, the apparent viscosity of HPAM reduced rapidly, whereas, its downtrend of HPBA began to flatten. Similar trend occur in the viscosity retention of polymer. HPBA had a steady retention rate (81.9%), which was larger than that of HPAM (32.4%), at shear time of 60s.

From FIGURE 2(c), both the HPDA and HPAM solutions showed a shear thinning behavior with increasing shear rate. Nevertheless, a higher viscosity retention rate was obtained for the HPBA solution. All the results showed that HPBA had a great thickening efficiency and excellent anti-shearing performance.

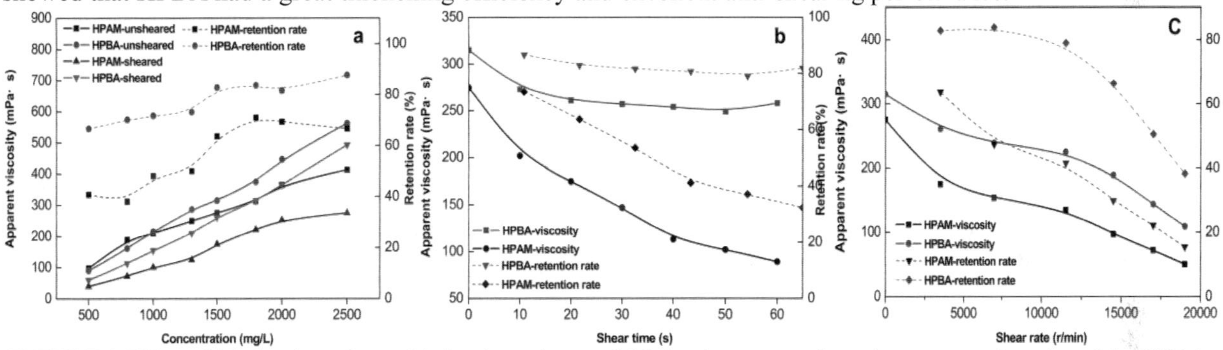

FITURE 2 Effect of (a) the shear force, (b) the shear time, and (c) the shear strength on the apparent viscosity of the HPBA and HPAM solution.

Mobility Control Ability

To study and analyze the feasibility of HPBA to enhance oil recovery, we took the HPAM into comparison and investigated the mobility control ability of HPBA though indoor displacement experiments. The flow characteristic curves of HPBA and HPAM under various permeabilities were presented in FIGURE 3. According to data in FIGURE 3, the RF and the RRF were constructed, as shown in TABLE 2. It was obvious that HPBA exhibited eminent mobility control ability that was favorable for enhancing the oil recovery. The HPBA solution could achieve a much higher RF and RRF in comparison to those of HPAM solution under the same conditions. The result showed that HPBA had a stronger mobility control capacity that was favorable to enhanced oil recovery.

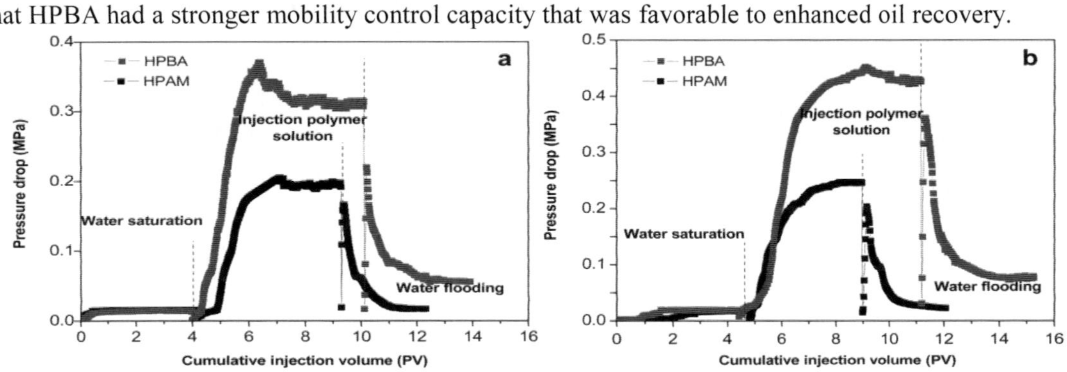

FITURE 3 Flow characteristic curves of the HPBA and HPAM solution at a permeability of about (a) 1.7 μ m^2, (b) 1.3 μ m^2

TABLE 2 RF and RRF of HPBA and HPAM solution

Polymer	Permeability (μ m^2)	Water saturation pressure (10^{-3} MPa)	Polymer solution pressure(10^{-3} MPa)	Water flooding pressure(10^{-3} MPa)	RF	RFF
HPBA	1.32	19.3	427.5	76.8	66.73	11.93
HPAM	1.38	18.4	246.9	22.5	40.26	3.67
HPBA	1.70	14.9	312.5	55.2	62.92	11.11
HPAM	1.61	15.8	196.1	17.2	37.23	3.37

CONCLUSIONS

A novel branched polymer (HPBA) was synthesized by aqueous free radical polymerization. The optimum synthetic conditions were determined as follows: initiator 0.3 %, reaction temperature 40 ℃, monomer concentration 24 %, of which n (AA): n (AM): n (BA) = 3: 7: 0.3 %. IR and ^{1}H NMR actually confirmed its chemical structure. The experiments of mechanical shearing model at different shearing condition (concentration of polymer, shear time, shear rate) convincingly proved that compared with HPAM which has been widely used in EOR, the branched structure and hydrophobic chains provided HPBA good thickening efficiency and excellent shear resistance. Thereby, HPBA had a better mobility control ability.

ACKNOWLEDGMENTS

This work was financial supported by National Natural Science Foundation of China(51404208), The Second Session of the "Young and Middle-aged Backbone Teachers" Training Program of Southwest Petroleum University, Scientific Research Project of Southwest Petroleum University, National College Students Innovation and Entrepreneurship Training Project(201510615008), China Scholarship Council.

REFERENCES

1. Z. Hu and G. Chen, *J. Mater. Chem. A*, **2**, 13593 (2014).
2. K. Li, M. Duan, H. Wang, J. Zhang and B. Jing, *RSC Adv.*, **5**, 17389-17395 (2015).
2. W. F. Pu, R. Liu, B. Li, F. Y. Jin, Q. Peng, L. Sun, D. J. Du and F. S. Yao, *RSC Adv.*, **5**, 88002-88013 (2015).
3. Y. Guo, J. Liu, X. Zhang, R. Feng, H. Li, J. Zhang, X. Lv and P. Luo, *Energy & Fuels*, **26**, 2116-2123 (2012).
4. B. Wang, Y. Chen, S. Liu, H. Wu and H. Song, *Colloid. Surface. A*, **287**, 170-174 (2006).
5. D. A. Z. Wever, F. Picchioni and A. A. Broekhuis, *Prog. Polym. Sci.*, **36**, 1558-1628 (2011).
6. A. M. Howe, A. Clarke and D. Giernalczyk, *Soft Matter*, **11**, 6419-6431 (2015).
7. A. Zaitoun, P. Makakou, N. Blin, R. S. Al-Maamari, A. A. R. AlHashmi and M. Abdel-Goad, *SPE J.*, **17**, 335-339 (2012).
8. L. Xue, U. S. Agarwal ,* and P. J. Lemstra, *Macromolecule*, **38**, 8825-8832 (2005).
9. H. Xin, D. Ao, X. Wang, Y. Zhu, J. Zhang and Y. Tan, *Colloid Polym. Sci.*, **293**, 1307-1316 (2015).
10. W. F. Pu, R. Liu, B. Li, F. Y. Jin, Q. Peng, L. Sun, D. J. Du and F. S. Yao, *RSC Adv.*, **5**, 88002-88013 (2015).
11. Y. L. Zhao, Q. Cai, J. Jiang, X. T Shuai, J. Z. Bei, C. F. Chen and F. Xi, *Polymer*, **43**, 5819-5825 (2002).
12. A. Pourjavadi, Z. M. Tehrani and S. H. Hosseini, *RSC Adv.*, **5**, 48586-48595 (2015).
13. Q. Zhu, F. Qiu, B. H. Zhu and X. Y Zhu, *RSC Adv.*, **3**, 2071-2083 (2013).
14. P. Zhang, X. Xu, M. Zhang, J. Wang, G. Bai and H. Yan, *Langmuir*, **31**, 7919-7925 (2015).
15. R. J. Kalbasi and F. Zamani, *RSC Adv.*, **4**, 7444-7453 (2014).
16. S. Veerapandian, S. Amudha, S. A. Suthanthiraraj, M. A. Rahman and A. S. Nasar, *RSC Adv.*, **5**, 31404-31409 (2015).
17. Lidija Slemenik Perše and Miroslav Huskic, *European Polymer Journal*, **76**, 188-195 (2016).
18. R. Ponnapati, O. Karazincir, E. Dao, R. Ng, K. K. Mohanty and R. Krishnamoorti, *Ind. Eng. Chem. Res.*, **50**, 13030-13036 (2011).

Development and Characterization of the Polymer Based on Modified Chitosan

Yanqin Liu [a] , Nanjun Lai [a,*], Zhongbin Ye [a,b]

[a] *School of Chemistry and Chemical Engineering, Southwest Petroleum University, Chengdu 610500, People's Republic of China*

[b] *State Key Lab of Oil and Gas Reservoir Geology and Exploitation, Southwest Petroleum University, Chengdu 610500, People's Republic of China*

* *e-mail: lainanjun@126.com*

Abstract. In order to obtain the environmental friendly polymer flooding agent , chemoselectively protecting the amino groups of chitosan has been developed to provide N-phthaloyl-chitosan that is indispensable as an organosoluble N-protected precursor for further controlled modification reactions of chitosan. And the derived 6-O-maleyl-2-N-phthaloyl chitosan has been prepared as a functional monomer , which then copolymerize with AA、AM and AMPS to synthetize the polymer based on modified chitosan . In this study, a preliminary study of some basic characterization and the feasibility of oil displacement of the copolymer solution is carried out. The copolymer showed positive characterization in shear resistance. Furthermore, this copolymer degraded with alpha amylase is biodegradable. Although the copolymer has low degree of polymerization, far less than the viscosity of HPAM, it also can obtain higher resistance factor, residual resistance factor and flow resistance in porous media., used for oil displacing agent may have potential advantages, and will have great significance to the production of oil field.

Keywords: Polymer oil displacement; Polymer Based on Modified Chitosan; Graft Copolymerization; Biodegradable Properties

61.41.+e

DEVELOPMENT OF MODIFIED CHITOSAN

Chitosan (CS) which contains a lot of $-NH_2$, is a rare natural polymer with a positive charge and has a unique structure and unique activity [1-2]. However, only in an acidic aqueous solution of chitosan and some organic solvents dissolved [3]. Firstly, the CS was phthalated, thus improving the performance of CS dissolved in some organic solvent, to facilitate the subsequent reactions [4]. And then double bond was introduced to get modified chitosan functional monomer [5-6] .Synthesis principle shown in **FIGURE 1**.

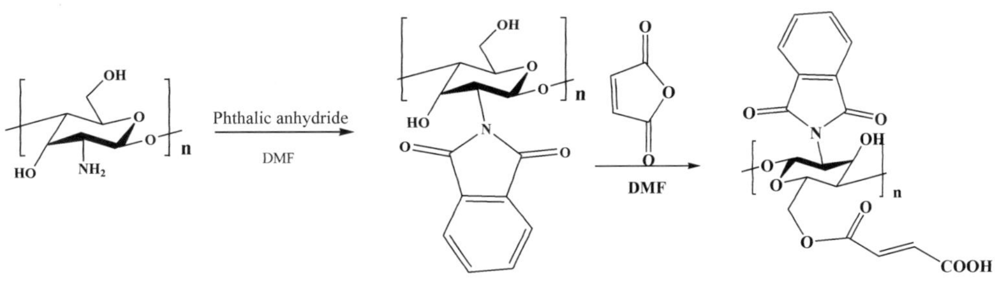

FIGURE 2. Synthesis Principle of the Modified Chitosan（PHCSMA）

The structure of PHCSMA was characterized by IR and ^{1}HNMR. The results were consistent with the proposed structure.

DEVELOPMENT OF THE POLYMER BASED ON MODIFIED CHITOSAN

Synthesis and Orthogonal Optimization Conditions of the Polymer Based on Modified Chitosan

Synthesis principle shown in **FIGURE 2**.

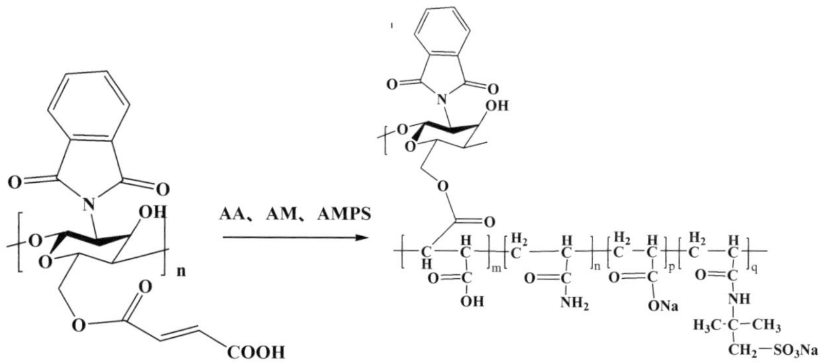

FIGURE 2. Synthesis Principle of the Polymer Based on Modified Chitosan(PLCS)

By aqueous solution polymerization, the modified chitosan functional monomer (PHMACS) and acrylic acid (AA), acrylamide (AM) and AMPS were copolymerized to obtain polymer based on modified chitosan; Through orthogonal experiment method, this polymer synthesis conditions were optimized, thereby determining the synthesis conditions were: monomer concentration 26%, PHCSMA and AMPS plus the amount of the total monomer mass 0.1%, AM: AA(mass ratio) = 1,0.1 % concentration $K_2S_2O_8$-$NaHSO_3$ (molar ratio1:1), the reaction temperature is 45 ℃. And by Ubbelohde viscometer, the polymer's viscosity average molecular weight was calibrated, about 2.94×10^6.

Structural Characterization of Polymer Based on Modified Chitosan

The structure of PLCS was characterized by IR and ^{1}HNMR. The results (**FIGURE 3**).were consistent with the proposed structure.

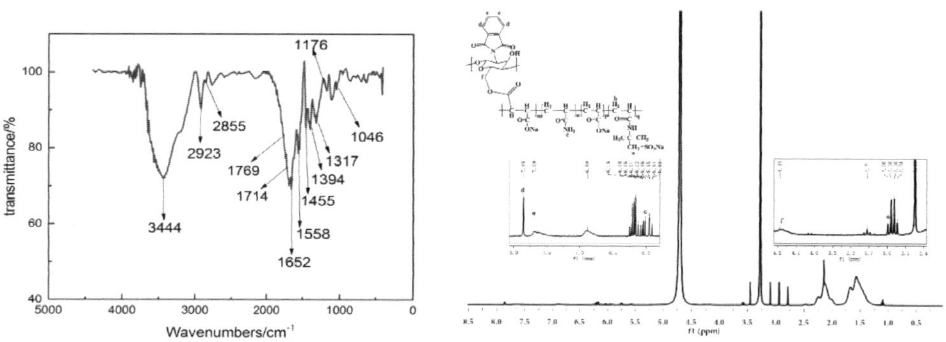

(a) (b)

FIGURE 3. IR Spectrum (a) of PLCS and ^{1}HNMR Spectrum (b) of PLCS

CHARACTERIZATION OF THE POLYMER BASED ON MODIFIED CHITOSAN

Thickening Performance

Efficient thickening properties of the polymer is an important property of the polymer. By Brookfiled DV-III viscometer having a different viscosity of the different concentration of the polymer solution at 60 ℃, the measurement results shown in **FIGURE 4.**

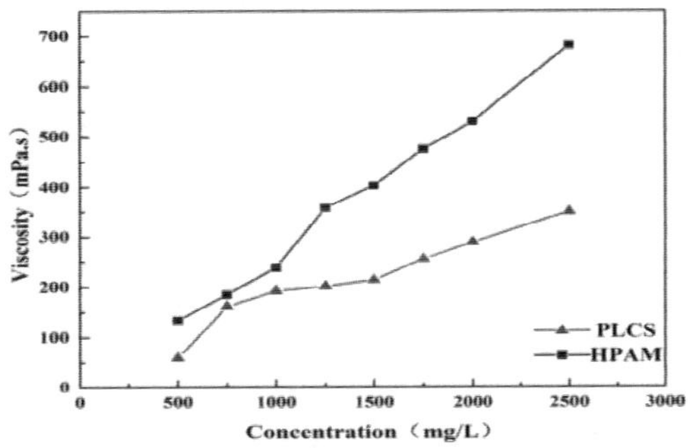

FIGURE 4. Thickening Performance of PLCS

Shear Resistance

Polymer was sheared 30 s by Wu Yin blender with 1 file (3000 r / min). At 60 ℃ by Brookfiled DV-III viscometer, viscosity of different concentration of the polymer solution before shearing and after shearing were investigated.The results are shown in **FIGURE 5.**

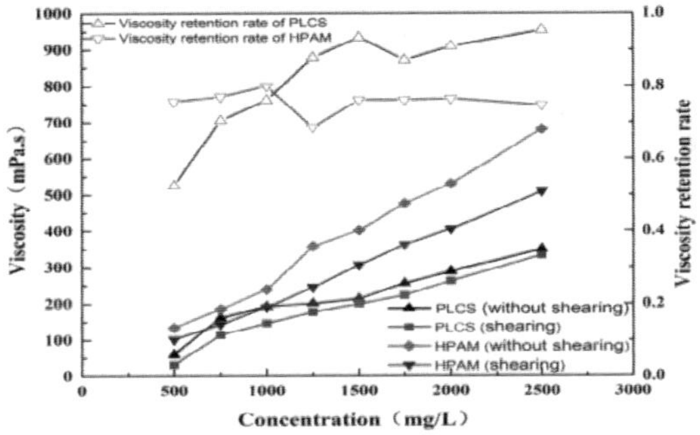

FIGURE 5. Shear Resistance of PLCS

Biodegradability

This paper studies the α-amylase (optimum temperature 50~70 ℃, the optimum pH 5.5~7.5) influences to the polymer's biodegradation. Effect of enzyme action on this polymer's M_η and viscosity were shown in **FIGURE 6 .**

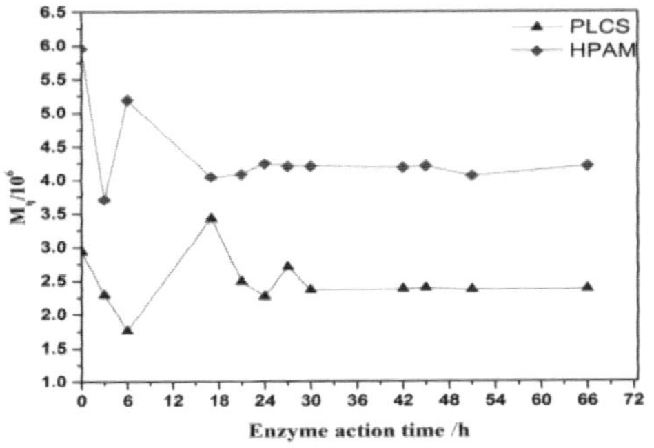

(a)

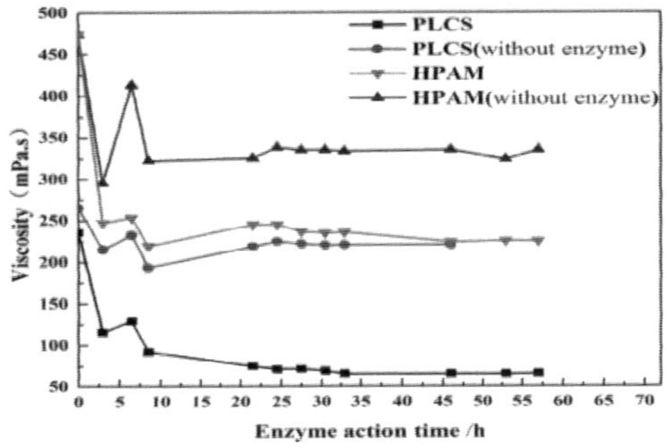

(b)

FIGURE 6. Biodegradation of PLCS

RF and RRF

The injection pressure curves of the copolymer solution in the permeability of 2.7 D and 1.6 D porous media are shown in **FIGURE 7** and **FIGURE 8,** respectively.

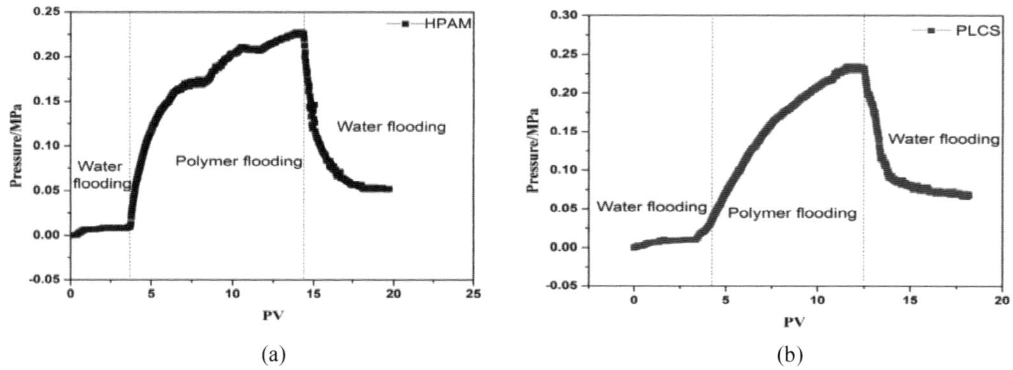

FIGURE 7. Injection Pressure Curve of HPAM and PLCS(2.7 D)

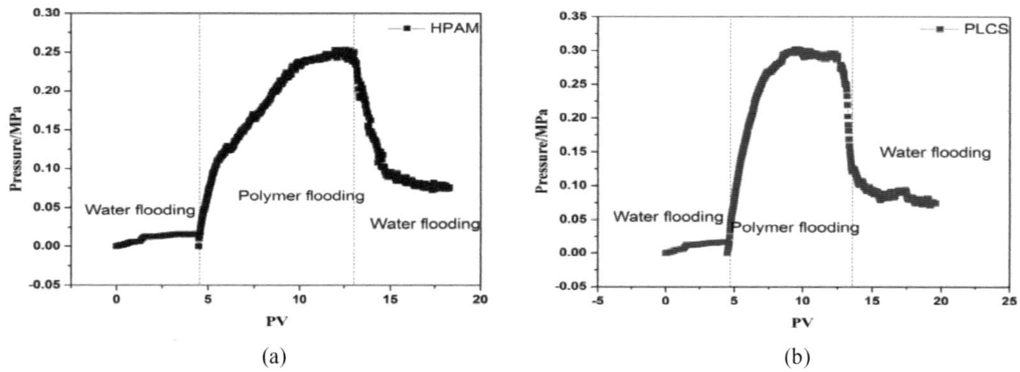

FIGURE 8. Injection Pressure Curve of HPAM and PLCS(1.6 D)

TABLE (1). RF and RRF of PLCS and HPAM

Polymer	Permeability (D)	Porosity (%)	Pressure of Water Drive (MPa)	Pressure of Polymer Flooding (MPa)	Pressure of Water Drive (MPa)	RF	RRF
HPAM	2.71	29.4	0.0096	0.2001	0.0513	62.4	16.3
PLCS	2.78	30.2	0.0098	0.2316	0.0674	70.9	20.9
HPAM	1.64	17.7	0.0121	0.2536	0.0828	62.9	14.4
PLCS	1.67	18.1	0.0127	0.3023	0.0943	71.4	17.1

Conclusion

In this study , polymer based on modified chitosan's thickening performance and shear resistance were investigated , its thickening effect is poor, but shear resistance is better, at a concentration of 2000 mg / L, viscosity of the solution retention rate of 95.3%, under conditions of high shear rate,this polymer can still be able to maintain a higher viscosity retention rates;

After 46 h enzyme action, the drop of viscosity average molecular weight of the polymer is not obvious, but the apparent viscosity of the solution decreased from 263.0 mPa • s to 65.2 mPa • s, the viscosity retention rate only 24.8%, indicating that it has good biodegradability;

Though core flow experiment [7] , indicating that this polymer was able to establish a higher resistance factor and residual resistance factor in porous media, flow resistance can be effectively established to improve sweep efficiency of the displacement fluid, used in oil displacing agent may have potential advantages, and will have great significance to the production of oil field.

ACKNOWLEDGMENTS

We acknowledge the support of key program for Undergraduate Extra-curricular Open Experiment (no. KSP15124) of Southwest Petroleum University.

REFERENCES

1. K.Kurita, T, KojimaY,and Nishiyama, , *Macromolecules. Letters* 33, 4711-4716(2000).
2. Mohamed E. I. and E.I. Rabea, *Biomacromolecules. Letters* 5,589-595(2004).
3. A.Q.Wang, *Chitin Chemistry,*Beijing: Science Publishing Company, 2008.
4. Keisuke K., Hiroyuki I., Yuya Y, et al, *Biomacromolecules. Letters* 3, 1-4 (2002).
5. Bo. Lu, Xu. Liu, Zhijun Huang, et al, *Carbohydrate Polymers. Letters* 87,1453-1459(2012).
6. K. Sarkara, A. Chatterjeeb, G. Chakraborti et al, *Carbohydrate Polymers. Letters* 98,596-606(2013).
7. B. S. Shiranand and A. Skauge, *Energy Fuels Letters* 27, 1223-1235(2013).

Research of Flooding Polymer with a Certain Degree of Surface Activity

Nanjun Lai [a,b], Shitao Li[a]

[a] *Chemistry and Chemical Engineering Institute of Southwest Petroleum University (Chengdu, Sichuan 610500)*

[b] *State Key Laboratory of Oil and Gas Reservoir Geology and Exploitation (Southwest Petroleum University, Chengdu, Sichuan 610500)*

Abstract. As the ordinary polymer flooding could not give consideration to both swept volume and washing oil efficiency and the polymer/surfactant dual-compound system has a problems of collaboration because of different migration velocity between different components, started with the relationship between molecular structure and performance, the can-be-polymerized surface activity monomer is designed, which is characterized by stronger capacity to reduce oil-water interfacial tension. The above monomer can be synthesized with acrylamide and acrylic acid into a new-type surface-activity polymers for oil flooding and further more with the help of Fourier transform infrared spectrum (FTIR), hydrodynamic radius (R_h), the structures and molecular weight are characterized. The studies show that there are much better tack ability and surface activity; and the chromatographic fractionation effects are solved, and furthermore the oil-water interfacial tension can be lowered to15.07mN/m. Polymer flooding experiments show that the oil recovery of the surface active polymer is higher than that of ordinary hydrolyzed polyacrylamide and close to polymer/surfactant dual-compound system because of its better viscosifying performance, lower oil/water interfacial tension and no chromatographic separation effect .

Keywords: oil displacement agent; surface active polymer; interfacial tension; tack ability; enhance oil recovery
PACS: 61.30.Hn, 61.41.+e, 61.82.Pv, 68.35.bm, 68.08.De

INTRODUCTION

Improving the sweep efficiency of swept volume and oil displacement system is one of the basic principles of chemical flooding to enhance oil recovery technology. Polymer flooding reduces the residual oil saturation by improving the swept volume, which exacerbates the reservoir heterogeneity and makes the remaining oil more dispersed in the formation[1]. And it is not very valuable to increase polymer concentration to increase oil recovery after polymer flooding[2]. Combination flooding can improve swept volume and washing oil efficiency, but alkali-surfactant-polymer flooding are often accompanied by formation damage and excessive mulsification of produced liquid[3-5]. Although surfactant-polymer flooding solves a part of problems of alkali-surfactant-polymer flooding at the cost of emulsifier's contribution to the recovery, the growth of oil recovery is narrow (about 15%) after that[6-7]. Therefore, exploring a kind of increasing the viscosity and interfacial activity in one and reducing the chromatographic separation of oil displacement system after injected into formation, is a useful supplement to the chemical flooding technology.

Surface active polymer flooding system is a kind of both interface activity and emulsification, reinforced adhesion, and many other features of functional system[8]. Especially for oil reservoir which has been carried out by polymer flooding, surface active polymer flooding system is a kind of extremely potential applications of new type of oil displacement system[9-10]. Thus, based on the relationship between the monomers molecule structure and its performance, low oil-water interfacial tension monomers are synthesized, furthermore, a novel polymer, which has better viscosifying and temperature resistance and salt tolerance is copolymerized by polymerizable surface active monomers and acrylamide, acrylic acid under the conditions of compound initiator. The polymer overcomes the chromatographic separation effect, and its structure is characterized by the infrared spectroscopy.

SYNTHESIS OF THE SURFACEACTIVE POLYMER

According to the molecular design theory, in view of the viscosifying property and low interfacial tension, the monomers have long chain hydrophobic groups, soluble groups (sulfuric groups) and rigid groups (benzene groups) have been developed. One type of polymerizable surface active monomers, PSA, were synthesized under the conditions of the solvent and valid temperature. A novel polymer was copolymerized by changing polymerizable surface active monomers and acrylamide, acrylic acid under the conditions of initiator and low polymeric temperature.

Materials and Apparatus

This is the paragraph spacing that occurs when you use the Enter key. Materials: deionized water, acrylamide (AM), acrylic acid (AA), ammonium persulfate, sodium bisulfite, sodium hydroxide, calcium chloride, magnesium chloride, potassium chloride, analytical reagent, made in Chengdu Kelong Chemical Reagent Co., polymerizable surface active monomers (PAS), home made; polyacrylamide (10.0×10^6, solid content 20.5 %), industrial product, France SNF.

Apparatus: grinding jars, constant temperature water bath, balance, Fourier transform infrared spectrometer (WQF-520), Brookhaven BI-200SM wide-angle dynamic/static light scattering instrument, DV-III Brookfield viscometer, KRUSS interface parameters measurement system et al.

Synthesis of the Surface Active Polymer for Oil Flooding

A certain amount of deionized water, sodium carbonate, AM, AA and PAS were added to the bottle and stirred until dissolving completely, sodium hydroxide adjusts pH to 7, The polymeric temperature was controlled at 40℃, nitrogen gas was purged for 15 minutes, after which the initiator was added and purged 5 minutes, and the bottle was sealed. After 5 hours, and after refining, drying, the white powdered polymer was obtained.

RESULTS AND DISCUSSION

Structural Characterization and Molecular Weight Analysis of the Surface Active Polymer

The structure of the surface active polymer (purified by ethanol) was characterized by Fourier transform infrared spectroscop, and its molecular weight was analyzed by wide-angle dynamic/static light scattering instrument. The results were shown in FIGURE 1 and 2.

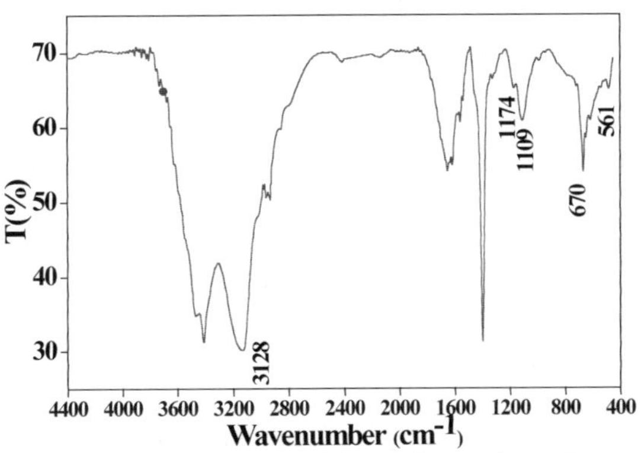

FIGURE 1 Infrared spectrum of the surface active polymer

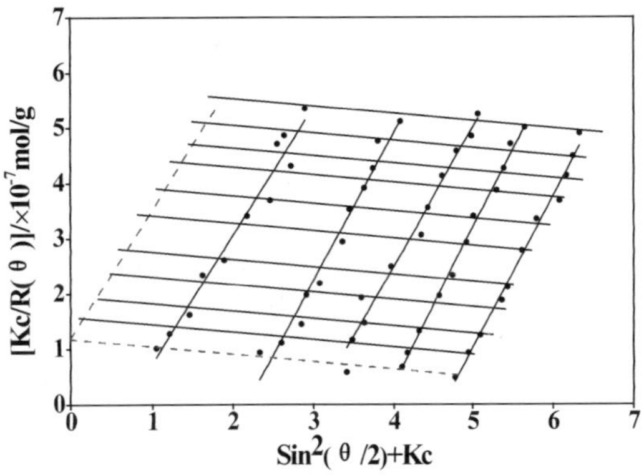

FIGURE 2 Berry figure of the surface active polymer

FIGURE 1 shows that the imaging site of 3128 cm-1 has corresponding absorption peak of benzene, and that of 561 cm-1 and 670 cm-1 have characteristic stretching vibrational absorption peak of -C=C- of benzene and bending vibration absorption peak of C-H of benzene respectively, and that of 1174 cm-1 and 1109 cm-1 have characteristic stretching vibrational absorption peak of -SO3-, indicating that PSA are copolymerized.

According to double extrapolation, weight-average molecular weight of the surface active polymer is 4.2×10^6 g/mol, and mean square radius of gyration of that is 170 nm.

Properties Evaluation of theSurface Active Polymer

Viscosity property

Surface active polymer is still a kind of polymer, which improve oil recovery still by increasing the displacement agent viscosity and expanding swept volume of displacement fluid, so it is necessary to contrasts its viscosity with ordinary polymer's. Viscosifying property of different polymers in deionized water were investigated, and the results are shown in FIGURE 3.

FIGURE 3 shows that during 500~1500 mg/L of concentration the surface active polymer has little lower viscosity than that of hydrolyzed polyacrylamide, while its viscosity is higher than that of hydrolyzed polyacrylamide when concentration over 1500 mg/L.

Interfacial tension

The oil/water interfacial tension of different polymers in deionized water was measured by the interface parameters measurement system, KRUSS, the results were shown in FIGURE 4. The results show that the ability to reduce oil/water interfacial tension of the surface active polymer is close to that of SDBS-HPAM system during concentration 1500~2000 mg/L. And the interfacial tension of the surface active polymer can be low to about 15 mN/m when 1500 mg/L.

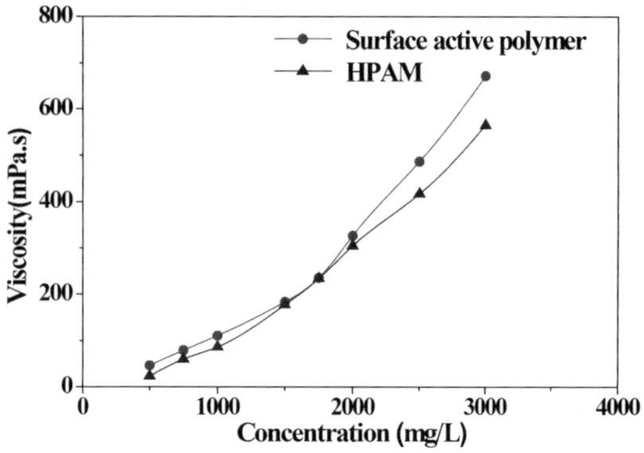

FIGURE 3 Viscosifying property of polymers

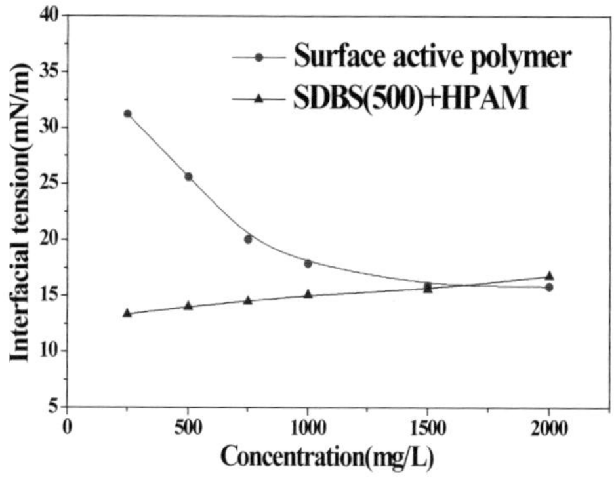

FIGURE 4 Oil/water interfacial tension of polymers

POLYMER FLOODING EXPERIMENT

Materials

Polymer solutions of 1500 mg/L HPAM, 1500 mg/L HPAM+500 mg/LSDBS, surface active polymer 1500 mg/L were prepared by deionized water. Polymer flooding effect was investigated with the sand packs.

Experiment Procedure

(1) At 70℃, the sand packs were drawn near vacuum by the pump, saturated by simulated formation water, and obtained permeability, The data of water quality were shown in Table 1 below. (2) At 70℃, the sand packs were saturated by oil, and obtained oil saturation. (3) At 70℃, the sand packs floods to certain water cut to reach water recovery. (4) At 70℃, the polymer oil displacement agent was injected to the 70℃ until the water flooding recovery was 95%, injected pore volume times were all 0.3, the results were shown in Table 2.

TABLE 1 Water quality data of simulated formation water

Constituent	Na^+、K^+	Ca^{2+}	Mg^{2+}	HCO_3^-	SO_4^{2-}	Cl^-	Salinity
Content（mg/L）	2049	80	28	400	90	3080	5257

TABLE 2 Polymer flooding results

Polymer	Permeability ($\times 10^{-3}\,\mu m^2$)	Porosity (%)	Oil recovery(%)		Oil recovery increment/% (%)
			Water flooding	Polymer flooding	
HPAM	3020	35.27	39.23	54.56	15.33
Surface active polymer	3040	35.45	40.11	65.67	25.56
HPAM+SDBS	2960	34.82	38.52	67.12	28.60
HPAM	980	33.65	35.25	47.53	12.28
Surface active polymer	1020	33.82	35.62	54.86	19.24
HPAM+SDBS	1010	33.86	34.91	57.00	22.09
HPAM	470	33.10	31.66	41.34	9.68
Surface active polymer	420	32.62	31.28	47.52	16.24
HPAM+SDBS	450	32.81	31.03	50.27	19.24

The results show that the novel polymer flooding has higher oil recovery than that of water flooding, at the same mass fraction, the oil recovery of the novel polymer is 5.2% higher than that of hydrolyzed polyacrylamide. Analysis illustrates that the novel polymer solution has higher viscosity, and stronger mobility control ability, which effectively enlarges the displacing volumes, besides, it has certain oil sweeping ability, so the oil recovery rises significantly.

CONCLUSIONS

The surface active polymer oil flooding agent has low oil/water interfacial tension (15 mN/m). Due to copolymerization of polymerizable monomers, AM and AA, the chromatographic separation is radically overcome. The polymer has good viscosifying property and it is also good at reducing oil/water interfacial tension. Experiments have proved that the surface active polymer has both greater mobility control capacity and better oil sweeping efficiency.

ACKOWNLEDGMENT

This work was financially supported by the National Natural Science Foundation of China (Grant No. 51404208), the Open Fund (Grant No. PLN1433) of State Key Laboratory of Oil and Gas Reservoir Geology and Exploitation (Southwest Petroleum University), the National Undergraduate Training Programs for Innovation and Entrepreneurship (Grant No. 201510615008).

REFERENCES

1. T.S. Urbissinova,J. Trivedi,E. Kuru.*Effect of Elasticity During ViscoelasticPolymer Flooding: A Possible Mechanism of Increasing the Sweep Efficiency*. SPE Journal of Canadian Petroleum Technology . 2010.
2. Yang Chengzhi. *Hua xue qu you ti gao shi you cai shou lv (In Chinese)*. Beijing: Petroleum Industry Press, 1999:132-145.
3. Wu Wenxiang, Zhang Yufeng, Hu Jinqiang. *Effect of pre-slug of Polymer/Surfactant upon the recovery efficiency of ASP compound flooding*. Journal of Northeast Petroleum University, 2005, 29(6):98-100.
4. Ying Guo, Weidong Liu, Linghui Sun, et al.*Chromatographic separation of components of ASP flooding system for conglomerate reservoir in Xinjiang Oil Fields*.Oilfield Chemistry, 25 (4) (2008), pp. 362–365.
5. Wang D,Xia H,Yang S,et al.*The influence of visco-elasticity on micro forces and displacement efficiency in pores,cores and in the field*. SPE EOR Conference at Oil & Gas West Asia . 2010.
6. WANG Yu，CHEN Quan-Sheng，LIN Li-Li. *Further EOR after Binary Compound Flooding by Application of Active Polymer*. Oilfield Chemistry, 2014, 31(3):414-418.

7. YE Peng, LU Xiang-Guo, LI Qiang. *Effect of Chromatographic Fractionation on the Solution Performance of Polymer Surface-Active Agent and the Concentration Detection of its Produced Solution*. Oilfield Chemistry, 2011, 28(2):158-162.
8. Feng Rusen, Qiao ZHihua, Wang He. *Study on properties of polymer surfactant aqeous solution*. Oilfield Chemistry, 2013, 30(1):64-69.
9. Shi L, Liu W D, Tong Z X. *Rheologic Behavior of Aqueous Solutions of Surface Active Polymer SLH- I for EOR[J]. Oilfield Chemistry*, 2008, 25(1):42-45.
10. LIU Tian, LIN Zhi-yu, LI Cui-qin. Study of aqueous solution properties of new type polymer surfactant. Chemical Engineer, 2015, 29(6):4-6.

Radiation Effect of Poly (Butylene Succinate)

Lu Yongjun, Liu Jiangwei, Wang Wenzhao, Jiao Yang, Zeng Xinmiao
Beijing Key Laboratory of Radiation Advanced Materials
Beijing Research Center for Radiation Application

Abstract: Radiation cross-linking and long chain branching of poly (butylene succinate) (PBS) was induced by Co-60 γ -ray with poly functional monomer as sensitizers respectively. The results show that PBS can be cross-linked by the addition of trimethylolpropane trimethacrylate (TMPTMA). The tensile strength and elongation at break of the cross-linked PBS was investigated. The PBS can be cross-linked or long chain branched with 1,6-hexanediol diacrylate (HDDA) as sensitizer. The mechanical properties and differential scanning calorimeter (DSC) were investigated.

Key words: Poly (butylene succinate); Radiation cross-linking; Long chain branching; crystallization temperature.

1. Introduction

Poly (butylene succinate) (PBS) is one of the most important biodegradable plastics (BDP). The melting point T_m of PBS is greater than 100℃ that can be used as package for hot food. General BDP like poly (lactic acid), which glass transition temperature is about 60℃, cannot be used for hot food[1, 2]. As the chain of PBS is linear, the melting strength is low and is difficult to be used as foam. When PBS is used as membrane, the tear strength is low. A cross-linked polymer cannot flow when heated, that is the melt strength get higher. The chain changed from a linear one to a net, which may improve the tear strength. Long chain branching can improve the melting strength of polymer through the change of chain structure.

In this study PBS was cross-linked by Co-60γ -ray to improve tear strength for used in membrane[3]. To improve the melting strength, both cross-linking and the long chain branching of PBS (LCB-PBS) were studied [2, 4].

2. Experimental

2.1 Main materials

Two commercially available PBS were used as based resins in this study. One is poly (butylene succinate) (PBS1), the other is a copolymerization product with a third monomer adipic acid –poly (BS-co-BA) (PBS2) and the amount about 3~5%.

In the study of PBS2, two poly functional monomers (PFM) were used as sensitizers. One is Trimethylolpropane trimethacrylate (TMPTMA) for cross-linking. The other is 1,6-Hexanediol diacrylate (HDDA) for cross-linking and long chain branching. The two PFM are products of Sartomer Corporation.

2.2 Mechanical properties

PBS and sensitizers was mixed in a twin screw extruder and molded into specified sheet in a machine with pressure at 130℃. The sheets were sealed in a plastic membrane to keep the air out. Then the sheets were irradiated by $Co^{60}\gamma$ -ray to a specified dose.

The dumbbell shape of PBS sheet, which was prepared according to GB/T 528, was used for mechanical property measurements. Mechanical properties were analyzed at a speed of 50 mm/min.

The right angle sample with no cut, which was prepared according to GB/T 529, was used for tear strength measurement. The tear strength was analyzed at a speed of 50mm/min.

2.3 Gel content

Gel content of the polymers was determined by extraction. The extraction used chloroform as solvent at room temperature. The extraction time is 48hr.

2.4 Differential scanning calorimeter (DSC)

The melting point T_m and crystallization temperature T_c of each sample (5mg) were measured in the temperature range 20~160℃ using a Mettler DSC1 differential scanning calorimeter at a heating rate of 10℃/min under a flow of nitrogen. The sample was heated to 160℃ and kept 5 minutes then cooled to 20℃ at a rate of 10℃/min. The sample was kept at 20℃ for 5 minutes and then heated again at a rate of 10℃/min. The T_c and T_m were taken as the peak temperature of endothermic peak respectively.

3. Results and discussion

3.1 Radiation cross-linking of PBS1

The PBS1 cross-linked with the irradiation ofγ -ray and the gel content increase with the increase of absorbed dose (see Fig1). The gel content increases linearly with absorbed dose under 20 kGy; with the absorbed dose increase, the gel content keep nearly constant and slightly decrease in a high absorbed dose (80 kGy or higher).

According to the gel content of PBS, we calculate the sol fraction of each sample. The sol fraction ($s+\sqrt{s}$) and reciprocal of absorbed dose -1/D relation under 20kGy was shown in Fig2.

The sol fraction $s+\sqrt{s}$ increase linearly with 1/D which indicates it's a random cross-linking. In more detail, the minimum absorbed dose to form gel is 2.3 kGy for PBS, which means PBS is very easy to cross-link by the irradiation of γ -ray.

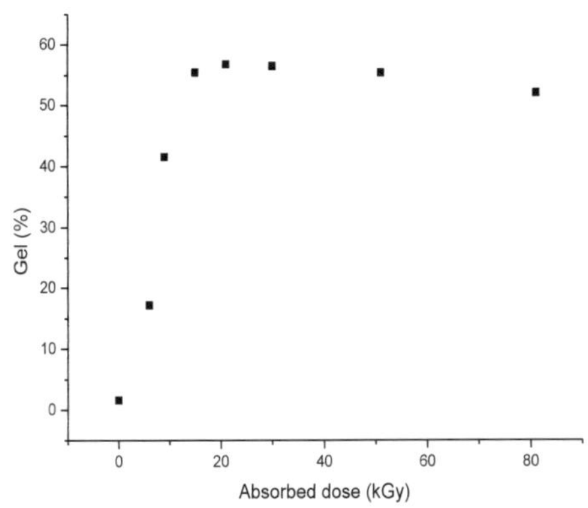

Fig1 The gel content of cross-linked PBS1

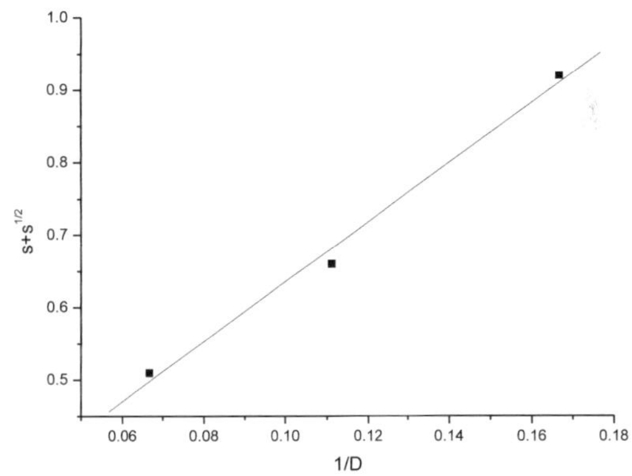

Fig2 The relationship of $s+\sqrt{s}$ and 1/D of PBS1

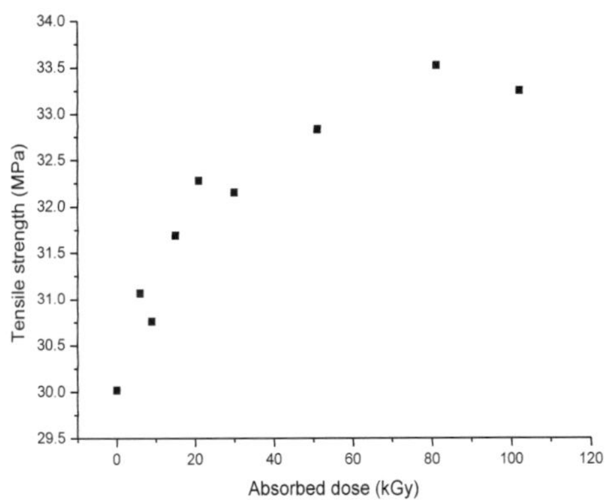

Fig3 The tensile strength of cross-linked PBS1

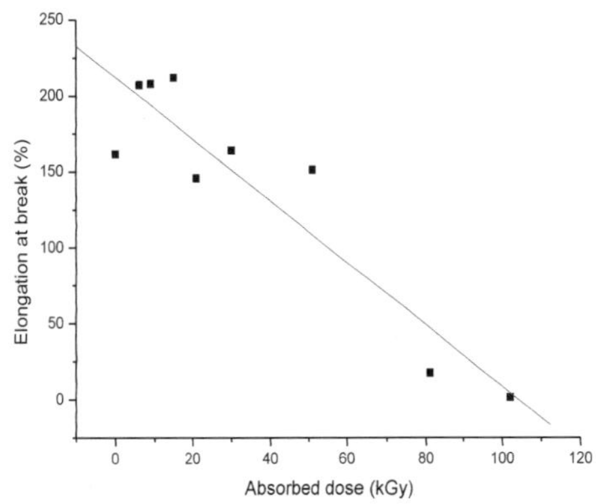

Fig4 The elongation at break of cross-linked PBS1

The relationships between the tensile strength and elongation at break with absorbed dose were shown in Fig3 and Fig4. The results indicate that with the increase of the absorbed dose, the tensile strength increase, and the elongation at break decrease almost linearly.

The tensile strength increase rapidly at low absorbed dose and increase rate decrease as the absorbed dose get higher. The tensile strength even decrease at very high absorbed dose which means there is degradation to some degree.

3.2 Cross-linking of PBS2

In the experiment of gel test, gel fraction less than 5% generally think as there is no gel. PBS2 is irradiated by γ -ray directly, the results see Fig5. We can safely draw a conclusion that the addition of adipic acid changes the structure of the polymer chain. PBS2 cannot be cross-linked by radiation directly.

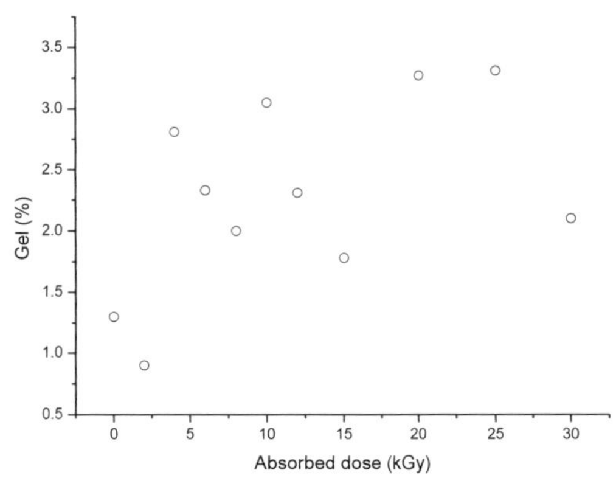

Fig5 Radiation effect of PBS2

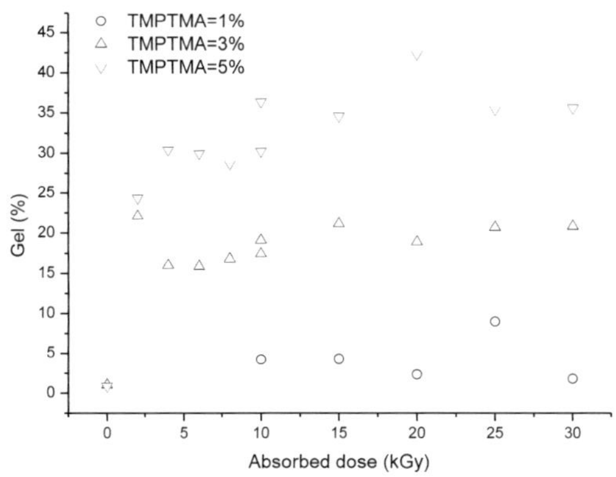

Fig6 Radiation effect of PBS2 with TMPTMA

To promote the cross-linking of PBS2, TMPTMA is added to PBS2. The radiation effect of PBS2/TMPTMA is shown in Fig6. PBS2 can be cross-linked with the addition of TMPTMA. When the content of TMPTMA is low (TMPTMA=1phr, phr = per hundred resin, same as below), PBS2 cannot be cross-linked. When the content of TMPTMA increase (TMPTMA=3~5phr), PBS2 can be cross-linked. The gel content reach a high level with a low absorbed dose and nearly keep constant with the increase of absorbed dose.

The tensile strength and elongation at break of PBS2 with TMPTMA=5phr was shown in Fig7 and Fig8. The properties both decrease with the increase of absorbed dose.

From the above we know that PBS2 is degrade with irradiation of γ -ray and can be cross-linked by γ -ray with TMPTMA as sensitizer.

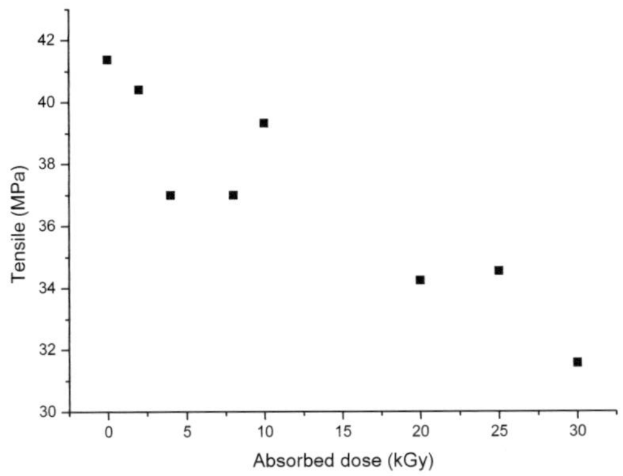

Fig7 Tensile strength of PBS2/TMPTMA

TMPTMA=5phr

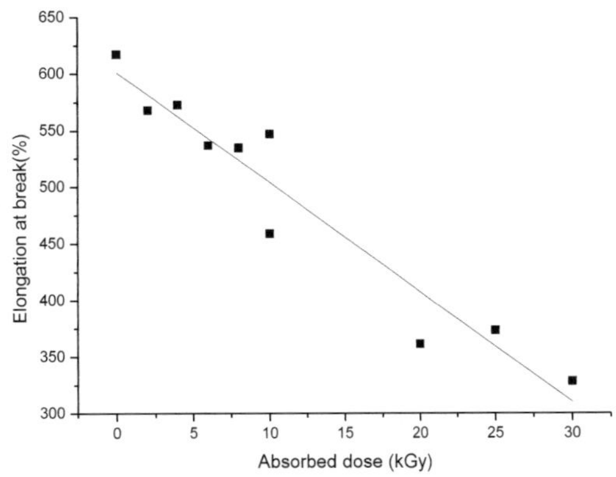

Fig8 Elongation of PBS2/TMPTMA

TMPTMA=5phr

3.3 Long chain branching of PBS2

As PBS1 can be cross-linked easily at a low absorbed dose, we discuss the long chain branching of PBS2 here. The HDDA is mixed with PBS2 and then irradiated by γ -ray.

The PBS2 can be cross-linked or long chain branched with different content of HDDA. The effect of different amount of HDDA and absorbed dose on the properties of PBS2 see Fig9 ~ Fig12.

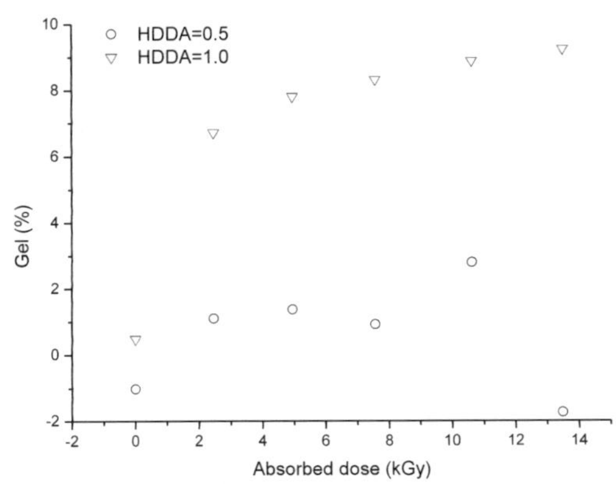

Fig9 The gel content of PBS2 with HDDA

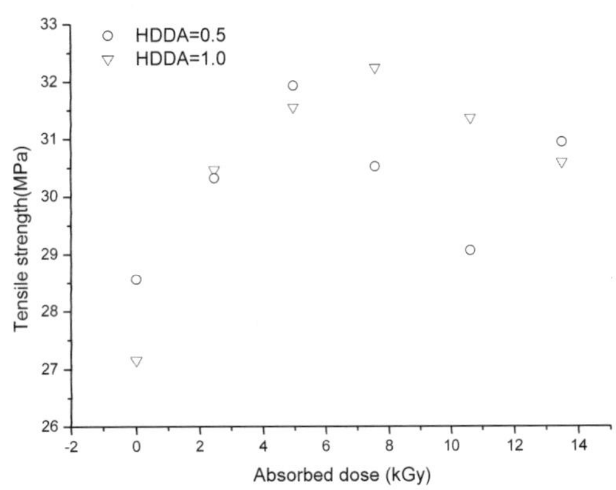

Fig10 The tensile stength of PBS2 with HDDA

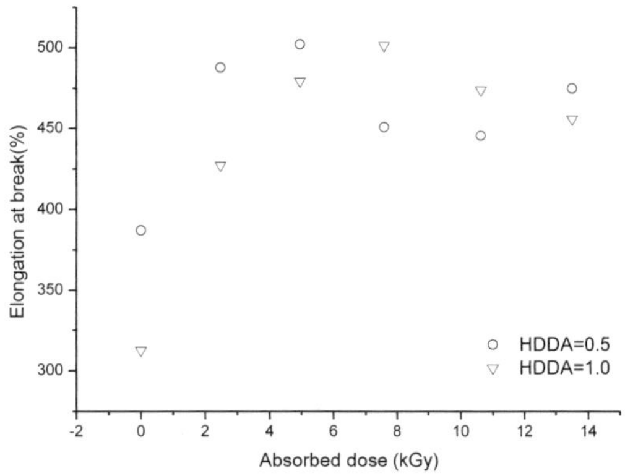

Fig11 The elongation at break of PBS2 with HDDA

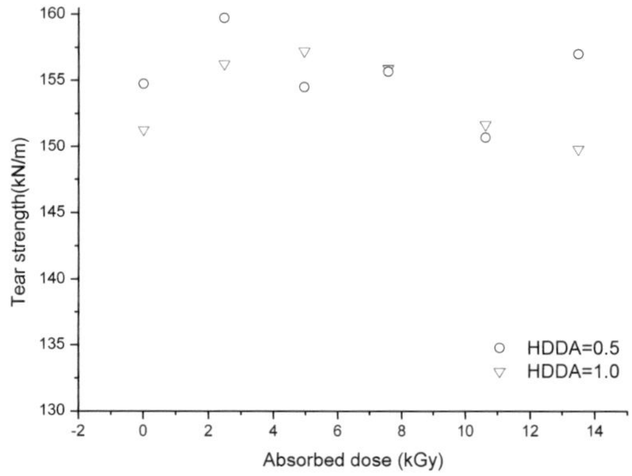

Fig12 The tear strength of PBS2 with HDDA

From Fig9 we know that there is no gel when the HDDA content is low (HDDA = 0.5phr); When HDDA is high (HDDA = 1phr), the gel content of PBS2 increase with the increase of absorbed dose. Firstly, the gel content increase rapidly and then increase rate get slowly with the absorbed dose is higher.

The tensile strength and elongation at break have the same trend with the increase of absorbed dose. Both increase soon firstly and then slightly decrease with the increase of absorbed dose. The tear strength nearly keeps constant with the increase of absorbed dose.

Whether cross-linking or long chain branching, the chain structure of PBS2 changed so as to the temperature of crystallization (T_c) and the melting point (T_m) changed (see Fig13 and Fig14).

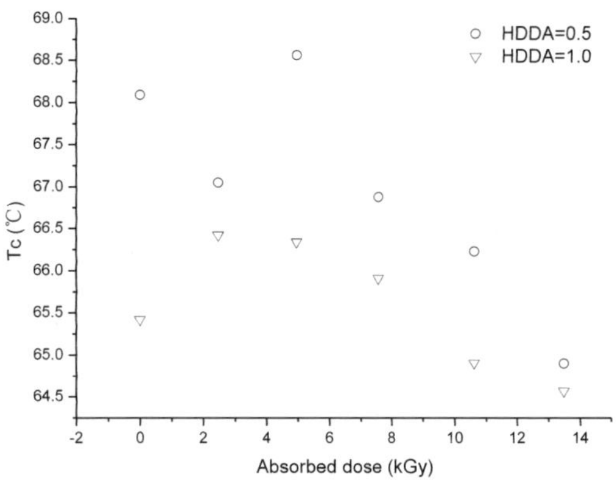

Fig13 the T_c of PBS2 with HDDA

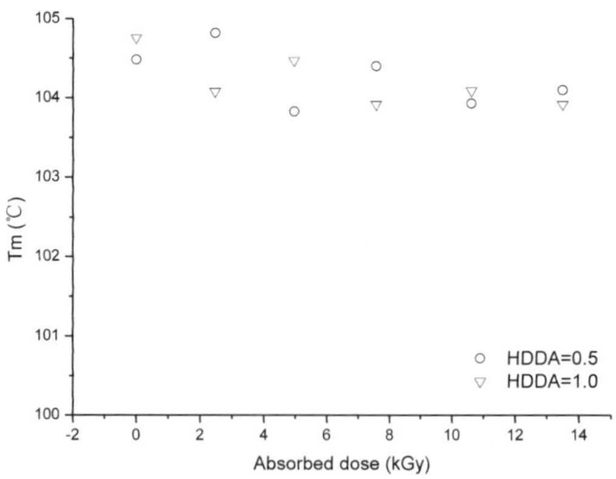

Fig14 the T_m of PBS2 with HDDA

As the chain changed from line to a branching shape or a net shape, the movement of chain was limited. So the T_c and T_m decrease with the increase of absorbed dose.

4. Conclusions

According to the above results and discussion, we can draw some conclusions:

1) PBS1 can be cross-linked by irradiation. The tear strength improved after radiation cross-linking;
2) PBS2 cannot be cross-linked directly by irradiation and can be cross-linked by the addition of poly functional monomer like TMPTMA. The tensile strength and elongation at break decreases with the increase of absorbed dose;
3) PBS2 can be long chain branched by bi-functional monomer like HDDA. The crystallization property of LCB-PBS2 changed with the amount of HDDA and absorbed dose.

5. References

[1] Ji Jun-hui, Developments and Applications of Fully Biodegradable Plastics. Plastics, 2007,36(2),37.

[2] Jianye Liu, Lijuan Lou, Wei Yu, Ruogu Liao, Runming Li, Chixing Zhou, Long chain branching polylactide: Structures and properties. Polymer, 2010,51,5186.

[3] Sun Chongxiao, Improving Tear Strength Resistance of Biodegradable poly (butylene succinate) by modification, [MD], 2008, Beijing University of Chemical Technology.

[4] Xia Qing, Research on PBS composite modified with hyperbranched polymer,[MD], 2013,Shaanxi University of Science and Technology.

Degradation of bisphenol A polycarbonate on outdoor exposure in western areas of China

Gao Weibin[1,2], Yang Minjiao[1], Han Shimin[1], Jiang long[1], Li Guangxian[1*], Dan Yi[1*]

1.State Key Laboratory of Polymer Materials Engineering of China(Sichuan University), Polymer Research Institute of Sichuan University, Chengdu 610065, China;
2. Changzhou vocational Institute of Engineering, Changzhou 213164, China

Abstract. To obtain better understanding of structure and properties of bisphenol A polycarbonate (PC) during outdoor exposure at the environments of western areas, Jiangjin, Lhasa, and Yuli of China, and to determine the effect of weathering on its fracture resistance, changes in structure and tensile properties of PC were studied. Structure and properties of PC after aging were investigated by scanning electron microscopy, fourier transform infrared and uv-visible absorption spectra in this paper. During outdoor weathering, the ultimate elongation and ultimate tensile stress decreased showing that PC became less ductile and more susceptible to fracture. The experimental results also show that the occurrence of surface microcracking of PC is attributed to a combination of a decrease in ultimate tensile stress and ductility of surface layer due to UV radiation, and alternating stresses due, primarily, to temperature changes. This paper reports associated studies carried out to help explain the mechanism of surface microcracking and also to determine the effect of weathering on the resistance of PC to fracture.

Keywords: Bisphenol A polycarbonate; weathering; outdoor exposure; property

1. INTRODUCTION

Bisphenol A polycarbonate (PC) is a widely used polymer in outdoor engineering because of its excellent physical properties such as transparency, high tensile strength, impact resistance and high modulus of elasticity [1-5]. Some outdoor applications of PC require a deep knowledge about the durability and predictability of the properties under different environmental conditions and over long periods to cover the whole lifetime of the object. Weather and radiation factors that contribute to weather creep or loss of properties in PC materials include temperature variations, moisture, sunlight, oxidation, and other environmental factors. The results of exposing PC products to these conditions are discoloration, loss of mechanical strength, embrittlement, and loss of electrical insulation and resistance properties [6-9]. The degree to which a particular PC product degrades depends on its susceptibility to each of the above factors. The destructive effect of the weather varies with geographic location, season, time of day, cloud cover, and exposure orientation because of variations in the critical weather factors with these conditions. Different weather conditions can also be encountered in individual years. Therefore, outdoor tests in one location during a specific time interval cannot be expected to provide information on either absolute durabilitiy or stability rankings of materials under various service conditions. Tests are commonly carried out in environments that have the most severe conditions, such as subtropical or desert climates, in order to determine resistance to the worst conditions [10]. This paper reports associated studies carried out to help explain the effect of weathering on the resistance of PC in different areas.

2. MATERIAL AND METHODS

2.1 Test specimens

The PC material was supplied as solid pellets from Chongqing Changfeng Chemical Factory (Chongqing city, China). In order to get the aged samples, the PC pellets were dried in a vacuum oven at

*Corresponding author. Fax: 86 028 85402465.

E-mail addresses: danyichenweiwei@163.com (Dan Yi); guangxianli@scu.edu.cn (Li Guangxian)

120℃ for 24 h before injection molding, then PC specimens were obtained by injection molding in K2 TEC40-111 equipment (Ferromatik Milacron, Germany). The barrel temperatures were 260℃, 260℃, 255℃, and 240℃. Mould temperature was 50℃. The specimens were made according to requirement of Chinese National Standard (GB/T 1040.1-2006) and ISO527, The tensile strength test specimens were in dumbbell shape with 150 mm long by 10 mm wide and 4 mm thick.

2.2 Outdoor exposure

In order to obtain data on the natural weathering of PC, sits of exposure best suited to this purpose should have a high UV intensity and/or relative humidity. In our research, the outdoor experimental locations were set in western areas of China. In general, the climate in those places was worse than that in other areas of China, and we selected three cities (Jiangjin, Lasa, and Yuli) where have typical climatic features in western areas of China. Jiangjin provides high humidity and relatively high temperatures; Lasa, on the other hand, provides intense sunlight; Yuli is often selected for intense sunlight, wide temperature fluctuations, and low humidity. The average climatic data of these areas according to three test stations are shown in Figure 1, Figure 2 and Figure 3.Specimens were mounted facing south uses specimen racks inclined at 45° with a southern exposure and then tested for retention of properties after being exposed to natural weather elements. Because of seasonal and year-to-year climatological variability of outdoor conditions, repeated testing during different seasons and over a period of at least 1～2 years is usually recommended. In this paper the experimental period was from January 24, 2007 to January 24, 2009. During the exposure time, all the meteorological data were measured and saved (temperature, humidity, solar radiation, etc.).The aged samples were collected at different exposure days for 3 months or 6 months. After the aging experiment, the aged samples were scratched from the surface of specimens after mechanical tests.

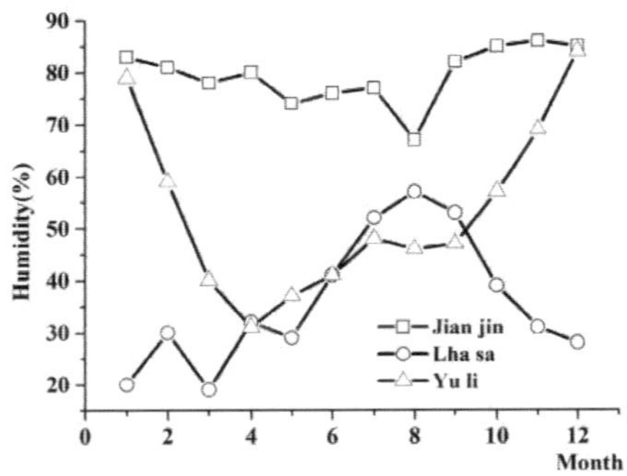

FIGURE 1 Humidity distribution of monthly in three zones

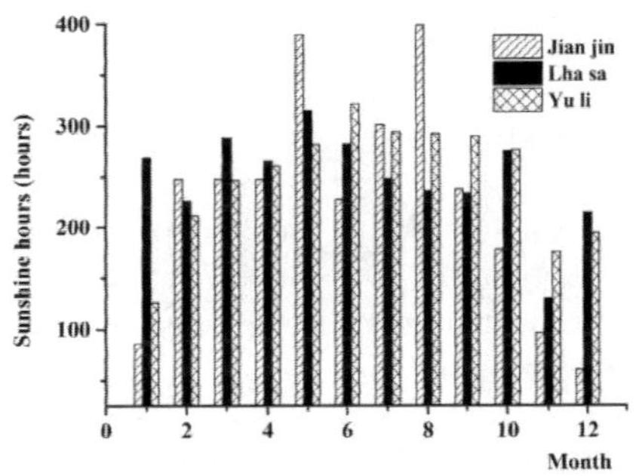

FIGURE 2 Sunshine hours in three zones

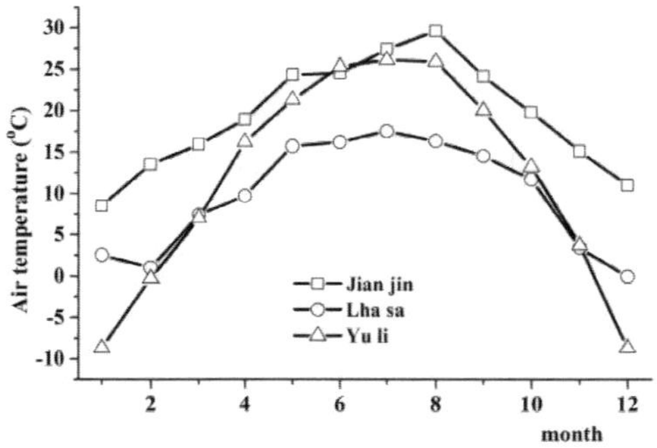

FIGURE 3 Temperature distribution of three zones

2.3 Mechanical performance testing

The stress-strain behaviors of specimens aged by accelerated aging test and outdoor exposure were measured on the CMT4204 floor-standing electromechanical universal testing machine (Sans Group Company, China). The rate of extension was 50 mm/min. In order to ensure the reproducibility of test results, not only specified production of specimens and sufficient uniformity of test climate (temperature and humidity) have to be guaranteed, but also the moisture content of the specimen. In this paper, these specimens were conditioned using standard conditioning procedures in a standard laboratory atmosphere (50% relative humidity, at 23℃) for 88 hours according to ISO 291. Five specimens of each aging period were tested and the sample thickness was measured with a pachymeter with 0.025 mm of exactitude. Finally, the test results were averaged.

2.4 Scanning electron microscopy observation

Scanning electron microscopy (SEM) analysis was performed on a JSM-5900LV (JEOL, Japan) to determine fracture surface appearance of PC. The samples were secured onto copper mounts using silver paint and gold coating in a polaron high resolution sputter coater to obtain enough contrast in the SEM micrographs. The energy of the electron beam was 20 Kv.

2.5 Fourier transform infrared spectra measurement

Fourier transform infrared (FTIR) spectra were recorded on a Nicolet 560 (Thermo Fisher Scientific inc., US) spectrometer at a resolution of 2 cm^{-1} in the frequency range from 4000 cm^{-1} to 400 cm^{-1}. The analysis was performed on thin films obtained from $CHCl_3$ solutions cast onto KBr plates by spontaneous evaporation of the solvent at room temperature.

2.6 UV-vis spectroscopy measurement

UV-vis spectra were recorded on a UV-2300 UV-vis spectrophotometer (Shanghai Tianmei Company, China). The scans were carried out in the range of 190～500 nm at the scanning rate of 200 nm/min. PC samples were dissolved in dichloromethane and the concentration of PC solution was 0.0001 g/ml. The pure dichloromethane was used as a reference.

3. RESULTS AND DISCUSSION

3.1The change of stress–strain behavior

The main change in the property of PC after aged are summarized as the change of stress–strain behavior, the evolution of morphological characteristics, and the change of chemical structure. The stress–strain behavior is a property that measures the ability of a material to withstand rough treatment and is related to toughness of the material. When PC is subjected to tensile loading it exhibits both elastic and plastic deformation. The former is characterized by complete recovery of strain when the load is removed; while the latter by irreversible yielding and flow under load. The high impact strength of PC is attributed to its ability to undergo large plastic deformation. Before aged PC is strong and tough, showing higher ductility than other amorphous polymers, with the formation of extended plastic deformation zones and shear bands. However, the occurrence of surface microcracking of PC on exposure to outdoor is attributed to a combination of a decrease in ultimate tensile stress and ductility of the surface layer due to UV radiation and alternating stresses due, primarily, to temperature changes. Cyclic variation of temperature and humidity in natural and artificial weathering impose on the surface material a type of stress fatigue. Such stress fatigue is caused by the non-uniform dimensional changes that result from thermal and moisture content gradients between the surface and the bulk of the resin and from inhomogeneities and defects. At the same time, many photochemical aging processes initiated by solar radiation cause embrittlement of the surface, enhancing the tendency to crack under tensile stresses during the drying period. This reduction in the tensile stress and elongation to failure is due to the reduction in the molecular weight of the surface layer [11]. These are illustrated in figure 4.

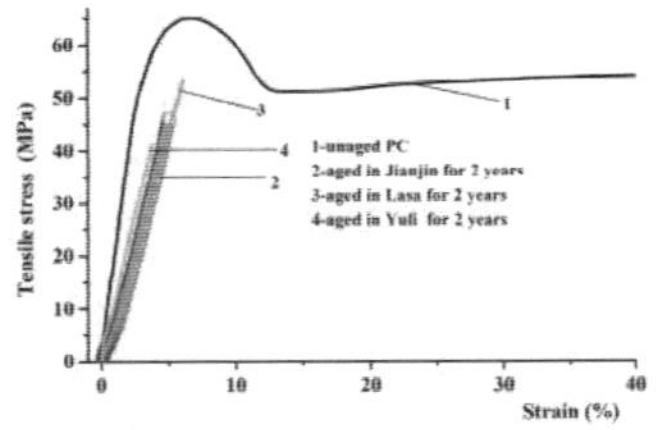

FIGURE 4 Stress-strain curves of PC after aging outdoor at three zones

3.2 Evolution of morphological characteristics with aging

The aging behavior of the outdoor exposed samples was further investigated by means of SEM analysis. The SEM micrographs (Figure 5) show that the aging process causes surface roughness. The surface microcracking appears on samples after outdoor exposed can be speculated that cracks are likely due to some aging on the PC surface due to decay of activity of species. Also, the cracks grow from the surface inward and are V-shaped, indicating that the stresses involved have a gradient · Changes in appearance and mechanical properties result from modification of the chemical structure of the material by its complex interaction with the environmental elements, primarily solar radiation, heat/cold, moisture (solid, liquid, and vapor), and oxygen.

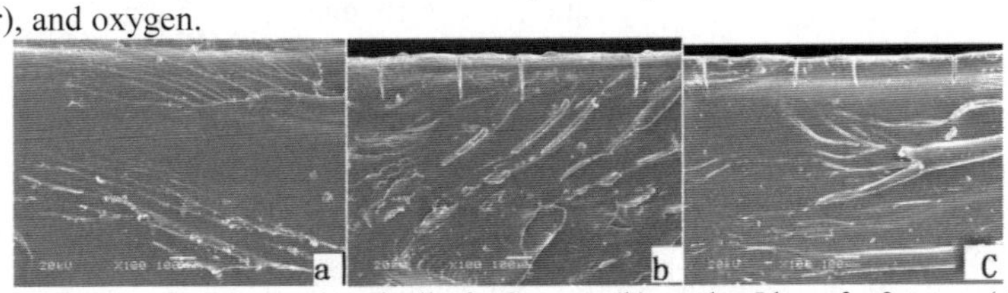

FIGURE 5 The SEM photos of PC (a) aged at Jianjin for 2 years; (b) aged at Lhasa for 2 years; (c) aged at Yuli for 2 years

3.3 The change of chemical structure

It is worthwhile to study the change of chemical structure occurred to the PC samples after accelerated weathering test and outdoor exposed. Figure 6 illustrates the FTIR absorption spectra of PC samples after outdoor exposed in three zones for 1 year and 2 years, respectively. With the aging time increasing, absorption bands form at different wavelengths. In this study, the degradation of PC were characterized by an increase of the FTIR absorption band at 1160-1240 cm^{-1}, the largest shifts involve the broad C–O stretching at 1235cm^{-1} and carbonate C–O, stretch at 1194cm^{-1} and carbonate C–O stretch at 1160cm^{-1}, respectively. The aging process is expected to change the conformation through internal rotation about the various single bonds in PC. In the previous work [12-14], a linear relation was found between the scaling factors, derived from UV and IR absorption, and changes in irradiation intensity, too.

The interpretation of the UV-visible absorption spectra had previously centred on the wavelength assignments of absorbing species produced by a photo-Fries rearrangement mechanism. These spectra correspond to a highly oxidized surface structure, as revealed by the data previously presented [15]. Peaks centred at 300~320 nm have been designated as arising from dihydroxybenzophenone and phenyl salicylate species, respectively. A "negative" peak forms at 300~320 nm. This is an indicative of "bleaching", i.e. the loss of chromophoric species in the exposed samples. The increase in absorption at ~285 nm is attributed to phenolic groups. Increases in absorption at these wavelengths are clearly seen from the data in Figure7.The data presented here are consistent with the literature [16] in the interpretation of a photo-Fries rearrangement, although, in the surface regions at least, extensive photooxidation has been shown to occur.

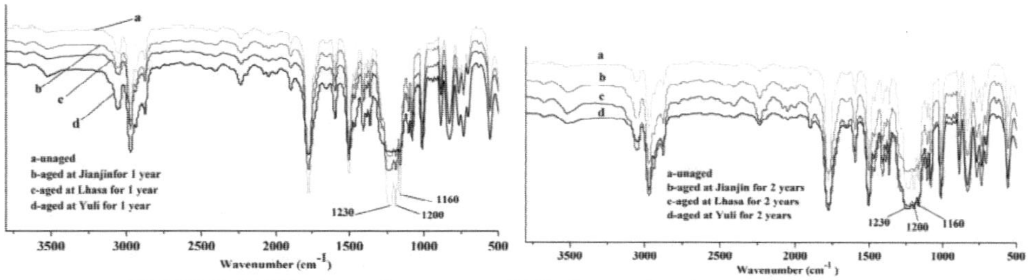

FIGURE 6 FTIR absorption spectra of PC after aged at three zones for 1 year

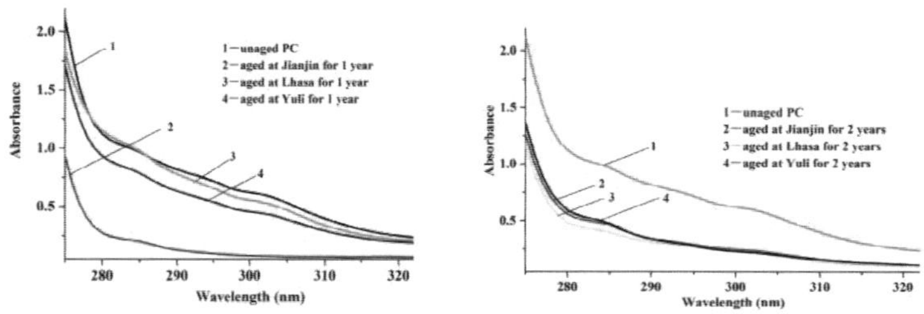

FIGURE 7 UV-visible absorption spectra of PC before and after aged at three zones

4. CONCLUSIONS

During outdoor weathering, PC is subjected not only to solar radiation, but also to other weathering factors such as temperature and moisture. Weathering causes PC to become less ductile and more susceptible to fracture when stressed.in other words, the reduction in the tensile stress and elongation to failure is due to the microcracking of the surface layer. Occurrence of surface microcraking of PC on exposure to outdoor is attributed to the combination of effects due to UV radiation and the presence of alternating thermal stresses.

ACKNOWLEDGMENTS

The authors gratefully acknowledge the National Natural Science Foundation of China (No.: 50533080 and 51133005) for financial support of this research.

REFERENCES

1. C.C. White, K.T. Tan, D.L. Hunston, et al. *Polymer Degradation and Stability*, **96**:1104-1110(2011).
2. Olivier Haillant, David Dumbleton, Allen Zielnik. *Solar Energy Materials & Solar Cells*, **95**:1889-1895(2011).
3. Bok Nam Jang, Charles A. Wilkie. *Polymer Degradation and Stability*, **86**: 419-430(2004).
4. Yong Ge,Tao Wang, Yue Yan, et al. *Applied Mechanics and Materials*, **377**: 255-260(2013).
5. Yong Xian Zhao, Le Zhao, Yang Yang Zhao, et al. *Advanced Materials Research* , **690-693**: 1590-1593(2013).
6. Mojtaba Haghighi-Yazdi,Pearl Lee-Sullivan. *Mechanics of Time-Dependent Materials*, **17** :171-193(2013).
7. Olivier Haillant. *Solar Energy Materials & Solar Cells*, **95**:1284-1292(2011).
8. Mojtaba Haghighi-Yazdi, Jacky K.Y. Tang, et al. *Polymer Degradation and Stability*, **96**: 1858-1865(2011).
9. Gao Weibin, Han Shimin, Yang Minjiao, Jang long, Dan Yi. *Polymer Degradation and Stability*, **94**:13-17(2009).
10. Pan Yonghao, Yang Mingjiao, Han Shimin, Gao Weibin, Dan Yi. *Journal of Applied Polymer Science*, **125**: 2128-2136(2012).
11. Oxana Y. Rodionova, Alexey L. Pomerantsev. *Journal of Applied Polymer Science*, **95** :1275-1284(2005).
12. J. W. McPherson editor. Reliability Physics and Engineering. New York: Springer International Publishing Switzerland,105-115(2013).
13. Diepens M, Gijsman P. *Polymer Degradation and Stability*,**94**:34-8(2009).
14. Marjolein Diepens, Pieter Gijsman. *Polymer Degradation and Stability*, **96**:649-652(2011).
15. I.B. Rufus, H. Shah, C. E. Hoyle. *Journal of Applied Polymer Science*, **51**: 1549-1558(1994).
16. Marjolein Diepens, Pieter Gijsman.*Polymer Degradation and Stability*, **92**:397-406(2007).

AUTHOR INDEX

Altan, M.C. 99
Altstadt, V. 64
Anderson, J.P. 99
Bhattacharya, S.N. 19
Boyer, S.A.E. 172
Buahom, P. 176
Chen, H. 84
Chen, J. 49
Chen, X.Y. 181
Cheng, S. 29
Costa, F.S. 19
Dan, Y. 242
Deceur, S. 192
Desplentere, F. 192
Ding, Y. 54
Dogan, C. 109
Feng, M. 114
Gao, W. 242
Gong, P. 176
Gonzalez-Gutierrez, J. 205
Guloglu, G.E. 99
Guo, X. 220
Gupta, R.K. 19
Hamidi, Y.K. 99
Han, S. 242
Hanana, F.E. 59
Harkin-Jones, E. 104
Haudin, J.M. 172
He, G. 29
He, H. 24, 29
Holzer, C. 205
Hu, X. 173
Huang, J. 14, 151
Huang, Y. 114, 214
Ishihara, S. 79
Jian, R. 54
Jiang, L. 242
Jiang, W. 151
Jiao, Y. 235
Jiao, Z. 54
Kajiyama, T. 124
Kaliaguine, S. 181
Kanai, T. 69
Kara, D. 109
Kawaguchi, T. 210
Keller, J.H. 64
Kikutani, T. 89
Kimmer, D. 74
Kitano, T. 119, 124
Kuga, S. 214
Kukla, C. 205
Lai, N. 5, 10, 220, 225, 230
Li, G. 242
Li, H. 84
Li, S. 230
Liu, J. 235
Liu, X. 114, 156, 161, 166
Liu, Y. 225
Long, S. 94, 133, 138, 143, 147
Lu, Y. 235
Luo, D. 186
Luo, X. 133, 138
Ma, X. 84
Minato, A. 79
Mosenbacher, A. 200
Mukaida, J. 124
Nie, X. 151
Ning, J. 24
Nishitani, Y. 119, 124
Nofar, M. 109
Oguz, H. 109
Ohshima, M. 79
Ovali, D. 109
Owada, S. 39
Ozen, Z.T. 109
Park, C.B. 176
Peng, M. 94
Peng, W. 138
Perumal, D.V. 19
Pinter, G. 200
Primetzhofer, A. 200
Qin, Q. 89
Ren, X. 34, 186
Rodrigue, D. 14, 59, 181
Saha, P. 74
Saito, T. 210
Sambaer, W. 74
Satoh, I. 210
Schuschnigg, S. 205
Sedlacek, T. 74
Shentu, B. 129
Shi, R. 98
Shinohara, T. 119
Spork, M. 205
Sun, L. 14
Takarada, W. 89
Takatsugi, S. 44
Takebe, T. 69
Tan, C. 24
Tan, J. 84
Tang, L. 5
Tong, L. 156, 166
Tran, M.P. 176
Wang, G. 176
Wang, L. 79
Wang, W. 167, 235
Wang, X. 94, 133, 138, 143, 147
Wang, Y. 49
Wang, Z. 161
Wei, R. 156, 161, 166

AUTHOR INDEX

Weng, Z. 129
Wu, M. 214
Wu, S. 167
Wu, Y. 129
Xiang, D. 104
Xie, P. 54
Xing, Y. 29
Yamada, Y. 124
Yamaguchi, M. 1
Yamanaka, T. 124
Yang, J. 94, 133, 138, 143, 147, 147
Yang, M. 242
Yang, W. 54, 84
Yang, Y. 24
Ye, J. 98
Ye, Z. 5, 10, 225
Yokoi, H. 39
You, Y. 166
Yu, T. 94
Zatloukal, M. 74
Zeng, X. 235
Zhai, S. 176
Zhang, G. 94, 133, 143, 147
Zhang, L. 98, 173
Zhang, M. 133, 138
Zhang, Q. 133
Zhang, X. 143
Zhang, Y. 10, 24
Zhao, M. 214
Zhou, X. 167
Zhu, Y. 151